Reviews of Environmental Contamination and Toxicology

VOLUME 193

Reviews of Environmental Contamination and Toxicology

Editor
David M. Whitacre

VOLUME 193

Springer

ISBN 978-1-4419-2511-4
e-ISBN: 978-0-387-73163-6

Springer
New York: 233 Spring Street, New York, NY 10013, USA
Heidelberg: Postfach 10 52 80, 69042 Heidelberg, Germany

ISSN 0179-5953

Printed on acid-free paper.

springer.com

Foreword

International concern in scientific, industrial, and governmental communities over traces of xenobiotics in foods and in both abiotic and biotic environments has justified the present triumvirate of specialized publications in this field: comprehensive reviews, rapidly published research papers and progress reports, and archival documentations. These three international publications are integrated and scheduled to provide the coherency essential for nonduplicative and current progress in a field as dynamic and complex as environmental contamination and toxicology. This series is reserved exclusively for the diversified literature on "toxic" chemicals in our food, our feeds, our homes, recreational and working surroundings, our domestic animals, our wildlife and ourselves. Tremendous efforts worldwide have been mobilized to evaluate the nature, presence, magnitude, fate, and toxicology of the chemicals loosed upon the earth. Among the sequelae of this broad new emphasis is an undeniable need for an articulated set of authoritative publications, where one can find the latest important world literature produced by these emerging areas of science together with documentation of pertinent ancillary legislation.

Research directors and legislative or administrative advisers do not have the time to scan the escalating number of technical publications that may contain articles important to current responsibility. Rather, these individuals need the background provided by detailed reviews and the assurance that the latest information is made available to them, all with minimal literature searching. Similarly, the scientist assigned or attracted to a new problem is required to glean all literature pertinent to the task, to publish new developments or important new experimental details quickly, to inform others of findings that might alter their own efforts, and eventually to publish all his/her supporting data and conclusions for archival purposes.

In the fields of environmental contamination and toxicology, the sum of these concerns and responsibilities is decisively addressed by the uniform, encompassing, and timely publication format of the Springer triumvirate:

Reviews of Environmental Contamination and Toxicology [Vol. 1 through 97 (1962–1986) as Residue Reviews] for detailed review articles concerned with any aspects of chemical contaminants, including pesticides, in the total environment with toxicological considerations and consequences.

Bulletin of Environmental Contamination and Toxicology (Vol. 1 in 1966) for rapid publication of short reports of significant advances and discoveries in the fields of air, soil, water, and food contamination and pollution as well as methodology and other disciplines concerned with the introduction, presence, and effects of toxicants in the total environment.

Archives of Environmental Contamination and Toxicology (Vol. 1 in 1973) for important complete articles emphasizing and describing original experimental or theoretical research work pertaining to the scientific aspects of chemical contaminants in the environment.

Manuscripts for *Reviews* and the *Archives* are in identical formats and are peer reviewed by scientists in the field for adequacy and value; manuscripts for the *Bulletin* are also reviewed, but are published by photo-offset from camera-ready copy to provide the latest results with minimum delay. The individual editors of these three publications comprise the joint Coordinating Board of Editors with referral within the Board of manuscripts submitted to one publication but deemed by major emphasis or length more suitable for one of the others.

Coordinating Board of Editors

Preface

The role of *Reviews* is to publish detailed scientific review articles on all aspects of environmental contamination and associated toxicological consequences. Such articles facilitate the often-complex task of accessing and interpreting cogent scientific data within the confines of one or more closely related research fields.

In the nearly 50 years since *Reviews of Environmental Contamination and Toxicology* (formerly *Residue Reviews*) was first published, the number, scope and complexity of environmental pollution incidents have grown unabated. During this entire period, the emphasis has been on publishing articles that address the presence and toxicity of environmental contaminants. New research is published each year on a myriad of environmental pollution issues facing peoples worldwide. This fact, and the routine discovery and reporting of new environmental contamination cases, creates an increasingly important function for *Reviews*.

The staggering volume of scientific literature demands remedy by which data can be synthesized and made available to readers in an abridged form. *Reviews* addresses this need and provides detailed reviews worldwide to key scientists and science or policy administrators, whether employed by government, universities or the private sector.

There is a panoply of environmental issues and concerns on which many scientists have focused their research in past years. The scope of this list is quite broad, encompassing environmental events globally that affect marine and terrestrial ecosystems; biotic and abiotic environments; impacts on plants, humans and wildlife; and pollutants, both chemical and radioactive; as well as the ravages of environmental disease in virtually all environmental media (soil, water, air). New or enhanced safety and environmental concerns have emerged in the last decade to be added to incidents covered by the media, studied by scientists, and addressed by governmental and private institutions. Among these are events so striking that they are creating a paradigm shift. Two in particular are at the center of ever-increasing media as well as scientific attention: bioterrorism and global warming. Unfortunately, these very worrisome issues are now super-imposed on the already extensive list of ongoing environmental challenges.

The ultimate role of publishing scientific research is to enhance understanding of the environment in ways that allow the public to be better informed. The term "informed public" as used by Thomas Jefferson in the age of enlightenment conveyed the thought of soundness and good judgment. In the modern sense, being "well informed" has the narrower meaning of having access to sufficient information. Because the public still gets most of its information on science and technology from TV news and reports, the role for scientists as interpreters and brokers of scientific information to the public will grow rather than diminish.

Environmentalism is the newest global political force, resulting in the emergence of multi-national consortia to control pollution and the evolution of the environmental ethic. Will the new politics of the 21st century involve a consortium of technologists and environmentalists, or a progressive confrontation? These matters are of genuine concern to governmental agencies and legislative bodies around the world.

For those who make the decisions about how our planet is managed, there is an ongoing need for continual surveillance and intelligent controls, to avoid endangering the environment, public health, and wildlife. Ensuring safety-in-use of the many chemicals involved in our highly industrialized culture is a dynamic challenge, for the old, established materials are continually being displaced by newly developed molecules more acceptable to federal and state regulatory agencies, public health officials, and environmentalists.

Reviews publishes synoptic articles designed to treat the presence, fate, and, if possible, the safety of xenobiotics in any segment of the environment. These reviews can either be general or specific, but properly lie in the domains of analytical chemistry and its methodology, biochemistry, human and animal medicine, legislation, pharmacology, physiology, toxicology and regulation. Certain affairs in food technology concerned specifically with pesticide and other food-additive problems may also be appropriate.

Because manuscripts are published in the order in which they are received in final form, it may seem that some important aspects have been neglected at times. However, these apparent omissions are recognized, and pertinent manuscripts are likely in preparation or planned. The field is so very large and the interests in it are so varied that the Editor and the Editorial Board earnestly solicit authors and suggestions of underrepresented topics to make this international book series yet more useful and worthwhile.

Justification for the preparation of any review for this book series is that it deals with some aspect of the many real problems arising from the presence of foreign chemicals in our surroundings. Thus, manuscripts may encompass case studies from any country. Food additives, including pesticides, or their metabolites that may persist into human food and animal feeds are within this scope. Additionally, chemical contamination in any manner of air, water, soil, or plant or animal life is within these objectives and their purview.

Manuscripts are often contributed by invitation. However, nominations for new topics or topics in areas that are rapidly advancing are welcome. Preliminary communication with the Editor is recommended before volunteered review manuscripts are submitted.

Summerfield, North Carolina D.M.W.

Table of Contents

Rev Environ Contam Toxicol 193:1–29

Remediation Technologies for Organochlorine-Contaminated Sites in Developing Countries

Alberto Bezama, Rodrigo Navia, Gonzalo Mendoza, and Ricardo Barra

Contents

I. Introduction

Despite its importance in human life, until recently the relationship between soils and human health has been undervalued, especially in least developed countries. Currently, a holistic approach has been incorporated to identify best practices in soil science, defining it as "the task of all people concerned with the soil to direct their interest, not just towards the physical, chemical, and biological aspects, but also to those environmental, economic, social, legal, and technical aspects that affect soil use" (Abrahams 2002; Fent 2003). Considering this definition, the European Union (EU) as well as most developed countries have recognized organochlorine-contaminated

Communicated by Lilia Albert

A. Bezama (✉), G. Mendoza, R. Barra
Environmental Sciences Center EULA—Chile, University of Concepción, Barrio Universitario s/n, Concepción, Chile.

R. Navia
Department of Chemical Engineering, University of La Frontera, Av. Francisso Salazar 01145, Temuco, Chile.

sites as potential threats to the human health, threats that take different forms, such as their influence on water (e.g., drinking water resources), soil, and air as well as their interrelationships, which can directly affect human health (EC 2002; EP 2002; Bezama et al. 2004). Moreover, economic expansion and industrial growth are linked with growing lack of "greenfields" (a term that defines all areas without previous history of development): the supply of new building sites is limited and must contend with other competing uses, such as housing, recreation, nature, traffic, or agriculture (De Sousa 2001; Tedd et al. 2001). Thus, cleaning and reusing contaminated sites can be a meaningful alternative to address this issue, because most contaminated sites are located in metropolitan centres and are, therefore, prime candidates for urban development (Lorber et al. 2004).

The identification of sites that pose a potential risk to human health and ecosystems, the verification of their actual pollution level, and assessment of the involved risks are the first steps when managing a contaminated land (Bezama et al. 2007a). Occurrence and distribution of soil problems in Europe are influenced by the diversity, distribution, and specific vulnerability of soils across the continent, coupled with physical aspects such as geology, relief, and climate (EEA 1999). About 550,000 sites across the EU have been identified as definitively or potentially contaminated, and the best estimate is that 1.3 million contaminated areas will be registered, although there is still a lack of information about the type and size of these contaminated areas. An estimated overview of the contaminated sites situation in Europe can be observed in Fig. 1 (Bezama 2006).

Organochlorine compounds represent an important fraction of the pollutants present in the identified contaminated sites in Europe, especially considering those originating from industrial and agricultural activities, which together are approximately 75% of the identified potentially contaminated sites (EEA 1999, 2003). For example, from the approximately 70,000 sites suspected to be contaminated in Austria, 33,549 have been registered as potentially industrial contaminated sites. Of these registered sites, 163 have already been investigated, evaluated, and classified. The data show that ~32% of the registered contaminated sites were contaminated with organochlorine compounds, whereas in 2002 these were found as main pollutants in ~29% of the 163 evaluated sites (FEA 2002).

As in most South American countries, the magnitude of the Chilean contaminated site problem has yet to be established. To date, two studies have been conducted for the identification and preliminary risk assessment of sites under suspicion of contamination, considering the associated human health and environmental risks (Fundación Chile 2004; Bezama et al. 2007a). The latter work is a case study in an industrial Region in South Central Chile that is the second most important nationally in social and economic terms. This preliminary investigation, based on historical databases of industrial activities and without a more accurate identification process due to limited funding, identified nearly 510 sites as suspect of

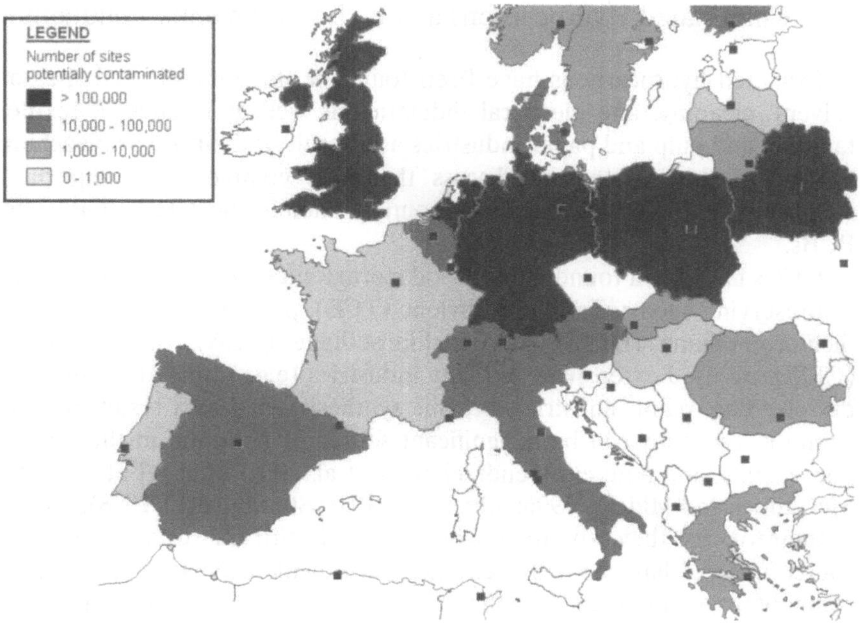

Fig. 1. Summary of the identified contaminated sites across Europe (Bezama 2006) (Adapted from EEA 1999; EEA 2003; FEA 2002; Eurostat et al. 1995, Andersen 2000.)

contamination in the first stage, from which approximately 10% were evaluated as dangerous to human health. As expected, a large number of these sites have an industrial origin. According to Seguel (2002), the total amount of industrial wastes generated in this Chilean Region amounts to about 350,000 t/yr. Most of the generated waste corresponds to the forestry sector (including sawmills, pulp and paper, cardboard, chipboards, and other companies, comprising ~21% of the regional waste generation) and to the iron and steel industry (17%). Considering these industrial activities, it is possible that chlorinated compounds are pollutants in this Region's contaminated sites.

The objective of this study is to define the technical considerations to be used when dealing with chlorinated-compound-contaminated sites, especially focusing on sites contaminated with chlorinated hydrocarbons (CHCs), chlorophenols (CPs), and polychlorinated biphenyls (PCBs), considering their occurrence in a developing economy such as Chile. These definitions are then followed by the determination of an integral site remediation technology that is based on the European experience but which could be easily transferable to the reality of developing countries in South America, specifically considering the Chilean case.

II. Chlorinated Hydrocarbons as Key Environmental Pollutants

Chlorinated hydrocarbons have been found in old contaminated sites of solvent, cleaning, and electrical industries as well as in sites of textile, tanning, and pulp and paper industries as a result of their use as bleaching agents. In this review, we discuss the most common and important subgroups among chlorinated compounds, namely the CHCs, CPs, and PCBs.

CHCs have been found in old wood storage sites, where they were used as preserving agents. Trichloroethylene (TCE), perchloroethylene (PCE), dichloromethane (DCM), 1,1,1-trichloroethane (TCA), or chloroform (CFM) are used extensively in many industries (e.g., in metal degreasing, dry cleaning, paint industries, organic synthesis), and as a result of their common use, they can be a significant source of pollution in the atmosphere, soils, and aquifers (Jendrzejewski et al. 2001; Michael et al. 1991; McCulloch and Midgley 1996; Stiber et al. 1998; Sturchio et al. 1998). Owing to their toxicity, they can threaten both public health and sustainable bioresource use and have actually generated worldwide concern as toxic environmental contaminants (Bidelman and Olney 1974; Tanabe et al. 1994; Guzzella et al. 2005; de Mora et al. 2004).

The principal industrial uses of CPs are in wood, textile, and leather preservation. An additional important source for the introduction of these compounds into the environment has been their application as pesticides and the degradation products of common pesticides. Moreover, CPs are by-products in the chloro-bleaching process of pulp and paper mills (Ohlenbusch et al. 2000), an industry of great relevance to Chile; while world pulp and paper production is estimated as 300 million t/yr, the Chilean production reaches about 4.4 million t/yr (INFOR 2006), a number expected to increase in the future because of the installation of several large pulp and paper mills in South-Central Chile.

The Cluster Rule (EPA 1998) introduced in the United States regulates the level of the 12 chlorophenols presented in Fig. 2 (McKague and Taylor 2001). The most important chlorophenols present in groundwater streams in the U.S. and Germany are 2,4,5-trichlorophenol, 2,4,6-trichlorophenol, 2,4-dichlorophenol, 2-chlorophenol, 3-chlorophenol, and pentachlorophenol (Kerndorff 1996). Moreover, these 6 compounds are on the list of the most common contaminants present in groundwater streams at contaminated sites. Finally, although 2,4,5-trichlorophenol, 2,4,6-trichlorophenol, 2,4-dichlorophenol, 2-chlorophenol, and pentachlorophenol are in the third group of priority contaminants according to EPA (1998), 3-chlorophenol and 2,6-dichlorophenol could join the first or the second group as their toxicity is not yet well established (Kerndorff 1996).

Another important group of organochlorine compounds are the PCBs. Synthesis of these compounds was first described in 1881, and their commercial production began at the end of 1920 (UNEP 2002), specifically

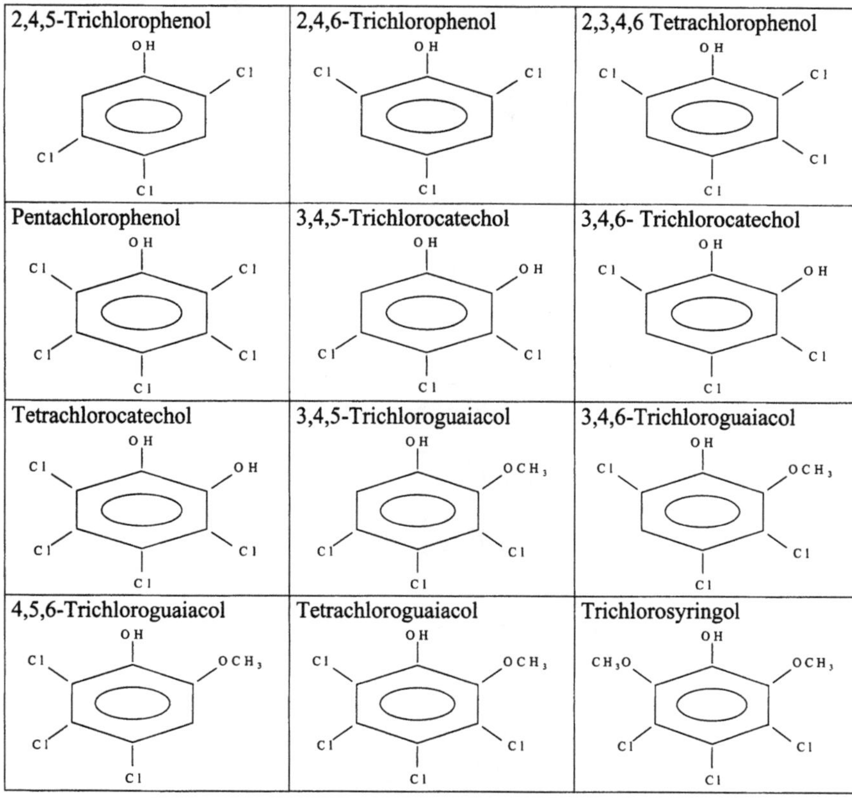

Fig. 2. Chlorophenols regulated by the Cluster Rule.

between 1929 and 1939. One of the most important characteristics of the PCBs is the wide variability in their physicochemical properties (vapor pressure, water solubility, and partition coefficients) as a consequence of their great number of congeners (209), varying from very low molecular weight/high volatility compounds (such as monochlorobiphenyl) to not very volatile/high molecular weight compounds (such as decachlorobiphenyl), which determines their behavior and mobility in different environmental compartments (Table 1). Such characteristics have resulted in international standards to regulate use of these compounds in all their life-cycle stages and in all their applications, e.g., recently in the European food industry (EC 2006).

Total PCB global production has been estimated to reach 1.3 million ts (Breivik et al. 2002). In Chile, as in other developing countries, PCBs have been used for more than 30 yr in many different industrial applications (Barra et al. 2004), principally as dielectric fluid, whose use was prohibited in 1982, in new transformers and condensers. However, their use in old devices and later storage may present a high environmental risk. In July

Table 1. Properties for polychlorinated biphenyls (PCB) congeners.

Formula	Name (chloro-biphenyl)	Number of isomers	IUPAC numbers	Molecular weight	Log K_{ow}	% Cl	Vapor pressure (Pa)	Solubility in water (g/m^3)	Number of identified isomers[a]
$C_{12}H_9Cl$	Mono	3	1–3	188.7	4.3–4.6	18.79	0.9–2.5	1.21–5.5	3
$C_{12}H_8Cl_2$	Di	12	4–5	233.1	4.9–5.3	31.77	0.008–0.6	0.06–2.0	12
$C_{12}H_7Cl_3$	Tri	24	16–39	257.5	5.5–5.9	41.3	0.003–0.22	0.015–0.4	23
$C_{12}H_6Cl_4$	Tetra	42	40–81	291.9	5.6–6.5	48.65	0.002	0.0043–0.01	41
$C_{12}H_5Cl_5$	Penta	46	82–127	326.4	6.2–6.5	54.3	0.0023–0.051	0.004–0.02	39
$C_{12}H_4Cl_6$	Hexa	42	128–169	360.8	6.7–7.3	58.9	0.0007–0.012	0.004–0.0007	31
$C_{12}H_3Cl_7$	Hepta	24	170–193	395.3	6.7–7	62.7	0.00025	0.000045–0.00001	18
$C_{12}H_2Cl_8$	Octa	12	194–205	429.8	7.1	66.0	0.0006	0.0002–0.0003	11
$C_{12}HCl_9$	Nona	3	206–208	464.2	7.2–8.16	68.7	—	0.00018–0.00012	3
$C_{12}Cl_{10}$	Deca	1	209	498.6	8.26	71.1	0.00003	1×10^{-6}–1×10^{-7}	1

[a]Refers to the number of isomers identified in commercial mixtures.

Source: Guitart et al. (1995); de Voogd et al. (1990); Ritter et al. (2005).

Table 2. PCB amounts for Chilean regions.

	Amounts of PCB in the Region (L)		
Region	In use	Stored	Total (L)
I	699	22	721
II	62,014	156,408	218,442
III	10,767	56,267	67,034
IV	7,745	681	8,426
V	12,132	250	12,382
R.M.	77,235	5,005	82,240
VI	44,944	742	45,686
VII	173	173	173
VIII	80,645	20,378	101,023
IX	450	0	450
X	60	0	60
XI	0	0	0
XII	30,314	2,616	32,930
Total	327,005	242,542	569,547

Source: CONAMA (2005b).

2004, Chile ratified the Stockholm Convention, which established different measurements to reduce and to eliminate POPs, including the prohibition of use and reduction of by-product liberation. This agreement has served as a diagnostic and assessment mechanism for these compounds at the national level in Chile. The Chilean PCB Register, carried out in 2004, which considered only dielectric fluids for condensers and transformers, shows the following inventories for Chile in each region, both for current use as well as for stored quantities (Table 2).

In the area of air quality, Barra et al. (2005) report that few PCB measurements have been performed in Chile. Recent reports for Santiago de Chile indicated values from 1.04 to 1.75 ng/m^3, which makes it comparable to other urban areas of the world. PCB levels in air particles smaller than 2.5 μm diameter ($PM_{2.5}$) in Temuco, Chile, ranged from 0.67 to 1.7 $\mu g/m^3$, while in Santiago the levels ranged from 1.15 to 2.7 $\mu g/m^3$. Unexpectedly, differences observed between these two cities were not high, in spite of differences in population density, indicating that the atmosphere in these two Chilean cities is fairly well polluted with PCBs, even when compared with other urbanized areas of the world (Mandalakis and Stephanou 2002).

Furthermore, there are very few officially recognized contaminated sites, and these are located mainly in heavily populated industrial areas as Santiago and Concepción, Chile. However, these official numbers grossly underestimate the real situation because of the existence of illegal and/or

Table 3. Distribution of the average sum of PCB concentrations in mussels throughout different Chilean regions.

Geographic distribution	Number of sampling points	Average sum of PCBs concentrations (ng/g d.w.)
North	4	25.3
North Center	3	37.2
South Center	3	12.4
South	3	15.3
South End	3	198.5

Source: Mendoza et al. (2006).

nonreported contaminated sites throughout the entire country (Barra et al. 2006). Available data are scarce and fragmentary; however, it is clear that PCB contamination is widespread through the country, as even the concentrations reported for background levels are lower than in the Northern Hemisphere (Barra et al. 2006; Borghini et al. 2005). PCB values in sediments from a highland lake have been reported, with quite low levels, reaching maximum values of 2 ng/g dw, but with trends indicating an increase in recent times (Barra et al. 2004). The most recent experience (Mendoza et al. 2006) consisted of the spatial distribution analysis of PCBs in mussels for the length of the Chilean coast. In this study, statistical analysis of congeneric PCB composition indicated five groups according to their molecular weight (number of chlorines), where the lighter congeners were observed in areas corresponding to high latitudes with total PCB values of 298 ng/g dry weight (Table 3).

III. Implementing Remediation Technologies for Organochlorine-Contaminated Sites

Remediation can be defined as the improvement of a contaminated site, to prevent, minimize, or mitigate damage to human health or the environment. It involves the development and application of a planned approach that reduces the availability of contaminants to receptors of concern (CSM 1997). The selection of an appropriate technology is therefore a crucial activity in a remedial project, where technical, economic, social, legal, and environmental aspects must be considered and evaluated (DEPA 2002; Bezama 2006). In general, it is possible to define two different types of remediation processes: *in situ* and *ex situ*. *In situ* processes are carried out in the same place where the pollutant is present, while *ex situ* processes are carried out in on-site or off-site plants or facilities. On-site refers to the place directly at the contaminated site, whereas off-site indicates that the place of a remediation facility is located outside the contaminated site (Hamby 1996; Guerin 1999; Bezama et al. 2004).

Any conventional *in situ* remediation scheme, such as bioremediation or soil vapor extraction, can be applied to the excavated soil, and because soil is usually handled in a closed system, process operational control is greatly simplified. Aggressive remediation schemes, such as surfactant or acid leaching of soil, can be applied without posing a threat to the environment. Often this cannot be done *in situ* for fear that mobilized contaminants will escape containment or will contaminate previously uncontaminated soil. Additionally, the treatment process can be more easily monitored, and unsuccessful processes may be abandoned with little environmental risk because the pile remains encapsulated until satisfactory results are attained. Furthermore, gradual, long-term remediation techniques can be used to provide low-cost, low-risk remediation that would not be feasible *in situ* (Lante 1991). This approach is similar to using *in situ* "natural attenuation" (Lin and Puls 2003; Kao et al. 2001; Youcai et al. 2002) but allows for greater mass transport control while the process continues.

In *ex situ* soil remediation, because wastes are removed from their original locations, much of the original site can be quickly recovered. This method can be a convincing advantage for redeveloping the affected areas where a small portion of the site can be dedicated to gradual remediation while the remainder of the site is redeveloped, or where one small site can host the remediation for several similar locations (Mesania and Jennings 2000). On the other hand, in contrast to *ex situ* remediation, *in situ* technologies can be used at a site with little disruption to ongoing operations. *In situ* treatments require neither heavy equipment for excavation nor large aboveground surface areas for treatment technology equipment facilities. Because *in situ* treatment does not involve movement of contaminated materials in the location, it minimizes both human and environmental exposure to contaminants. In contrast, transferring subsurface contaminated materials to the surface increases exposure risk for the same receptors (Volkwein et al. 1999).

In addition, in comparison with *ex situ* technologies, *in situ* technologies may be a more cost-effective and less intrusive means of remediating contaminated soils, sediments, and groundwater (Dupont 1993; Lin et al. 1996). Excavating contaminated material as well as operating and maintaining facilities for *ex situ* treatments often result in higher costs for treatment on the surface. On the other hand, to use an *ex situ* approach, one must remove contaminated soil from its original location and treat it above ground (Mesania and Jennings 2000; Van Deuren et al. 2002; Bezama et al. 2007b).

To determine the remediation processes that could be useful for organochlorine-contaminated sites, the physicochemical properties of some relevant chlorinated compounds present in the soil–aquifer environment need to be examined. In this case, compounds that are considered to be representative for this analysis because of their toxicity and ubiquitous presence are PCP, 2,4,6-trichlorophenol, and 2-chlorophenol (Table 4).

Table 4. Physicochemical properties of some chlorophenols.

Contaminant	K_{ow}	Water solubility* (mg/L)	Specific gravity	Boiling point °C (760 mmHg)	pK_a	Vapor pressure (mmHg)
PCP	$1.0 * 10^5$	$1.4 * 10^1$	1.9278	311.0	4.7	0.0002
TCP	$7.4 * 10^1$	$8.0 * 10^2$	1.49	244.5	6.2	400[a]
CP	$1.5 * 10^1$	$2.9 * 10^4$	1.241	174.5	8.5	1[b]

PCP, pentachlorophenol; TCP, 2,4,6-trichlorophenol; CP, 2-chlorophenol.
[a]At 12.1°C.
[b]At 25°C.
*10^6 = infinite solubility.
Source: Adapted from Shiu et al. (1994).

PCP properties are of special importance for Chile because the intensive forestry activity in the South Central region results in a large number of sawmills and wood-related enterprises, where the Potentially Contaminated Sites are highly probable from PCP use, especially in those companies operating before 1999, when the Chilean government banned PCP use. Until that date, more than 500 t PCP was imported and used for wood impregnation, where sodium salt is the most common form used.

In Table 4, it can be observed that nonionic PCP has low solubility and very low vapor pressure, which means, looking at its K_{ow} value, that PCP has a great affinity for organic matter and will probably remain adsorbed onto the organic part of the soil in the unsaturated zone. It is possible to assume that nonionic PCP has a low Henry's law constant, so that it will be very difficult to find PCP in the vapor phase. Regarding nonionic TCP, this compound would be preferentially found in the vapor phase of the vadose zone because of its high vapor pressure, moderated water solubility (Table 4), and high Henry's law constant. The K_{ow} for TCP is small, and TCP can be assumed to adsorb lightly on organic matter of the soil. Finally, it is possible to assume that nonionic CP would be preferentially dissolved in the aqueous phase of the vadose zone because of its high solubility value, low vapor pressure (see Table 3), and moderated Henry's law constant. The K_{ow} value for CP is also small, and it is possible that this pollutant could lightly adsorb onto soil organic matter. For the unsaturated zone, the situation would be as described in Fig. 3a, assuming as well an unsaturated zone pH smaller than pK_a and that all compounds are present in their nonionic form.

All these considerations correspond to the pollutants in their original form, not dissociated, although they may change depending on the pH of the water and soil system, affecting each compound's dissociation capacity in dependence with its pK_a value. These dissociated species may present different behaviors in the surrounding environment and will have different

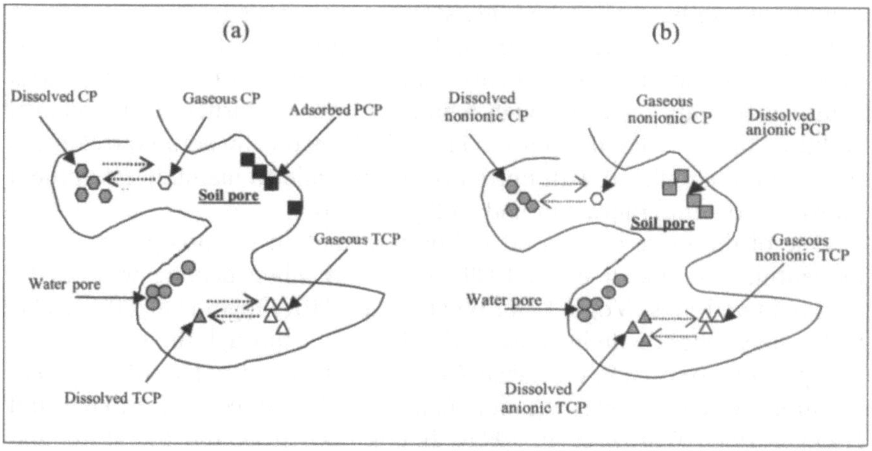

Fig. 3. Possible contaminants (chlorophenols) distribution in the vadose zone: (a) $pH < pK_a$ and (b) $pH = 7$.

solubilities, Henry's law constants, and vapor pressures. Thus, soil and water pH would become an important factor when choosing or designing a remediation process. If we consider that pH controls the presence of the dissociated species in a soil–water environment, the dissociation curve for each pollutant and its dependence with pH needs to be known. Taking as example a pH value of 7 in the soil/water system, PCP would be preferentially in the anionic form and its solubility will increase from 1.4×10^1 (nonionic PCP) to 10^4 mg/L (anionic PC-phenolate). In contrast, the situation for TCP and CP should be quite different, because at pH 7 TCP would be present at about 50% in the anionic form and 50% in the nonionic form and CP should be mainly present in the nonionic form in the aqueous phase. Therefore, there might be only small quantities of nonionic TCP and CP that remain adsorbed onto the soil pores (Fig. 3b).

A. Soil

A widely used remediation practice for soils polluted with hydrocarbons is bioremediation. In recent years, bioremediation has been increasingly selected to treat organochlorine compounds, because it is more publicly acceptable as it relies on natural processes to treat contaminants (EPA 2000; GeoSyntec Consultants 2005). Bioremediation utilizes microorganisms to degrade organic contaminants in soil, sludge, and solids, either excavated or *in situ*. Microorganisms break down contaminants by using them as a food source or cometabolizing them with a food source. Aerobic processes require an oxygen source, and the end products typically are carbon dioxide and water, although metabolites can also be found as intermediates (Zink and Lorber 1995). In *ex situ* processes, optimal aerobic conditions can be achieved adding structure materials (e.g., perlite) to an optimal surface for bacterial growth. In these processes, the soil is arranged in a pile and aeration is normally achieved using a vacuum pump. The contaminated air passes through a biofilter and is returned to the pile. In addition, the percolated water is collected, cleaned in an aerated treatment plant, and recirculated with nutrients into the pile to maintain its optimal moistore content (Eiermann and Bolliger 1996).

One of the more often used soil bioremediation technologies is *in situ* bioventing (Stehmeier et al. 1999), especially when dealing with CHC-contaminated soils (Vogel 1996; Whyte et al. 2001; Gogoi et al. 2003). This technology relies on an increase in the flow of air through the vadose zone that provides oxygen in the subsurface to optimize natural aerobic biodegradation, which becomes the dominant remedial process (Eiermann and Bolliger 1996; Warith et al. 1999). It is known from the literature that aerobic biological remediation processes are highly dependent on pollutant bioavailability, oxygen, and nutrients (nitrogen and phosphorus). As a result, in the ideal bioventing condition, the pollutants, oxygen, and nutrients are all dissolved in the water phase, which is very difficult to achieve. Therefore, the main problem for biological *in situ* treatment is that usually long process times are required (years), the microorganisms must be very specific, and the products of the biochemical reactions involved should be determined to be innocuous. Another application is the soil–air extraction *in situ* process, which relies on extraction of the air in the vadose zone through a vacuum pump. This process is suitable for high vapor pressure (>1500 Pa) contaminants in the soil matrix. The outlet air stream should be further treated in a biofilter or an adsorption system (Hamby 1996; Bezama 2006).

Another interesting technological option that has gained much attention recently is soil washing (Semer and Reddy 1996; DEPA 2002; Haapea and Tuhkanen 2006). Soil washing is an *ex situ* technology that relies on water as the fluid medium and energy to separate pollutants in nonsaturated media (Mann 1999; Norris et al. 1999). Soil washing liberates pollutants from the soil pores with mechanical energy, creating a pollutant suspension or an emulsion in the wash fluid (Tokunaga and Hakuta 2002; Urum et al. 2005). An ultimate pollutant separation is required to recover a concentrated pollutant and a clean wash fluid. Pure water is normally used, but some alternatives include addition of organic solvents, complex agents, and high-pressure soil wash (Chu and Chan 2003; Conte et al. 2005). Use of this technology is normally *ex situ*, but there are also some cases of *in situ* application, depending principally on soil characteristics.

When determining *in situ* or *ex situ* remediation possibilities, organic matter content and soil permeability play a key role. Indeed, for soil with organic matter content <5%, *in situ* remediation technologies are applicable; for organic matter >5%, *ex situ* technologies should be applied. In cases of intermediate organic matter values, the pollutants adsorption onto soil will be the limiting mechanism determining whether an *in situ* or *ex situ* technology should be applied (Okx and Stein 2000).

Moreover, a permeability factor (K_f) > 0.2 m/d allows *in situ* remediation, and *ex situ* technologies are most suitable for K_f < 0.2 m/d. An *ex situ* biological treatment is also feasible, but it will depend particularly on the soil properties and the needed remediation time. In terms of time duration, *ex situ* remediation processes last around 1 yr, whereas the remediation time ranges normally between 1 and 5 yr years for *in situ* technologies (Bezama et al. 2004).

B. Groundwater

Since the late 1970s, the on-site "pump and treat" process has been used for groundwater remediation. Groundwater is pumped, treated normally in a filter/adsorbent system, and returned to the groundwater layer (Voudrias 2001). Sand filters, granular activated carbon adsorbents, biofilters, and advanced oxidation processes have commonly been used as treatment systems. The main disadvantages of this technology are related to its high operational costs and the use of water as the carrier fluid. Still, the tendency is toward increased use of *in situ* groundwater remediation.

In situ bioremediation has potential to remove contaminants present in the groundwater plume, but it is limited by low ambient temperatures from the aquifers. Mineralization processes from the contaminants could be seriously affected by this factor, and biological on site treatment of contaminated groundwater could even require heating (Valo et al. 1990). There has been very little experience with *in situ* biological treatment of contaminated groundwater, as *ex situ* systems are more controllable, predictable, and less complex than *in situ* treatments (Hamby 1996).

Activated carbon, minerals, waste-derived adsorbents, and zero-valent metals are the adsorbent materials most used, with an interesting application as a reactive wall for contaminant compound adsorption. Activated carbon is mainly used in drinking and wastewater treatment facilities as well as in groundwater remediation, and it can adsorb a wide range of organic and inorganic pollutants present in water systems (USEPA 1998). Its amphoteric character governs adsorption onto activated carbon. In fact, in response to pH changes, the carbon surface develops coexisting electric charges of opposite sign, whose prevalence depends on the solution's chemistry. Therefore, attractive or repulsive electrostatic interactions between the adsorbate and the adsorbent must be taken into consideration (Radovic et al. 2000).

Activated carbon can adsorb heavy metals such as Cr(III), Cr(VI), Mo, Co, Ni, Cu, Zn, Cd, Hg, Pb, U, Au, and As as well as phosphates (Radovic et al. 2000), phenols (Streat et al. 1995), substituted phenols and benzenes (Mollah and Robinson 1996; Aksu and Yener 2001), dyes (Walker and Weatherley 1999), natural organic matter (NOM) (Radovic et al. 2000), and other organics such as trihalomethanes, amines, chlorinated organic compounds, alcohols, carboxylic and fatty acids, and amino acids (Radovic et al. 2000). Because of its wide range of adsorption spectra, activated carbon is considered a non selective adsorption material, although the high cost of this adsorption material [between 2.5 and 9.00 (US)/kg] has driven the search for other cheaper and selectively adsorbing materials. Research and development on the use of natural soils as abundant, cheap, and selectively natural adsorbents has been mainly focused on special soils. Such soils have been recently investigated for their pollutant adsorption and remediation capacity. As they are normally present abundantly in the environment, they could consequently present interesting cost advantages. Historically, the soil's porosity and pore-size distribution were of interest, principally for their effects on water retention and flow, gas advection and diffusion, and nutrient transport. With increased concern about environmental pollution, soil pore structure is now recognized as a critical factor in pollutant sorption. Sorption has an underlying influence on pollutant transport, chemistry, and biological activity (Navia 2004).

Soil is composed of individual minerals and organic matter grains that are cemented together to form particles. In turn, these particles may agglomerate to form higher-ordered structures (Yaron et al. 1996). In short, the association of its mineral and organic parts controls a soil system's porosity where soil water has a strong effect. From the point of view of potential interactions with various pollutants, the constituents of the soil's solid phase should be grouped according to their surface area. The fate of pollutants is affected by all the components of the soil solid phase. Still, soil constituents with low surface area could principally affect the transport of the pollutants as solutes, as immiscible with water liquids, or as vapors. The soil solid phase can also indirectly induce organic pollutant degradation in the soil medium through its effects on the water/air ratio in the system and consequently on the soil's biological activity. The group of constituents with high surface area controls pollutant transport, retention, and release on and from the soil surface as well as their surface-induced chemical degradation (Yaron et al. 1996).

Regarding waste-derived materials, extensive research regarding non-conventional, low-cost adsorption materials has been undertaken mainly to identify the possible reuse of some organic and inorganic wastes. Inorganic waste materials act as pure adsorbents, whereas organic waste materials could act as adsorbents as well as biosorbents because of the microbiological flora present in them.

Fly ash from thermal power stations and dried activated sludge were used to adsorb and biosorb 2- and 3-chlorophenol, respectively, with satisfactory results (Aksu and Yener 2001). Furthermore, chlorophenols and nitrophenols were found to adsorb strongly onto a cheap carbonaceous material obtained from the waste slurry generated in fertilizer plants (Gupta et al. 2000), while cheaper activated carbons based on coconut shell, wood, coal, straw, and tires were tested successfully for the removal of phenol and *p*-chlorophenol from contaminated water (Streat et al. 1995).

Bagasse fly ash, a sugar industry waste, has been recently investigated for the removal of some specific toxic and carcinogenic compounds such as pesticides based on a chlorobiphenyl structure (Gupta and Ali 2001). This same residue is also capable of removing Cd and Ni from wastewater streams (Gupta et al. 2003).

Low-cost adsorbents obtained from organic residues have also been developed to remove hexavalent chromium, which is often found in wastewater discharges from electroplating, metal finishing, and chrome preparation, and is considered to be highly toxic with a potential carcinogenic effect. At acidic pH, sawdust, sugar cane bagasse, sugar beet pulp, and corn cob, which are naturally occurring cellulosic waste materials, are able to adsorb Cr(VI) present in contaminated water (Sharma and Foster 1994). Moreover, Cu and Cd ions have been found to adsorb successfully onto bone char (Ko et al. 2000), whereas Al, Ca, Cd, Cu, Fe, Mg, Ni, Pb, and Zn ions (especially Pb) were efficiently removed from an acidic leachate by cocoa shells and, to a lesser degree, by cedar bark (Meunier et al. 2002).

With respect to colored effluents, these wastewaters are not only aesthetically displeasing, but they also impede light penetration, thus upsetting biological processes within a stream. Additionally, many dyes are toxic to some organisms and may cause direct destruction of aquatic communities, requiring some form of advanced treatment. Adsorption of acid and basic dyes present in aqueous solutions onto low-cost adsorbents such as bagasse pith, peat, corn cob, bean waste, and sugar-industry mud (Magdy and Daifullah 1998), as well as onto low-cost sewage sludge-based activated carbon (Rozada et al. 2003), have produced very successful results.

Zero-valent metals are one of the most promising adsorbent materials, with an interesting application for pollutant adsorption in *in situ* groundwater remediation, and are partially displacing on-site groundwater treatment processes like "pump and treat." In this process, the groundwater is pumped, treated normally in a filter/adsorbent system, and returned to the ground, although the tendency is toward increased use of *in situ* remediation (Zorzi and Hammer 1998). Fe^0, Al^0, Zn^0, Ni^0, Cu^0, Pd^0/C, and their combinations have been successfully tested for organochloride removal from groundwater. Moreover, it has been demonstrated that zero-valent iron is capable of producing reductive dechlorination of a great spectra of chlorinated organic compounds (Table 5).

Some unknown Fe^0-treatable CHCs have recently been carefully studied. Particularly, PCB reductive dechlorination may also occur in the presence of zero-valent iron while extracting PCBs from soil and sediments with subcritical water (Yak et al. 1999). From the chlorophenol group, the electrochemical dechlorination of 4-chlorophenol to phenol was determined to occur rapidly on palladized carbon cloth or palladized graphite electrodes

Table 5. Organochlorine compound treatability with Fe^0 as reactive wall.

Treatable organochlorine compounds	
Methanes	Trichloromethane. tetrachloromethane
Ethanes	1,1-Dichloroethane; 1,1,2-trichloroethane; 1,1,1-trichloroethane; 1,1,1,2-tetrachloroethane; 1,1,2,2-tetrachloroethane; hexachloroethane
Ethenes	Vinyl chloride; 1,1-dichloroethene; *trans*-1, 2-dichloroethene; *cis*-1,2-dichloroethene; trichloroethene; tetrachloroethene
Propanes	1,2-Dichloropropane; 1,2,3-trichloropropane
Other	Hexachlorobutadiene
Nontreatable organochlorine compounds	
Methanes	Chloromethane; dichloromethane
Ethanes	Chloroethane; 1,2-dichloroethane
Organochlorine compounds with unknown treatability	
Chlorobenzenes	—
Chlorophenols	—
Certain pesticides	—
PCBs	—

Source: Adapted from USEPA (1998); Gillham and O'Hannesin (1994).

(Cheng et al. 1997). At laboratory scale, complete dechlorination of pentachlorophenol was observed in a Pd/Mg reactive wall, while the dechlorination of less substituted chlorinated phenols by different metallic systems was found to be more likely to be achieved (Morales et al. 2002).

Moreover, reductive dechlorination of lindane was achieved in aqueous solution using a Zn-modified carbon cloth cathode (Kulikov et al. 1996). Therefore, it is stated that in general zero-valent metal barriers could be used for CHC dechlorination. The principal disadvantages of activated carbon are its price and its consequent thermal regeneration, whereas Fe^0 is an inexpensive material that should react and dechlorinate the CHCs present in groundwater (Navia et al. 2002).

IV. Definitions of an Integrated Approach

On May 19, 2005, Chile ratified the Stockholm Convention to prevent the damage caused by persistent organic pollutants (POPs) on the environment, including human health. With this ratification, Chile agreed to apply all the necessary measures to eliminate or restrict intentionally generated POPs as well as to appropriately eliminate the existent unused POPs, to minimize unintentional POP generation, and to establish a National Stockholm Convention Implementation Plan. Within this framework, Chile adopted as one of its priority lines the remediation of POP-contaminated environmental liabilities.

In Chile at present there is a generalized search for appropriate and advanced soil and groundwater remediation technologies, where the inclusion of regional conditions for this technology development is highly appreciated (De Palma 2002; Bezama et al. 2007c). In site remediation, the *ex situ* soil treatment plants are gaining importance in the European market (Schmitz and Andel 1997) and also when dealing with chlorinated organics (Deshpande et al. 1999; Khodadoust et al. 1999; Sheets and Bergquist 1999; Chu and Kwan 2003; Haapea and Tuhkanen 2006). Although cost restraints can be a factor, as mentioned earlier, in the case of CHC- and/or PCB-contaminated soils and groundwater, a certain degree of accuracy in monitoring and controlling the handling and degradation of pollutants is required. Therefore, in this work we propose an *ex situ* soil and groundwater remediation scheme (Fig. 4) consisting of a soil washing process followed by a further wastewater treatment with a filtration and an adsorption process. In addition, characterization of the waste fractions is made to define the possible final treatment alternatives.

A. Soil Washing

First, contaminated soil should be sieved. The large size soil fraction, which is normally less contaminated, should be immediately disposed. The contaminated soil fraction should be washed under optimal temperature, pH, and tensoactive concentration. With these optimized values, the final pollutant distribution in the different soil-size fractions should be determined. It is expected that the pollutants should concentrate in the fine soil fraction <63 μm (Reinert et al. 1999), which will be characterized and treated further in a windrow composting system. As guide values for the chlorinated organics concentration in cleaned soil, it is possible to use the Austrian Norm for soil remediation S2088-2 (OeN 2000), where limit values and prevention values are established (Table 6). For possible soil landfilling after the cleaning process, the Austrian Landfill Policy (List 2001) should be applied, differentiating between four landfill types, and each one has limit values for pollutants in the soil (Table 7).

B. Wastewater Treatment

The wastewater stream from the soil washing process will be contaminated with CHCs. This stream should be cleaned by adding flocculating compounds through a press filter and a further activated carbon adsorption process. The filtration process will treat the suspended compounds, while the activated carbon system should adsorb the dissolved pollutants. The cleaned wastewater should be returned to the washing system as process water, following an integrated process concept. The Austrian Water Law will be used for first-limit wastewater values. The emission limit values will follow the indication for emission into open canalization (Table 8).

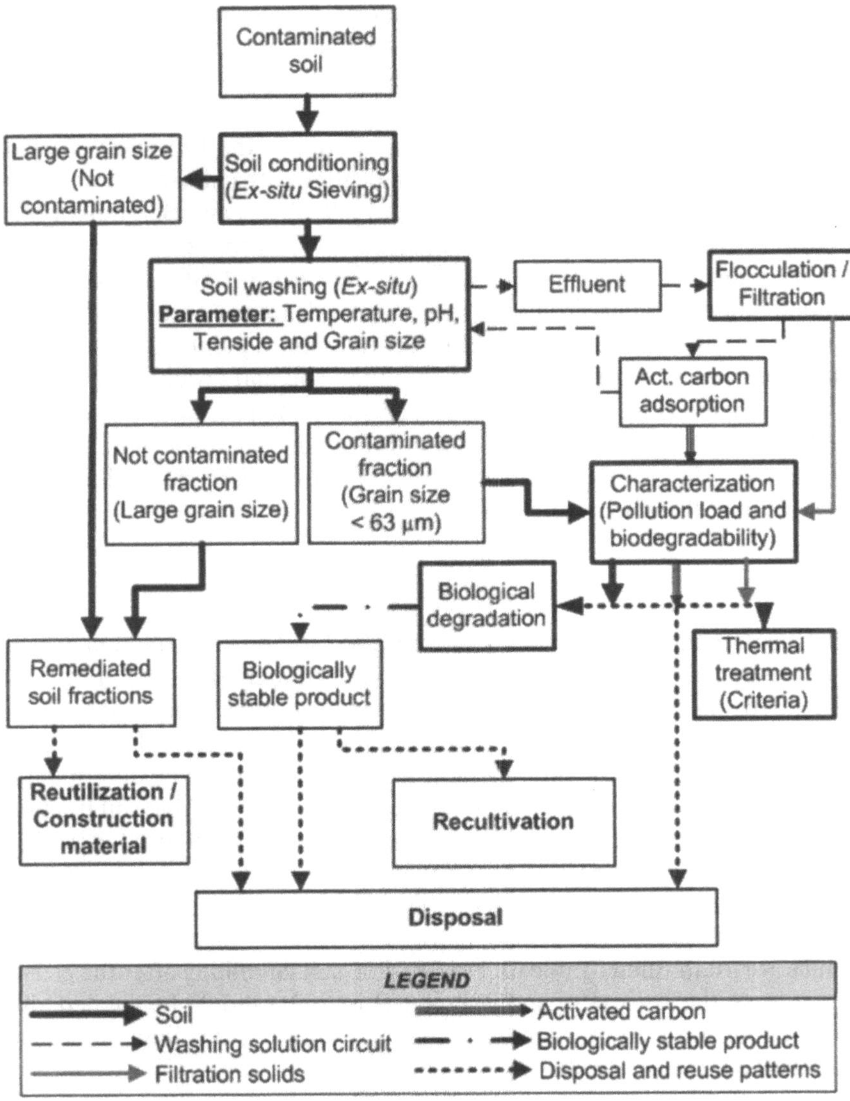

Fig. 4. Overview of the proposed integral concept for the remediation of organochlorine compound-contaminated soils.

Table 6. Limit and prevention values for polyaromatic hydrocarbons (PAH) and PCB presence in soil.

Pollutant	Limit value (mg/kg dry soil)	Prevention value (mg/kg dry soil)
PAH	1	50
PCB	0.2	1

Source: OeN (2000).

Table 7. Limit values for selected pollutants depending on landfill type.

	Landfill type (differentiated by the source of the wastes)[a]			
Pollutant	Excavated soil (mg/kg dry mass)	Demolition waste (mg/kg dry mass)	Residual waste (mg/kg dry mass)	Mass waste (mg/kg dry mass)
Total content				
PAH	0.5	2	t.b.d.	100
Hydrocarbons	20	100	5,000	20,000
TOC (as C)	20,000	30,000	30,000	50,000
POX (as Cl)	—	—	—	1,000
Elute				
Hydrocarbons	5	50	100	—
TOC (as C)	200	500	500	t.b.d.
EOX (as Cl)	0.3	3	t.b.d.	30
Surfactants (as TBS)	1	5	20	t.b.d.

t.b.d., to be determined if relevant; TOC, total organic carbon; POX, purgeable organic halogens; EOX, extractable organic halogens.

[a]Types of sanitary landfills defined according to FEA (2001); Jacobsen and Kristoffersen (2002).

Table 8. Wastewater limit values.

Organic pollutants	Limit value (mg/L)
AOX (as Cl)	0.5
Nonvolatile lipophilic compounds	100
Hydrocarbons	20
POX (as Cl)	0.1
Phenol index (as phenol)	10
BTX	0.1

AOX, adsorbable organic halogens; POX, purgeable organic halogens; EOX, extractable organic halogens; BTX, benzol, toluol, xylol.

Source: Rossmann (1993).

C. Characterization of Waste Streams

The different waste streams (soil fine fraction, filtration steps, solid waste, and spent activated carbon) will be tested for pollutant load level and biodegradability. The wastes with a pollutant load level below the Austrian Landfill Policy will be immediately landfilled, and the biodegradable

Table 9. Limit values for pollutants in compost.

Pollutant	Limit value (mg/kg dry mass)
PAH	2
PCB	0.2
Hydrocarbons	200

polluted fractions will be treated in a composting system. Possible reuse of the cleaned soil fractions in the construction market as well as the reuse of the stabilized material (from the windrow composting process) in recultivation purposes will be analyzed. For the nonbiodegradable fractions, a thermal processing criterion will be developed, focused especially on use at cement industry facilities.

D. Windrow Composting

Biological degradation of the organochlorine compounds should be performed in a windrow composting system. The fractions to be handled are the biodegradable, still-contaminated fine fraction from the soil washing process, the spent activated carbon, and the waste fraction from the wastewater treatment's filtration process. To optimize the relationship between the nutrients (i.e., carbon, phosphorus, and nitrogen) in the biological process, structural material and appropriate substrates (i.e., amendments) should be added. As a final product, a stabilized material is obtained that should at least comply with the limit values presented in Table 8 for its further landfilling. To search for additional uses, a complete characterization of the stabilized material shall be determined. For land applicability uses, the compost limit values are regulated by the Austrian Compost Policy as summarized in Table 9 (BLFUW 2001).

V. Discussion

Contaminated sites are becoming more important throughout the world, as their containment and/or remediation can be coupled with a reasonable further redevelopment of these sites, which is foreseen as one of the most important future environmental tasks. In the future, research on contaminated site remediation must focus on the application of combinations of already existing technologies in such a way as to appropriately degrade the most abundant pollutants in an economically and environmentally sound way. Particularly, organochlorine compounds represent an important fraction of the pollutants found at the identified contaminated sites in Europe, especially considering those originating from industrial and agricultural activities, being present in approximately 75% of the identified potentially contaminated sites. Therefore, the sites contaminated with such pollutants require innovative remediation solutions that make possible future reuse of these derelict areas.

This work has focused on the three most representative groups of organochlorine compounds, namely polychlorinated biphenyls (PCBs), chlorinated hydrocarbons (CHCs), and chlorophenols (CPs). A complete analysis has been established for the determination of an appropriate technology that could handle these compounds in an environmentally and technically sound manner but that could also be suitable for developing countries and especially for Chile. Such a concept has been defined as an integral approach to treat and manage the contaminated material that comprises soil excavation and subsequent mechanical sorting of the material. Two material streams can be defined from this process, where each one will undergo a different treatment according to their pollution loads. The defined treatment options comprise mechanical sorting and soil conditioning and a soil washing process coupled with either biological or thermal remediation, depending on the nature of the pollutants present in each project. Final treatment options comprise the disposal of the material in a sanitary/security landfill but also consider the recycling of the material in applications such as filling materials or for the recultivation of degraded areas, an aspect that can be considered only on a case-by-case basis.

In the case of developing countries such as Chile, such technological approaches could be easily implemented because they would not result in technological adversities or barriers in the already existing market. The technical concepts analyzed and presented in this review can be considered as suitable for application regardless of the country's economic situation and thus provide an interesting option when looking for sound environment recovery alternatives.

Summary

Chlorinated hydrocarbons represent an important aspect of the contaminated sites identified worldwide, especially when considering those originated from industrial or agricultural activities, which represent approximately 75% of the potentially contaminated sites identified in developed European countries.

Chlorinated hydrocarbons, such as trichloroethylene, perchloroethylene, dichloromethane, 1,1,1-trichloroethane, or chloroform, are used extensively in many industries (e.g., in metal degreasing, dry cleaning, paint industries, organic synthesis, etc.), and their common use can make them into a significant pollution source in the atmosphere, soils, and aquifers. Another important group of organochlorine compounds are polychlorinated biphenyls (PCBs). Among the characteristics of the PCBs, one of the most important is a wide variability in their physicochemical properties (vapor pressure, water solubility, and partition coefficients) because of their great number of congeners (varying from very slight compounds, such as monochlorobiphenyl, to the not very volatile compound decachlorobiphenyl), which determines their behavior and mobility in different environmental compartments.

Remediation can be defined as the improvement of a contaminated site to prevent, minimize, or mitigate damage to human health and/or the environment. Remediation involves the development and application of a planned approach that reduces the availability of contaminants to receptors of concern. The selection of an appropriate technology is therefore a crucial activity in a remedial project, where technical, economic, social, legal, and environmental aspects must be considered and evaluated. Conventional groundwater treatment methods (e.g., based on sorption onto granular activated carbon) are in many cases not very effective, in particular for some low sorption potential compounds. Permeable reactive barriers constructed of elemental iron have emerged as an effective passive remediation method for groundwater contaminated with a range of contaminants, mainly chlorinated hydrocarbons. Furthermore, soil washing is a state-of-the-art technique that transfers the contaminants from the soil to a liquid phase by desorption and solubilization. In these applications, extracting solutions (typically water with or without additives to improve contaminant removal) are infiltrated into the soil, and the loaded aquifer is then either treated *in situ* or recovered and treated on site. Nowadays there is a generalized search for appropriate and advanced soil and groundwater remediation technologies whereby the inclusion of the regional conditions to this technology development is highly appreciated.

The present work aimed to first present the technical aspects that need to be considered for an appropriate selection of remedial technologies for site remediation when dealing with organohalogenated compounds as the main pollutants. From these aspects, an integral site remediation strategy was defined, based on the European experience but easily transferable to the reality of developing countries in South America, specifically considering Chile. This strategy consists in the remediation of CHC-contaminated soil through a soil washing process followed by a further wastewater treatment consisting of a filtration and an adsorption process. In addition, a characterization of the waste fractions will be made to define the possible treatment alternatives.

Acknowledgments

This work was partially supported by FONDECYT N° 1050647 (R. Barra), and by The Research Directorate of the University of Concepción.

References

Abrahams PW (2002) Soils: their implications to human health. Sci Total Environ 291:1–32.

Aksu Z, Yener J (2001) A comparative adsorption/biosorption study of monochlorinated phenols onto various sorbents. Waste Manag 21:695–702.

Andersen JN (ed) (2000) Management of contaminated sites and land in Central and Eastern Europe. Danish Cooperation for Environment in Eastern Europe – Ministry of Environment and Energy, Copenhagen, Denmark.

Barra R, Cisternas M, Suarez C, Piñones O, Popp P (2004) PCBs and HCHs in a saltmarsh sediment record from south central Chile: use of tsunami signatures and ^{137}Cs fallout as temporal markers. Chemosphere 55:965–972.

Barra R, Colombo JC, Eguren G, Gamboa N, Jardim WF, Mendoza G (2005) Persistent organic pollutants (POP) in Eastern and Western South American countries. Rev Environ Contam Toxicol 185:1–33.

Bezama A (2006) A knowledge-based tool for contaminated sites management. Dissertation. Institute for Sustainable Waste Management and Technology, University of Leoben, Austria.

Bezama A, Navia R, Novak J, Staber W, Lorber KE (2004) Novel approaches for the management and redevelopment of contaminated sites. Österr Wasserwirtsch 56(11–12):139–144.

Bezama A, Agüero R, Barrera S, Márquez F, Lorber KE (2007a) Land register and evaluation of suspected contaminated sites in an industrial Chilean Region. Waste Manag. DOI: 10.1016/j.wasman.2007.02.029 (in press).

Bezama A, Szarka N, Wolfbauer J, Lorber KE (2007b) Development and use of a balanced scorecard system for supporting decision-making in contaminated sites remediation. Water Air Soil Pollu, 181(1–4):3–16.

Bezama A, Szarka N, Navia R, Lorber KE (2007c) Lessons learned for a more efficient technology transfer to South American countries in the fields of solid waste and contaminated sites management. Waste Manag Res 25:148–161.

Bidelman TF, Olney CE (1974) Chlorinated hydrocarbons in the Sargasso Sea atmosphere and surface water. Science 183:516–518.

Borghini F, Grimalt J, Sanchez-Hernandez J, Barra R, Torres C, Focardi S (2005) Organochlorine compounds in soils and sediments of the mountain Andes Lakes. Environ Poll 136:253–266.

Breivik K, Sweetman A, Pacyna J, Jones K (2002) Towards a global historical emission inventory for selected PCB congeners a mass balance approach: 1. Global production and consumption. Sci Total Environ 290(1–3):181–198.

BLFUW (Bundesministerium für Land- und Fortwirtschaft, Umwelt und Wasserwirtschaft) (2001) “Kompostverordnung.” Vienna, Austria.

Cheng IF, Fernando Q, Korte N (1997) Electrochemical dechlorination of 4-chlorophenol to phenol. Environ Sci Technol 31(4):1074–1078.

Chu W, Chan KH (2003) The mechanism of the surfactant-aided soil washing system for hydrophobic and partial hydrophobic organics. Total Environ 307(1–3):83–92.

Chu W, Kwan CY (2003) Remediation of contaminated soil by a solvent/surfactant system. Chemosphere 53:9–15.

CONAMA (Chilean National Environmental Commission) (2005a) Plan Nacional de Implementación para la Gestión de los Contaminantes Orgánicos Persistentes (COPs) en Chile; Fase I: 2006–2010 (National Implementation Plan for the Management of Persistent Organic Pollutants (POP) in Chile; Phase I: 2006–2010). Gobierno de Chile, Comisión Nacional de Medio Ambiente, Santiago.

CONAMA (Chilean National Environmental Commision) (2005b) Inventario Nacional de Bifenilos Policlorados en Chile.
Conte P, Agretto A, Spaccini R, Piccolo A (2005) Soil remediation: humic acids as natural surfactants in the washings of highly contaminated soils. Environ Pollut 135(3):515–522.
CSM (Contaminated Sites Management Group) (1997) Site Remediation Technologies: A Reference Manual. Federal Interdepartamental Contaminated Sites Management Working Group of Canada (ed). Environment Canada, Hazardous Waste Branch, Quebec, Canada.
de Mora S, Villeneuve J, Sheikholeslami MR, Cattini C, Tolosa I (2004) Organochlorinated compounds in Caspian Sea sediments. Mar Pollut Bull 48: 30–43.
DEPA (Danish Environmental Protection Agency) (2002) Guidelines on Remediation of Contaminated Sites. Environmental Guidelines No. 7, 2002c. Danish Ministry of the Environment, Copenhagen.
De Palma R (2002) Background paper. UNIDO Programme on Transfer of Environmentally Sound Technologies in CEE and "NIS." In: Ministry of Environment of the Slovak Republic and UNIDO (eds) International Forum on Strategies and Priorities for Environmental Industries, June 12–14, Bratislava (Slovakia).
De Sousa C (2001) Contaminated sites: the Canadian situation in an international context. J Environ Manag 62:131–154.
Deshpande S, Shiau BJ, Wade D, Sabatini DA, Harwell JH (1999) Surfactant selection for enhancing ex situ soil washing. Water Res 33 (2):351–360.
Dupont RR (1993) Fundamentals of bioventing applied to fuel contaminated sites. Environ Prog 12:45–53.
EC (Commission of the European Communities) (2002) Towards a Thematic Strategy for Soil Protection. COM (2002) 179 (final).
EC (Commission of the European Communities) (2006) Reglamento (CE) No. 199/2006 de la Comisión de las comunidades Europeas. Contenido máximo de determinados contaminantes en los productos alimenticios por lo que se refiere a dioxinas y PCB similares a dioxinas. Diario Oficial de la Unión Europea (ES), 04.2.2006, L32/34–38.
EEA (European Environmental Agency) (1999) Environment in the European Union at the turn of the century. Environmental assessment report no. 2. EEA, Copenhagen, Denmark.
EEA (European Environmental Agency) (2003) Europe's Environment: The Third Assessment. EEA, Copenhagen, Denmark.
Eiermann D, Bolliger R (1996) Industiestandort Gaswerk: Neueste Erkenntnisse aus einer biologischen Langzeit-Bodensanierung. Terra Tech 3:37–39.
EP (European Parliament and European Council) (2002) Laying down the Sixth Community Environment Action Programme. Decision No. 1600/2002/EC, 22 July 2002.
EPA (United States Environmental Protection Agency) (1998) National Emission Standards for Hazardous Air Pollutants for Source Category: Pulp and Paper Production; Effluent Limitations Guidelines, Pretreatment Standards, and New Source Performance Standards: Pulp, Paper, and Paperboard Category. 40 CFR Parts 63, 261, and 430 [FRL–5924–8] RIN 2040–AB53. Federal Register Notices for Final Air and Water Rules, Vol. 63, No. 72, pp. 18504–18751.

EPA (United States Environmental Protection Agency) (2000) Engineered Approaches to In Situ Bioremediation of Chlorinated Solvents: Fundamentals and Field Applications. Report EPA 542-R-00-008. U.S. Environmental Protection Agency, Office of Solid Waste and Emergency Response, Technology Innovation Office, Washington, DC.

Eurostat, European Environment Agency, United Nations Economic Commission for Europe, Organisation for Economic Co-operation and Development & World Health Organisation (1995) Europe's Environment: Statistical Compendium for the Dobris Assessment. Statistical Office of the European Communities, Luxembourg.

Fam S (1996) Vapor extraction and bioventing. In: In Situ Treatment Technology. CRC Press, Baton Rouge, p. 329.

FEA (Austrian Federal Ministry of Agriculture and Forestry, Environment and Water Management) (2001) Federal Waste Management Plan 2001. Federal Ministry of Agriculture and Forestry, Environment and Water Management, Division III/3, Vienna, Austria.

FEA (Austrian Federal Ministry of Agriculture and Forestry, Environment and Water Management) (2002) State of the Environment, vol 6. Umweltbundesamt, Vienna.

Fent K (2003) Ecotoxicological problems associated with contaminated sites. Toxicol Lett 140–141:353–365.

Fundación Chile (2004): Manual de procedimiento de identificación y priorización de Sitios Contaminados (Manual on the procedure for identifying and prioritizing contaminated sites). Area de Medio Ambiente y Metrología Química. Programa de Remediación Ambiental, Santiago.

GeoSyntec Consultants (2005) Bioaugmentation for Remediation of Chlorinated Solvents: Technology Development, Status, and Research Needs. White paper for the Environmental Security Technology Certification Program (ESTCP), U.S. Department of Defense. Arlington, VA.

Gillham RW, O'Hannesin SF (1994) Enhanced degradation of halogenated aliphatics by zero-valent iron. Ground Water 32(6):958–967.

Gogoi BK, Dutta NN, Goswami P, Krishna Mohan P (2003) A case study of bioremediation of petroleum hydrocarbon-contaminated soil at a crude oil spill site. Adv Environ Res 7:767–782.

Guerin TF (1999) Bioremediation of phenols and polycyclic aromatic hydrocarbons in creosote contaminated soil using ex-situ land treatment. J Hazard Mater B65:305–315.

Gupta VK, Ali I (2001) Removal of DDD and DDE from wastewater using bagasse fly ash, a sugar industry waste. Water Res 35(1):33–40.

Gupta VK, Srivastava SK, Tyagi R (2000) Design parameters for the treatment of phenolic wastes by carbon columns (obtained from fertilizer waste material). Water Res 34(5):1543–1550.

Gupta VK, Jain CK, Ali I, Sharma M, Saini VK (2003) Removal of cadmium and nickel from wastewater using bagasse fly ash, a sugar industry waste. Water Res 37:4038–4044.

Guzzella L, Roscioli C, Vigano L, Saha M, Sarkar SK, Bhattacharya A (2005) Evaluation of the concentration of HCH, DDT, HCB, PCB and PAH in the sediments along the lower stretch of Hugli estuary, West Bengal, northeast India. Environ Int 31:523–534.

Haapea P, Tuhkanen T (2006) Integrated treatment of PAH contaminated soil by soil washing, ozonation and biological treatment. J Hazard Mater, 136(2): 244–250.

Hamby DM (1996) Sites remediation techniques supporting environmental restoration activities: a review. Sci Total Environ 191:203–224.

INFOR (Instituto Forestal de Chile) (2006) Estadísticas Forestales Chilenas 2005. Boletín Estadístico N° 111. Unidad de Estudios Económicos y del Ambiente del Instituto Forestal, Sede Metropolitana INFOR, Santiago, Chile.

Jacobsen H, Kristoffersen M (2002) Case studies on waste management minimisation practices in Europe. Topic report 2/2002. European Environmental Agency, Copenhagen, Denmark.

Jendrzejewski N, Eggenkamp HGM, Coleman ML (2001) Characterisation of chlorinated hydrocarbons from chlorine and carbon isotopic compositions: scope of application to environmental problems. Appl Geochem 16:1021–1031.

Kao CM, Chen SC, Liu SC, Wang L (2001) Application of microbial enumeration technique to evaluate the occurrence of natural bioremediation. Water Res 35(8):1951–1960.

Kerndorff H (1996) Chemische und humantoxikologische Grundlagen. In: Chemie und Biologie der Altlasten, Fachgruppe Wasserchemie in der GDCh (eds). VCH, Weinheim, Germany.

Khodadoust A, Suidan M, Acheson C, Brenner R (1999) Remediation of soils contaminated with wood preserving wastes: crosscurrent and countercurrent solvent washing. J Hazard Mater 64(2):167–179.

Ko DCK, Porter JF, McKay G (2000) Optimised correlations for the fixed-bed adsorption of metal ions on bone char. Chem Eng Sci 55:5819–5829.

Kulikov SM, Plekhanov VP, Tsyganok AI, Schlimm C, Heitz E (1996) Electrochemical reductive dechlorination of chlororganic compounds on carbon cloth and metal-modified carbon cloth cathodes. Electrochim Acta 41(4):527–531.

Lante DW (1991) Techniken der Deponiesicherung und Deponiesanierung. Entsorgungspraxis 1-2:14–18.

Lin G, Sauer NE, Cutright TJ (1996) Environmental regulations: a brief overview of their applications to bioremediation. Int Biodeterior Biodegrad 38(1): 1–8.

Lin Z, Puls RW (2003) Potential indicators for the assessment of arsenic natural attenuation in the subsurface. Adv Environ Res 7:825–834.

Lorber KE, Bezama A, Novak J, Navia R (2004) A novel approach for the redevelopment of old landfill sites. In: Proceedings of the ISWA World Environment Congress and Exhibition, Oct. 17–21, Rome, Italy.

List W (2001) Abfallrecht Kodex. 10. Auflage, Orac Verlag, Vienna, Austria.

Magdy YH, Daifullah AAM (1998) Adsorption of basic dye from aqueous solutions onto sugar-industry-mud in two modes of operation. Waste Manag 18:219–226.

Mann MJ (1999) Full-scale and pilot-scale soil washing. J Hazard Mater 66: 119–136.

Mandalakis M, Stepahanou E (2002) Polychlorinated biphenyls associated with fine particles (PM2.5) in the urban environment of Chile: concentrations levels, and sampling volatilization losses. Environ Toxicol Chem 21(11):2270–2275.

McCulloch A, Midgley PM (1996) The production and global distribution of emissions of trichloroethene, tetrachloroethene and dichloromethane over the period 1988–1992. Atmos Environ 30:601–608.

McKague AB, Taylor DR (2001) Isomer specific syntheses of chlorinated catechols and guaiacols relevant to pulp bleaching. Chemosphere 45:261–267.

Mendoza G, Gutierrez L, Pozo-Gallardo K, Fuentes-Rios D, Montory M, Urrutia R, Barra R (2006) Polychlorinated biphenyls (PCBs) in mussels along the chilean coast. Environ Sci Pollut Res 13(1):67–74.

Mesania F, Jennings A (2000) Modelling soil pile bioremediation. Environ Model Softw 15:411–424.

Meunier N, Blais JF, Tyagi RD (2002) Selection of a natural sorbent to remove toxic metals from acidic leachate produced during soil decontamination. Hydrometallurgy 67:19–30.

Michael LC, Pellizzari ED, Norwood DL (1991) Application of the master analytical scheme to the determination of volatile organics in wastewater influents and effluents. Environ Sci Technol 25:150–155.

Mollah AH, Robinson CW (1996) Pentachlorophenol adsorption and desorption characteristics of granular activated carbon: I. Isotherms. Water Res 30(12): 2901–2906.

Morales J, Hutcheson R, Cheng IF (2002) Dechlorination of chlorinated phenols bay catalyzed and uncatalyzed Fe(0) and Mg(0) particles. J Hazard Mater B90: 97–108.

Navia R (2004) Environmental use of volcanic soil as natural adsorption material. Ph.D. thesis. Institute for Sustainable Waste Management and Technology, University of Leoben, Austria.

Navia R, Diez MC, Lorber KE (2002) Remediation of sites contaminated with chlorophenols. Abfallwirtschaftstagung DepoTech, Nov. 20–22, Leoben, Austria.

Norris G, Al-Dhahir Z, Birnstingl J, Plant SJ, Cui S, Mayell P (1999) A case study of the management and remediation of soil contaminated with polychlorinated biphenyls. Eng Geol 53:177–185.

OeN (Österreichisches Normungsinstitut) (2000) Altlasten – Gefährdungsabschätzung für das Schutzgut Boden. ÖNORM S 2088-2, Vienna, Austria.

Ohlenbusch G, Kumke MU, Frimmel FH (2000) Sorption of phenols to dissolved organic matter investigated by solid phase microextraction. Sci Total Environ 253:63–74.

Okx JP, Stein A (2000) An expert support model for in situ soil remediation. Water Air Soil Pollut 118:357–375.

Radovic LR, Moreno-Castilla C, Rivera-Utrilla J (2000) Carbon materials as adsorbents in aqueous solutions. In: Radovic LR (ed) Chemistry and Physics of Carbon, vol 27. Dekker, New York–Basel, p. 227.

Reinert G, Lotz W, Fiedler J, Bauer T, Heidingsfelder S, Neesse T, Keller U, Breiter R (1999) Trennschnitt 5 μm für die Bodenwäsche-biologische Reinigung der feinstfraktion. Terra Tech 3:49–51.

Ritter L, Solomon KR, Forget J (2005) Persistent Organic Pollutants. International Programme on Chemical Safety (IPCS) Within the Framework of the Inter-Organization Programme for the Sound Management of Chemicals (IOMC).

Rossmann H (ed) (1993) Das Österreichische Wasserrechtsgesetz. Verlag der Österreichischen Staatsdruckerei, Dritte Auflage, Vienna, Austria.

Rozada F, Calvo LF, García AI, Martín-Villacorta J, Otero M (2003) Dye adsorption by sewage sludge-based activated carbons in batch and fixed-bed systems. Bioresour Technol 87:221–230.

Schmitz HJ, Andel P (1997) Bodenbehandlungsanlagen: Die Jagd nach dem Boden wird härter. Terra Tech 5:17–31.
Seguel P (2002) Catastro de residuos industriales de la VIII Región (Register of industrial waste in the VIII Region). Bachelor of Engineering thesis. Universidad del Desarrollo, Concepción, Chile.
Semer R, Reddy KR (1996) Evaluation of soil washing process to remove mixed contaminants from a sandy loam. J Hazard Mater, 45(1):45–57.
Shiu WY, Ma KC, Varhanickova D, Mackay D (1994) Chlorophenols and alkylphenols: a review and correlation of environmentally relevant properties and fate in an evaluative environment. Chemosphere 29(6):1155–1224.
Sharma DC, Forster CF (1994) A preliminary examination into the adsorption of hexavalent chromium using low-cost adsorbents. Bioresour Technol 47: 257–264.
Sheets RG, Bergquist BA (1999) Laboratory treatability testing of soils contaminated with lead and PCBs using particle-size separation and soil washing. J Hazard Mater 66:137–150.
Stehmeier LG, Francis M, Jack TR, Diegor E, Winsor L, Abrajano TA (1999) Field and in vitro evidence for in-situ bioremediation using compound-specific 13C/12C ratio monitoring. Org Geochem 30:821–833.
Stiber NA, Small MJ, Fischbecl PS (1998) The relationship between historic industrial site use and environmental contamination. J Air Waste Manag Assoc 48:809–818.
Stoops G, Cheverry C (1992) New Challenges for Soil Research in Developing Countries: A Holistic Approach. Proceedings of the Workshops, funded by the European Community, Life Sciences and Technologies for Developing Countries (STD 3 Programme), Rennes, France.
Streat M, Patrick JW, Camporro-Perez MJ (1995) Sorption of phenol and parachlorophenol from water using conventional and novel activated carbons. Water Res 29(2):467–472.
Sturchio NC, Clausen JL, Heraty LJ, Huang L, Holt BD, Abrajano TA (1998) Chlorine isotope investigation of natural attenuation of trichloroethebe in an aerobic aquifer. Environ Sci Technol 32:3037–3042.
Tanabe S, Iwata H, Tatsukawa R (1994) Global contamination by persistent organochlorines and their ecotoxicological impact on marine mammals. Sci Total Environ 154:163–177.
Tedd P, Charles JA, Driscoll R (2001) Sustainable brownfield re-development—risk management. Eng Geol 60:333–339.
Tokunaga S, Hakuta T (2002) Acid washing and stabilization of an artificial arsenic-contaminated soil. Chemosphere 46:31–38.
UNEP (2002) PCB Transformers and Capacitors, from Management to Reclassification and Disposal. Geneva, Switzerland.
Urum K, Pekdemir T, Ross D, Grigson S (2005) Crude oil contaminated soil washing in air sparging assisted stirred tank reactor using biosurfactants Chemosphere 60(3):334–343.
USEPA (1998) Permeable reactive barrier technologies for contaminant remediation. EPA/600/R-98/125. EPA, Washington, DC.
Valo R, Häggblom MM, Salkinoja-Salonen M (1990) Bioremediation of chlorophenol containing simulated groundwater by immobilized bacteria. Water Res 24: 253–258.

Van Deuren J, Lloyd T, Chetry S, Liou R, Peck J (2002) Remediation Technologies Screening Matrix and Reference Guide, 4th Ed. U.S. Federal Remediation Technology Roundtable (FRTR). http://www.frtr.gov.

Videla S, Diez MC (1997) Experiences of wastewater treatment in Chilean forest industry. Water Sci Technol 35(2-3):221–226.

Vogel TM (1996) Bioaugmentation as a soil bioremediation approach. Curr Opin Biotechnol 7:311–316.

Volkwein S, Hurtig H, Klöpfer W (1999) Life cycle assessment of contaminated sites remediation. Int J Life Cycle Anal 4(5):263–274.

Voudrias EA (2001) Pump-and-treat remediation of groundwater contaminated by hazardous waste: can it really be achieved? Global Nest Int J 3(1):1–10.

Walker GM, Weatherley LR (1999) Kinetics of acid dye adsorption on GAC. Water Res 33(8):1895–1899.

Warith M, Fernández L, Gaudet N (1999) Design of in-situ microbial filter for the remediation of naphthalene. Waste Manag 19:9–25.

Whyte LG, Goalen B, Hawari J, Labbe D, Greer CW, Nahir M (2001) Bioremediation treatability assessment of hydrocarbon-contaminated soils from Eureka, Nunavut. Cold Regions Sci Technol 32:121–132.

Yak HK, Wenklawiak BW, Cheng IF, Doyle JG, Wai CM (1999) Reductive dechlorination of polychlorinated biphenyls by zerovalent iron in subcritical water. Environ Sci Technol 33(8):1307–1310.

Yaron B, Calvet R, Prost R (1996) Soil Pollution. Processes and Dynamics. Springer-Verlag, Berlin/Heidelberg.

Youcai Z, Luochun W, Renhua H, Dimin X, Guowei G (2002) A comparison of refuse attenuation in laboratory and field scale lysimeters. Waste Manag 22:29–35.

Zink G, Lorber KE (1995) Mass spectral identification of metabolites formed by microbial degradation of polycyclic aromatic hydrocarbons (PAH). Chemosphere 31(9):4077–4084.

Zorzi M, Hammer S (1998) In Situ-Filter-Reaktionswand zur Sanierung von KW und LHKW-kontaminierten Grundwässern. In: Hengerer D, Lorber KE, Nelles M, Wolber G (eds) Restabfallbehandlung, Deponietechnik, Entsorgungsbau und Altlastenproblematik, Balkema, Rotterdam, pp. 151–155.

Manuscript received March 22; accepted April 4, 2007.

Rev Environ Contam Toxicol 193:31–52

Chemistry and Fate of Triazolopyrimidine Sulfonamide Herbicides

Thomas W. Jabusch and Ronald S. Tjeerdema

Contents

I. Introduction

The triazolopyrimidine sulfonamide (TSA) herbicides were registered in the United States in 1993. Their mode of action is by inhibition of acetolactate synthase (ALS), an enzyme common to plants and microorganisms but not found in animals. ALS inhibitors include other herbicide families such as the sulfonylureas, imidazolinones, and pyrimidinyl thiobenzoates. In the 1970s, sulfonylureas were the first ALS inhibitors to be introduced to the market. Their discovery was greeted as a great achievement because of their ability to suppress weeds at extremely low application rates compared to previously used herbicides and with low toxicity risk to humans and wildlife. By 1991, the market value of ALS inhibitors had reached $1.3 billion (U.S.). However, some of the problems associated with their use include the induction of resistance in weed species to both ALS-inhibiting and alternative herbicides. ALS inhibitors have also been found to damage

Communicated by Ronald S. Tjeerdema.

Thomas W. Jabusch ✉
San Francisco Estuary Institute, 7770 Pardee Lane, Oakland, CA 94621, U.S.A.

Ronald S. Tjeerdema
Department of Environmental Toxicology, College of Agricultural and Environmental Sciences, University of California, One Shields Avenue, Davis, CA 95616-8588, U.S.A.

nontarget plants at residual levels that are below the detection limits of standard analytical methods (Saari et al. 1994; Whitcomb 1999).

TSA herbicides were first developed in the late 1980s by Dow Agro Sciences and are now widely applied for postemergence control of broadleaf weeds in food and forage crops (Table 1), commonly in rotation and/or mixtures with different mode of action herbicides. Flumetsulam was the first to be introduced to the U.S. market in 1993 to control weeds in corn and soybean fields. It was followed by cloransulam-methyl and diclosulam, which were developed for use with soybeans and peanuts for broadleaf weed control. Although neither metosulam nor florasulam are registered in the United States, they have been introduced in other countries for weed control with maize (Kleschick and Gerwick 1989; Ware and Whitacre 2004; PAN 2006). Penoxsulam, the newest addition, was registered in the U.S. in 2004 for the control of annual grasses, sedges, and broadleaf weeds in rice culture (USEPA 2004a). At this writing, the new analogue pyroxsulam is pending U.S. registration for control of annual grasses and broadleaf weeds with winter wheat (USEPA 2006).

II. Chemistry

A. Physicochemical Properties

Triazolopyrimidine sulfonamides are among the newer structural types of herbicides that have made important impacts upon the industry over the past 10–15 yr; they represent an increasingly science-based approach to pest control. They are selective ALS inhibitors, generally exhibit a broad activity spectrum with good crop selectivity and typically high herbicidal activity at low application rates of 10–50 g active ingredient (a.i.)/ha (Kleschick et al. 1992; Baskaran et al. 1996), and include seven compounds: flumetsulam, metosulam, cloransulam-methyl, diclosulam, florasulam, penoxsulam, and pyroxsulam (Fig. 1). Their physicochemical properties are summarized in Table 1.

Triazolopyrimidine sulfonamides exhibit acidic dissociation constants (pK_as); thus, their water solubilities are highly pH dependent (see Table 1). For example, the water solubility of penoxsulam ranges from 0.006 g/L at pH 5 to 1.46 g/L at pH 9 (Roberts et al. 2003). Generally, these compounds tend to be highly water soluble at basic and neutral pHs, and their solubility decreases at lower pHs. Octanol–water partitioning and soil sorption coefficients, as well as hydrolytic rates, for these compounds are also pH dependent. Log octanol–water partition coefficients (log K_{ow}) are low and indicate that these agents do not significantly partition into lipids or other organic solvent phases. The reported soil sorption coefficients (K_ds) for flumetsulam (5–182) and penoxsulam (0.2–5.1) indicate moderate to high mobility in soils (Kleschick et al. 1992; Jabusch and Tjeerdema 2005). Chemical degradation by hydrolysis is also pH dependent; the hydrolytic half-life ($t_{1/2}$)

Table 1. Physical, chemical, and toxicological properties of the TSA herbicides.

Property	Flumetsulam	Metosulam	Diclosculam	Florasulam	Cloransulam-methyl	Penoxsulam	Pyroxsulam
Chemical name	*N*-(2,6-Difluorophenyl)-5-methyl-[1,2,4]triazolo[1,5-*a*]pyrimidine-2-sulfonamide	*N*-(2,6-Dichloro-3-methylphenyl)-5,7-dimethoxy-[1,2,4]triazolo[1,5-*a*]pyrimidine-2-sulfonamide	*N*-(2,6-Dichlorophenyl)-5-ethoxy-7-fluoro-[1,2,4]triazolo [1,5-*c*]pyrimidine-2-sulfonamide	*N*-(2,6-Difluorophenyl)-8-fluoro-5-methoxy-[1,2,4]triazolo[1,5-*c*]pyrimidine-2-sulfonamide	Methyl 3-chloro-2-(5-ethoxy-7-fluoro-[1,2,4]triazolo[1,5-c]pyrimidine-2-sulfonamido)benzoate	2-(2,2-Difluoroethoxy)-*N*-(5,8-dimethoxy-[1,2,4]triazolo[1,5-*c*]pyrimidin-2-yl)-6-(trifluoromethyl)benzenesulfonamide	*N*-(5,7-Dimethoxy-[1,2,4]triazolo[1,5-*a*]pyrimidin-2-yl)-2-methoxy-4-(trifluoromethyl)pyridine-3-sulfonamide
Trade name(s)	Accent gold herbicide®, Broadstrike®, Frontrow®, Hornet®, Naf-280, Naf-281, Python®, Scorpion®		Strongarm®		Firstrate®, Frontrow®, Gangster®, Gauntlet®, Riverdale 565®	Gf-881®, Granite®, Grasp®	
Application	Corn, soybean, barley, wheat	Cereals	Soybean, peanut	Cereals, pasture	Soybean	Rice	Wheat, hay, straw
Application rate	10–80 g a.i./ha	30–70 g a.i./ha	Maximum global use rate:_52 g a.i./ha	Canada 5 g a.i./ha Europe 7.5 g a.i./ha	17.5–44 g a.i./ha	40 g a.i./ha	
CAS number	98967-40-9	139528-85-1	145701-21-9	145701-23-1	147150-25-4	219714-96-2	422556-08-9
Chemical formula	$C_{12}H_9F_2N_5O_2S$	$C_{14}H_{13}Cl_2N_5O_4S$	$C_{13}H_{10}Cl_2FN_5O_3S$	$C_{12}H_8F_3N_5O_3S$	$C_{15}H_{13}ClFN_5O_5S$	$C_{16}H_{14}F_5N_5O_5S$	$C_{14}H_{13}F_3N_6O_5S$
Molecular weight	325.3	418.3	406.1	359.3	429.8	483.4	434.4

Table 1. (*cont.*)

Property	Flumetsulam	Metosulam	Diclosculam	Florasulam	Cloransulam-methyl	Penoxsulam	Pyroxsulam
Vapor pressure	4×10^{-9} Pa @ 25°C	NA	7×10^{-10} Pa @ 25°C	NA	4×10^{-14} Pa @ 25°C	9.5×10^{-14} Pa @ 25°C	NA
pK_a	4.6	NA	4.1	NA	4.8	NA	NA
Water solubility	0.049 g/L at pH 2.5	NA	0.117 g/L at pH 5 0.124 g/L at pH 7 4.29 g/L at pH 9	NA	0.003 g/L at pH 5 0.184 g/L at pH 7	0.006 g/L at pH 5 0.41 g/L at pH 7 1.46 g/L at pH 9	NA
Log K_{ow}	0.21	NA	1.42 at pH 5 –0.047 at pH 7 –0.448 at pH 9	NA	1.12 at pH 5 –0.365 at pH 7 –1.24 at pH 9	NA	NA
Soil sorption coefficient K_d	5–182	NA	NA	NA	NA	0.14–5.05	NA
Soil $t_{1/2}$ (d)	30–246	22	48	0.7–85	9–263	5.2–12.8	NA
Aqueous photolysis $t_{1/2}$ (d)	NA	NA	101	NA	0.015	2.9–26.5	NA
Hydrolysis $t_{1/2}$ (d)	NA	NA	1007 at pH 5 241 at pH 7 1.9 at pH 9	NA	>365 at pH 5 118–231 at pH 7 3 at pH 9	NA	NA

Sources: Kleschick et al. 1992; Wolt et al. 1992; Frear et al. 1993; Baskaran et al. 1996; USEPA 1997; Parnell and Hall 1998; Jackson et al. 2000; Krieger et al. 2000a; Baron 2001; van Weesenbeeck et al. 2001; Zabik et al. 2001; Rouchaud et al. 2002b; Roberts et al. 2003; USEPA 2004b; Borges et al. 2005b; Jabusch and Tjeerdema 2005; Cambridge Soft Corporation 2006; Jabusch and Tjeerdema 2006a; PAN 2006; Wood 2006.

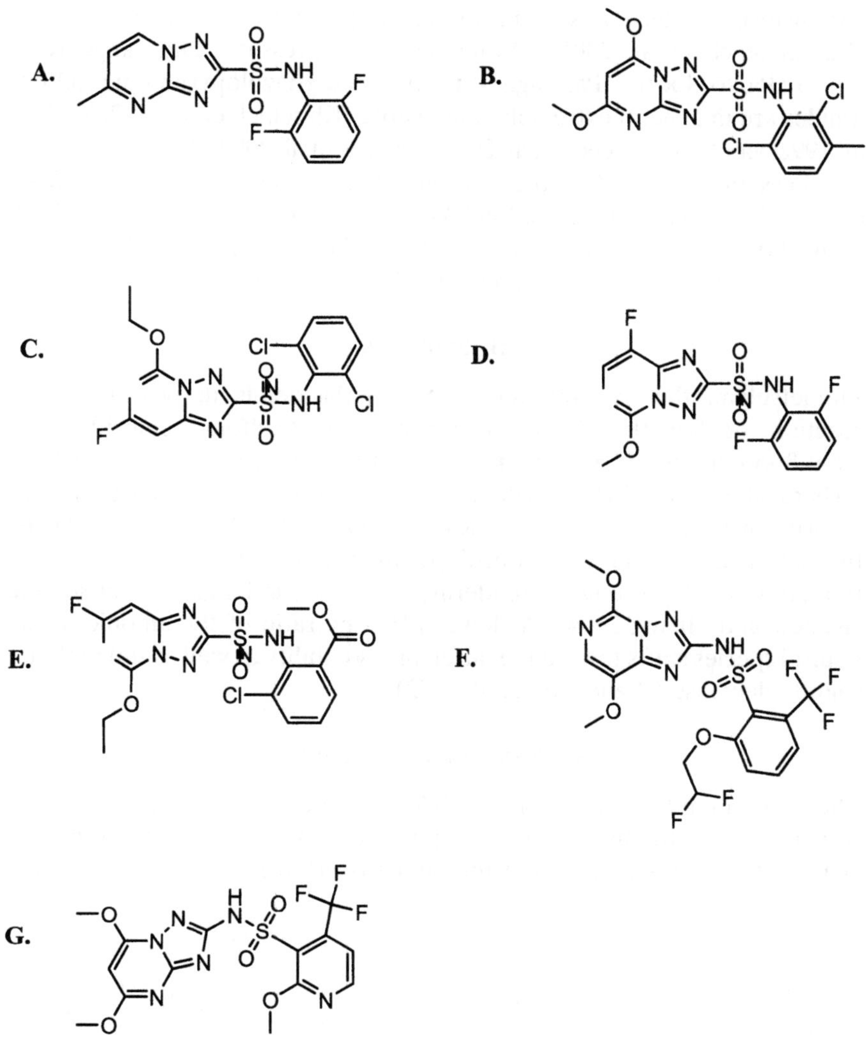

Fig. 1. Chemical structures of the triazolopyrimidine sulfonamides. A. Flumetsulam (*N*-(2,6-difluorophenyl)-5-methyl-[1,2,4]triazolo[1,5-α]pyrimidine-2-sulfonamide. B. Metosulam (*N*-(2,6-dichloro-3-methylphenyl)-5,7-dimethoxy-[1,2,4]triazolo[1,5-α]pyrimidine-2-sulfonamide). C. Diclosulam (*N*-(2,6-dichlorophenyl)-5-ethoxy-7-fluoro-[1,2,4]triazolo[1,5-*c*]pyrimidine-2-sulfonamide). D. Florasulam (*N*-(2,6-difluorophenyl)-8-fluoro-5-methoxy-[1,2,4]triazolo[1,5-*c*]pyrimidine-2-sulfonamide). E. Cloransulam-methyl (methyl 3-chloro-2-(5-ethoxy-7-fluoro-[1,2,4]triazolo[1,5-*c*] pyrimidine-2-sulfonamido)benzoate). F. Penoxsulam (2-(2,2-difluoroethoxy)-*N*-(5,8-dimethoxy-[1,2,4]triazolo[1,5-*c*]pyrimidin-2-yl)-6-(trifluoromethyl) benzenesulfonamide). G. Pyroxsulam (*N*-(5,7-dimethoxy-[1,2,4]triazolo[1,5-*a*] pyrimidin-2-yl)-2-methoxy-4-(trifluoromethyl)pyridine-3-sulfonamide).

for diclosulam decreases from 1,007 d at pH 5 to 1.9 d at pH 9 (van Weesenbeeck et al. 2001). Available vapor pressure data range from 4×10^{-9} Pa to 4×10^{-14} Pa, suggesting that the triazolopyrimidine sulfonamides are in general extremely nonvolatile (Kleschick et al. 1992; Wolt et al. 1992; van Weesenbeeck et al. 2001; Roberts et al. 2003). Soil degradation half-lives range from 0.7 to 246 d and photodegradation half-lives from 0.015 to 101 d (Kleschick et al. 1992; Wolt et al. 1992; USEPA 1997; Jackson et al. 2000; Krieger et al. 2000a; van Weesenbeeck et al. 2001; Zabik et al. 2001; Roberts et al. 2003; Jabusch and Tjeerdema 2005, 2006a).

B. Synthesis

Flumetsulam, the first marketed TSA herbicide, is derived from the reaction of 5-methyl-[1,2,4]triazolopyrimidine-2-sulfonyl chloride with 2,6-difluoroaniline in pyrimidine. The sulfonyl chloride is prepared from 2-(benzothio)-5-methyl-[1,2,4]triazolo[1,5-α]pyrimidine by reaction with chlorine in aqueous acid (Fig. 2; Kleschick et al. 1992). At near-neutral pHs, the sulfonamide bridge of triazolopyrimidine sulfonamide molecules is thought to be deprotonated, rendering the compound highly water soluble (Renew and Huang 2004). At lower pHs, the ratio of the anionic to the neutral species shifts toward the latter, and solubility as well as soil mobility slightly decrease (Wauchope et al. 2002).

C. Mode of Toxic Action

The general mode of action of the TSA herbicides is the inhibition of ALS, which catalyzes the first common step in the biosynthesis of the branched-chain amino acids (valine, leucine, and isoleucine) in plants and micro-

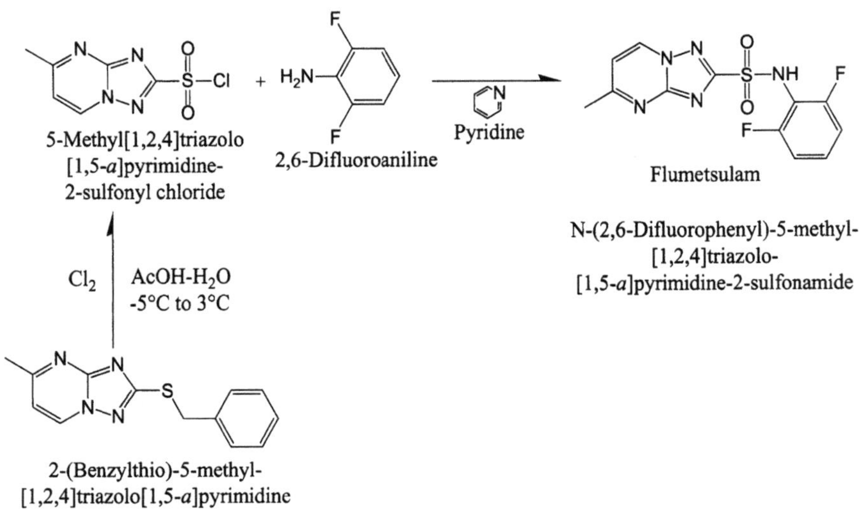

Fig. 2. Reaction scheme for the synthesis of flumetsulam.

organisms (Subramanian et al. 1989). There are to date four classes of ALS–inhibiting herbicides: the triazolopyrimidines, sulfonylureas, imidazolinones, and pyrimidyl-oxy-benzoates (Mourad and King 1992). They all bind to closely overlapping sites on the ALS molecule that are presumably separate from binding sites for substrates or feedback inhibitors such as valine (Landstein et al. 1993).

The herbicidal activities of both flumetsulam and metosulam were found to be temperature mediated. Elevating temperature after spraying from 1° to 20°C increased the activities of both flumetsulam and metosulam in the broad-leaved weed *Raphanus raphanistrum* by factors of 97 and 7, respectively, with 50% reduction rates (ED_{50}) for flumetsulam ranging between 19.5 (1°C) and 0.2 g a.i. ha^{-1} (20°C) and for metosulam between 0.55 (1°C) and 0.08 g a.i. ha^{-1} (Madafiglio et al. 2000).

ALS herbicide resistance has been observed worldwide, and the development of target site resistance resulting from repeated applications is probably the most significant barrier to the long-term successful use of TSA herbicides (Gerwick and Kleschick 1991; Rubin 1996; Whitcomb 1999). Almost all cases of resistance to TSA herbicides or other ALS inhibitors are the result of an altered ALS enzyme that is less sensitive to inhibitors (Schmitzer et al. 1993; Bernasconi et al. 1995). Resistance is commonly the result of a point mutation within discrete conserved domains of the *als* gene, resulting in an altered target site and, thus, target site resistance. There are different possible mutations within the ALS system, conferring a broad spectrum of resistance to ALS inhibitors (Shaner and Singh 1997; Sprague et al. 1997). In addition, varying degrees of cross-resistance between triazolopyrimidines, sulfonylureas, and pyrimidinyl-oxy-benzoates have been detected in different plants (Landstein et al. 1993; Foes et al. 1998; Ferguson et al. 2001; Hashem et al. 2001; Patzoldt et al. 2001). Patterns of resistance and cross-resistance appear to be specific to the particular point mutation and its position on the ALS enzyme (Whaley et al. 2006). In most instances, resistance has been inherited as a semidominant trait (Saari et al. 1994; Boutsalis and Powles 1995; Devine and Eberlein 1997; Patzoldt and Tranel 2002; Sibony and Rubin 2003; Zheng et al. 2005).

A different resistance mechanism, observed in *Alopecurus myosuroides*, is the rapid detoxification of the ALS-inhibiting herbicide (Kemp et al. 1990). Selectivity of florasulam in wheat has been determined to be primarily related to a differential rate of metabolism between wheat ($t_{1/2}$ of 2.4 hr) and broadleaf weeds ($t_{1/2}$ ranging from 19 to >48 hr) and, to a lesser extent, to uptake differences (deBoer et al. 2006). Differences in cloransulam absorption and translocation partially explained differences in susceptibility among some weed species but not others (Barnes and Oliver 2004).

Cloransulam and other ALS inhibitors have also been found to antagonize graminicides. This antagonism has not been fully explained. Proposed hypotheses suggest either a decline in graminicide absorption or prevention of the inhibiting action of the graminicide on meristematic growth (Barnwell and Cobb 1994).

III. Methods of Analysis

Because of the low use rates of TSA herbicides, ultratrace residue methodologies are required for their detection in environmental samples or to enforce tolerance levels in food crops (Maycock et al. 1995; Shackelford et al. 1996; Whitcomb 1999). Analytical methods with high sensitivity levels for the detection of TSA herbicides in environmental matrices include liquid chromatography with ultraviolet detection (LC-UV), thermospray liquid chromatography with mass spectrometry (TSP-LC-MS), capillary gas chromatography with mass spectrometry (GC-MS), LC tandem MS (LC-MS/MS), capillary electrophoresis with UV detection (CE-UV), coelectroosmotic CE-UV, and direct enzyme-linked immunosorbent assay (ELISA). All these methods are to varying degrees associated with some inherent difficulties but are able to detect one or several TSA herbicides in the ppb range after varying extraction, cleanup, and preconcentration procedures (Maycock et al. 1995; Baskaran et al. 1996; Shackelford et al. 1996; Krynitsky 1997; Parnell and Hall 1998; Krieger et al. 2000b; Laganà et al. 2000; Borges et al. 2005a–c).

LC-MS and CE were successfully employed for ultratrace analysis of TSA herbicides in environmental matrices in multiresidue determinations at the ppb range. CE is a high-efficiency separation method for pesticide analysis in complex environmental matrices, but its low detection sensitivity, which is normally in the ppm range, needs to be overcome; this can be achieved by sample preconcentration with a method called sample stacking. Krynitsky (1997) analyzed flumetsulam along with 12 sulfonylurea herbicides in runoff water by CE and LC/MS and achieved a limit of quantitation (LOQ) of 0.02 ppb with a sample preparation method that included acidification, extraction by reversed-phase SPE, and extract cleanup with a tandem sample stacking system consisting of a strong anion-exchange SPE cartridge stacked on an alumina cartridge. For the determination of five TSA herbicides (diclosulam, cloransulam-methyl, flumetsulam, metosulam, and florasulam) in soy milk, Borges et al. (2005c) developed a methodology combining SPE and CE-MS with sample stacking in normal stacking mode (NSM) and achieved limits of detection (LODs) down to 74 ppb. The same authors achieved LODs between 0.1 and 0.24 ppb for analysis of the same suite of TSA herbicides (diclosulam, cloransulam-methyl, flumetsulam, metosulam, and florasulam) in water samples with a methodology that employed CE-UV, SPE using C_{18} cartridges, and on-line preconcentration by sample stacking with matrix removal (SWMR; Borges et al. 2005b). The separation electrolyte was 24 mM formic acid and 16 mM ammonium carbonate at pH 6.4; recovery ranged between 55% and 110%. For the detection of TSA herbicides in soils, the authors combined a methodology using SPE (using C_{18} cartridges), on-line field-enhanced sample injection (FESI), and coelectroosmotic CE-UV and achieved LODs between 18 and 34 ppb (Borges et al. 2005a).

IV. Environmental Fate and Occurrence

The physicochemical properties and experimentally determined degradation rate constants of the TSA herbicides (see Table 1) indicate their moderate to low persistence in the environment, which is confirmed by available field dissipation studies. For instance, penoxsulam applied to flooded and nonflooded rice plots in Arkansas, California, Italy, and Spain yielded first-order $t_{1/2}$s of 3–7 d (Roberts et al. 2003). Field dissipation of cloransulam-methyl and diclosulam was best characterized by two-compartment modeling, presumably resulting from multiple concurrent degradation mechanisms (photolysis, hydrolysis, and aerobic soil metabolism; van Weesenbeeck et al. 1997; Zabik et al. 2001). Field dissipation rates were determined with initial $t_{1/2}$s ranging from 2.5 to 4.8 d for cloransulam-methyl (van Weesenbeeck et al. 1997) and from 13 to 43 d for diclosulam (Zabik et al. 2001).

Dissipation of flumetsulam in the 0–8 cm soil layer of cornfield soils also followed first-order kinetics with $t_{1/2}$s of about 41 d for crops grown on sandy-loam soil and 30 d for crops grown on loamy-sand soil (Rouchaud et al. 2002a). Flumetsulam degradation $t_{1/2}$s in five Mississippi soils of varying texture (loam, silt loam, and clay loam), organic matter content (1.2%–3.5%), and pH (6–7.6) were ranging between 20 and 46 d (Shaw and Murphy 1997). In this study, flumetsulam persistence decreased with increasing cumulative rainfall and decreasing organic matter content (within a range of 1.2%–3.5%). Soil pH had no measurable effect (Shaw and Murphy 1997). Lehmann et al. (1992) observed similar degradation rates in flumetsulam but found a dependency on both sorption and pH in 21 U.S. soils. They described degradation rates with a first-order model, using organic carbon and pH to estimate sorption from K_{oc} values for the neutral and anionic forms of flumetsulam (Lehmann et al. 1992). At applications rates of 50–100 g a.i. ha^{-1} on corn, flumetsulam residues in soils gave injury syndromes to sensitive cabbage crops planted 1 yr later (O'Sullivan et al. 1999). Also, there were no, or only very light, phytotoxicity symptoms observed in sensitive crops (sugar beet, lettuce) sown 5 mon after application of flumetsulam at a low application rate of 20 g a.i. ha^{-1}. Flumetsulam is used in the U.S. at the rate of 60–80 g a.i. ha^{-1} for the control of broad-leaved weeds and grasses in corn and soybean culture (Rouchaud et al. 2002b).

Soil mobility varies between different TSA herbicides. Residues of penoxsulam in soil were low and transient after application to flooded and nonflooded rice fields (Roberts et al. 2003). Cloransulam-methyl was retained in the soil surface layer (0–15 cm) after application to soybean crops despite strong leaching conditions at test sites (van Weesenbeeck et al. 1997). Modeling results with the PRZM 2 (Pesticide Root Zone Modeling) model suggest that cloransulam-methyl is potentially mobile in finer-texture soils with <1.5% organic matter. Flumetsulam concentrations in three Mississippi soils of varying texture and organic content were primarily limited to the upper 8cm of soil 84d after treatment (Murphy and Shaw 1997). During the cropping period, flumetsulam residues were detectable but low (near the GC-MS method detection limit of 0.3ppb) in the 8–15cm surface layer and were never detected in the 15–20cm layer. The results indicate that flumetsulam strongly adsorbed to soil and that leaching is not a significant dissipation route (Rouchaud et al. 2002a). The reported soil $t_{1/2}$ of flumetsulam ranges from 30 to 60d at pH 6–7 (2%–4% organic carbon; Gerwick and Kleschick 1991).

A runoff study combined with PRZM modeling revealed that less than 6% of diclosulam applied to peanut fields would be lost from severe rainfall, assuming worst case management practices. Of the total diclosulam runoff, 97% was transported off the field by water, with <3% associated with the sediment (van Weesenbeeck et al. 2001). Soil aging effects, resulting in increased adsorption, can contribute to the limited mobility of diclosulam (Yoder et al. 2000; Zabik et al. 2001), cloransulam-methyl (van Weesenbeeck et al. 1997), and florasulam in soils (Krieger et al. 2000a). The leaching potential of cloransulam-methyl to groundwater was modeled using conservative inputs for soil, weather, and application rate. The maximum predicted concentration in groundwater for the most vulnerable soil was 2.7µg/L (ppb), which is more than two orders below the U.S. EPA-defined health level of 1,000ppb (USEPA 1997; van Weesenbeeck and Havens 1999). The same study also suggests no significant risk for nontarget terrestrial plants, based on comparisons of estimated environmental concentration (EEC) and EC_{25} distributions obtained from conservative Tier 2 phytotoxicity studies (van Weesenbeeck and Havens 1999).

So far, few monitoring studies have attempted to determine ambient TSA herbicide concentrations. In 1998, a monitoring study was conducted to determine the occurrence of ALS inhibitors in rivers, reservoirs, and groundwater in the midwestern U.S. Flumetsulam, the only TSA herbicide measured, was detected in 83% of samples collected from 75 surface water (63%) and 25 groundwater sites (12%), at a median concentration of 0.01ppb and a maximum of 0.035ppb. These concentrations are not likely to be toxic to nontarget aquatic plants or of concern for human consumption, but do add to the overall burden of pesticides carried by midwestern rivers (Battaglin et al. 2000).

V. Photodegradation

Photolytic processes are considered important contributors to the degradation of TSA herbicides, both on soil and, particularly, in water (USEPA 1997; van Weesenbeeck et al. 1997; Krieger et al. 2000c; Zabik et al. 2001; USEPA 2004b; Jabusch and Tjeerdema 2006b). Photodegradation studies with florasulam yielded a $t_{1/2}$ of 3.3 d in a natural water system in summer at 51.5°N latitude compared to an estimated $t_{1/2}$ of 36 d based on photolysis experiments in sterile buffered solution. The much faster rate in the natural system suggests an important role for indirect photolytic processes in the overall photodegradation of florasulam (Krieger et al. 2000c). In contrast, degradation studies for penoxsulam in summer sunlight at 39°N yielded no significant differences between the rates determined in sterile buffered solution and rice field water. Degradation $t_{1/2}$s ranged from 3.3 to 3.7 d in field waters of different origin compared to a $t_{1/2}$ of 3.9 d determined in sterile buffered solution, indicating a minor role for indirect photolytic processes in controlling the overall degradation process and rate (Tjeerdema et al. 2005).

Proposed photodegradation pathways have been published for florasulam (Krieger et al. 2000c; Fig. 3) and penoxsulam (Jabusch and Tjeerdema 2006b; Fig. 4). In studies using radiolabeled florasulam in aqueous buffer, only one product (triazolopyrimidine sulfonic acid, TPSA; Fig. 3) was found at greater than 10% of the applied radioactivity (it attained 18% after 32 d; Krieger et al. 2000c). Analogous TPSA degradates were also found as products of aqueous photolysis for diclosulam (Concha et al. 1994) and penoxsulam (Fig. 4, pathway 2; Jabusch and Tjeerdema 2006b). However, the TPSA degradates were not detected in experiments using natural water systems (Krieger et al. 2000c; Jabusch and Tjeerdema 2006b).

Photodegradation of florasulam in natural waters is proposed to proceed by two main routes: (1) formation of 5-hydroxy florasulam by oxidation of the methoxy group on the triazolopyrimidine ring to a hydroxyl group, or (2) loss of the difluorophenyl moiety by cleavage of the phenyl carbon–nitrogen bond, followed by conversion of the methoxy group of the triazolopyrimidine ring to a hydroxyl group (Krieger et al. 2000c). Subsequent degradation of 5-hydroxy florasulam is by either (a) opening of the pyrimidine ring and subsequent loss of the difluorophenyl moiety, or (b) loss of the difluorophenyl moiety first, followed by opening of the pyrimidine ring.

Degradation studies with ^{14}C-labeled penoxsulam in a "merry-go-round" reactor suggest that aqueous photodegradation proceeds via three pathways: (1) cleavage of the sulfonamide bridge, (2) stepwise degradation of the triazolopyrimidine system and its substituents, and (3) photooxidation of the sulfonyl group. Seven major photoproducts were found (maximum levels ≥5% during the course of the study) and six were identified (see Fig. 4). Two of the identified photodegradation products, 2-ATP and TPSA, were present as 1.4% and 8.9%, respectively, of the initially applied radioactivity. However, these compounds are assumed to be either rapidly biodegraded when formed or not formed in significant amounts in natural waters (Krieger et al. 2000c; Jabusch and Tjeerdema 2006b).

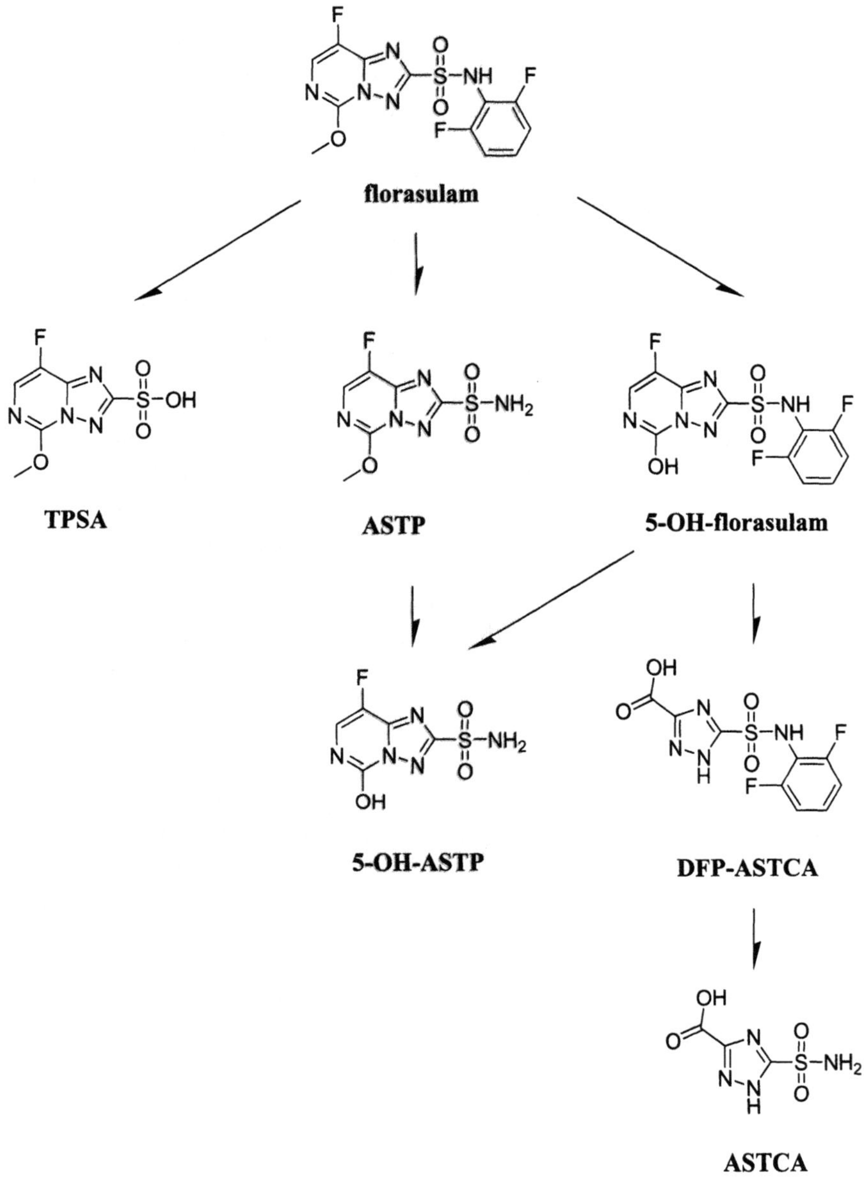

Fig. 3. Suggested photodegradation pathways for florasulam in water (adapted from Krieger et al. 2000c).

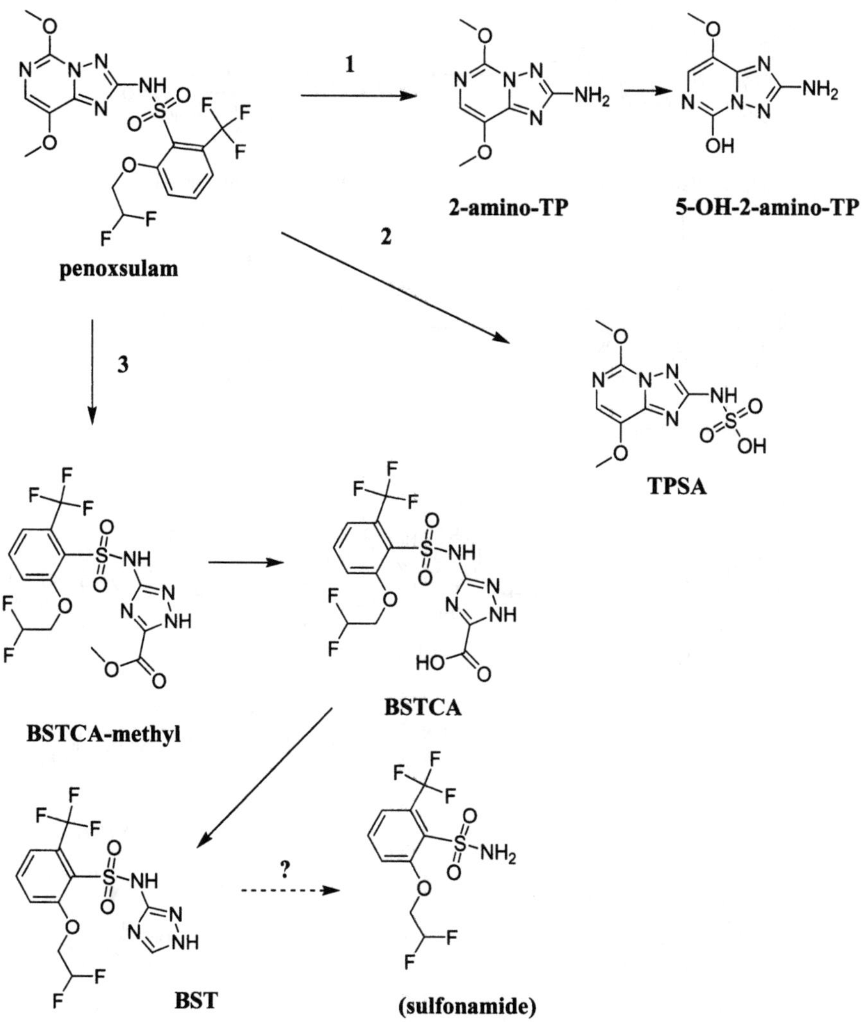

Fig. 4. Suggested aqueous photodegradation pathways for penoxsulam: (1) cleavage of the sulfonamide bridge, (2) photooxidation of the sulfonyl group, and (3) stepwise degradation of the triazolopyrimidine system and its substituents. BSTCA,3-(2-(2,2-difluoroethoxy)-6-(trifluoromethyl)pheuylsulfonamido)-1H-1,2,4-triazole-5-carboxylic acid (Adapted from Jabusch and Tjeerdema 2006b.)

VI. Metabolism in Microorganisms, Plants, and Animals

Aerobic soil metabolism of cloransulam-methyl, florasulam, and diclosulam is shown in Fig. 5. Degradation rates of the TSA herbicides and their degradation products appear to be strongly influenced by temperature, as well as to some degree by soil moisture content (Kleschick et al. 1992; Wolt et al. 1992, 1996; van Weesenbeeck et al. 1997; Cupples et al. 2000; Zabik et al. 2001; Rouchaud et al. 2002b). For example, soil $t_{1/2}$s of flumetsulam

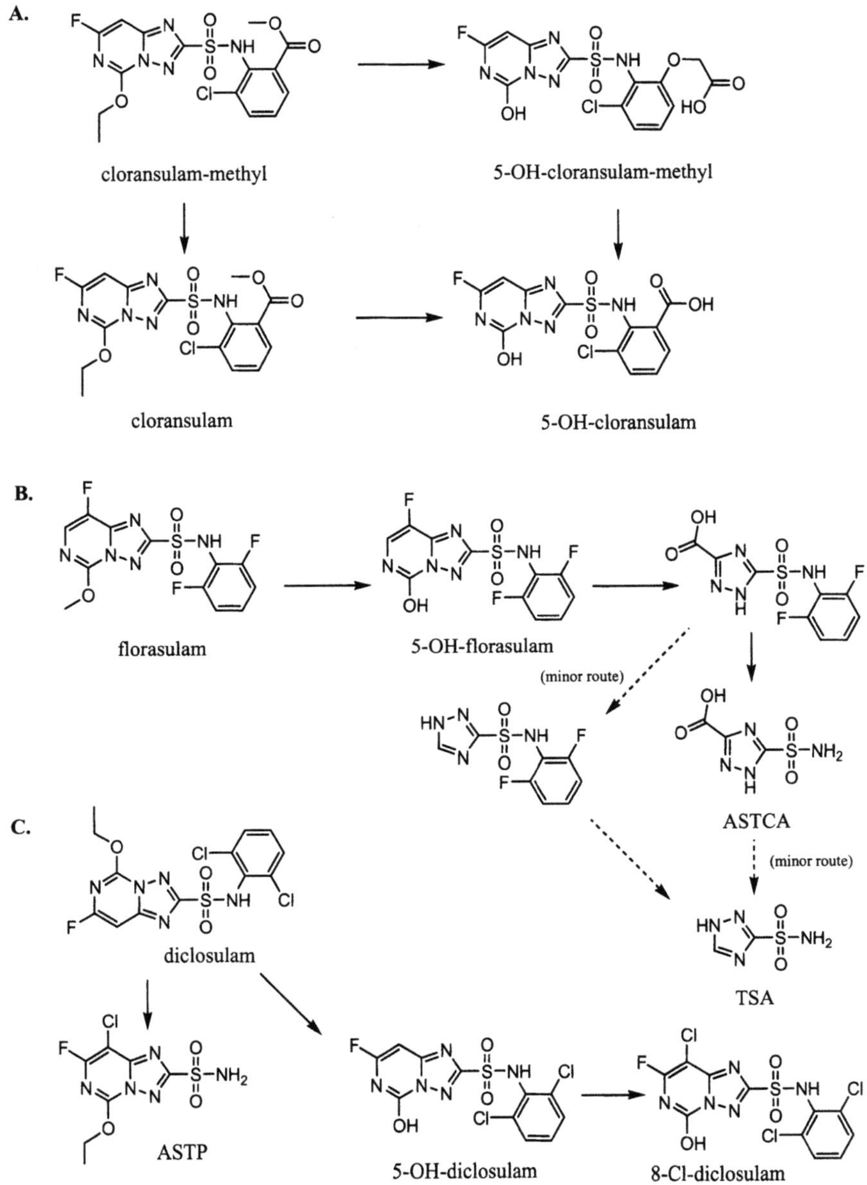

Fig. 5. Aerobic metabolism of triazolopyrimidine sulfoanimide (TSA) herbicides in soil. A. Cloransulam-methyl (adapted from Wolt et al. 1996). B. Florasulam (adapted from Jackson et al. 2000). C. Diclosulam (adapted from Yoder et al. 2000). ASTP, aminosulfouyl triazdopyrimidine.

in laboratory studies (both at 26°C) were found to range from 49 to 246 d, whereas those of florasulam ranged from 1 to 8.5 d at 20°–25°C, and from 6.4 to 85 d at 5°C (Rouchaud et al. 2002b); the $t_{1/2}$ of its major degradation product, 5-hydroxy florasulam, ranged from 8 to 36 d at 20°–25°C and 43 to 78 d at 5°C (Krieger et al. 2000a). Dissipation $t_{1/2}$s of diclosulam ranged from 16 to 54 d in U.S. and South American soils incubated in the dark at 20°C and 45% moisture-holding capacity and at 20°C at 75% 0.3 bar moisture (Yoder et al. 2000). Three soil metabolites reached levels of ≥10% of applied compound (5-hydroxy diclosulam, aminosulfonyl triazolopyrimidine (ASTP), and 8-chloro-5-hydroxy diclosulam). The terminal products were CO_2 and bound soil residues. The K_D values of the metabolites were slightly lower than that for diclosulam, and the sorptivity of both diclosulam and its metabolites were found to increase over time (Yoder et al. 2000).

The degradation kinetics of cloransulam-methyl in tubes simulating aerobic soil systems was observed to follow a biphasic pattern with a fast initial step followed by a slower step. In this study, analyzed metabolites had lower K_Ds and were less phytotoxic than the parent compound (Wolt et al. 1996). However, in a different study the influence of temperature on the mineralization rate (conversion to CO_2) was dependent upon the location of the radiolabel within the cloransulam-methyl molecule, suggesting that distinct groups of microorganisms degrade different parts of the molecule at different temperatures (Cupples et al. 2000). Also, mineralization rates for cloransulam-methyl were found to be moisture dependent, with direction of response (increase versus decrease) depending on soil type (Cupples et al. 2000).

The $t_{1/2}$ of penoxsulam in flasks simulating flooded rice field conditions (anaerobic conditions but open to the atmosphere) ranged from 2 to 13 d; faster degradation rates occurred in flasks with steeper redox gradients (Jabusch and Tjeerdema 2006a). The suggested anaerobic pathway for penoxsulam is shown in Fig. 6. In a strongly anaerobic test system (sulfate-reducing or methanogenic) isolated from the atmosphere, flumetsulam was transformed abiotically to flumetsulam hydrate with a $t_{1/2}$ of 186 d. The transformation was reversed upon aeration, and the $t_{1/2}$ of flumetsulam hydrate was 2 d versus 40 d for flumetsulam (Wolt et al. 1992).

In plants, TSA herbicides are metabolized via *para*-hydroxylation of the aniline ring, followed by conjugation with glucose (Fig. 7; Frear et al. 1993; deBoer et al. 2006). Microsomal cytochrome P450-dependent monooxygenases were identified as the enzymes responsible for flumetsulam hydroxylation (Frear et al. 1993). Metabolism of florasulam by broadleaf weeds was so slow that metabolite characterization was not possible, but comparison of HPLC data suggests hydroxylation as the major pathway (deBoer et al. 2006). There is very little information on the metabolism of TSA herbicides in animals. Lactating goats, when given a daily oral dose of radiolabeled cloransulam-methyl for 5 consecutive d, rapidly excreted most of the compound within 23 hr of receiving the last dose, with 99.9% of the

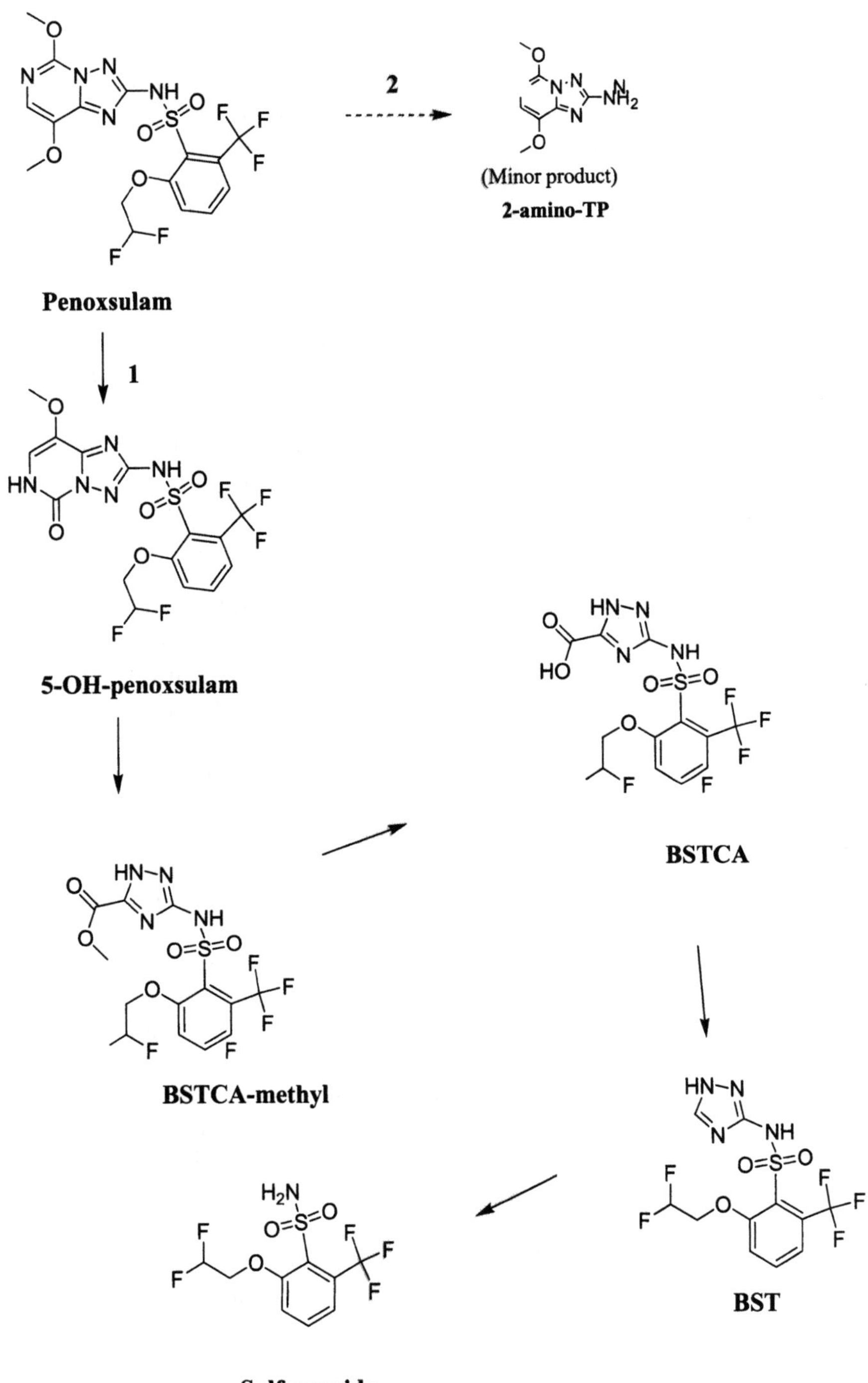

Fig. 6. Suggested anaerobic metabolism of penoxsulam in flooded rice field soils. (Adapted from Jabusch and Tjeerdema 2006a.)

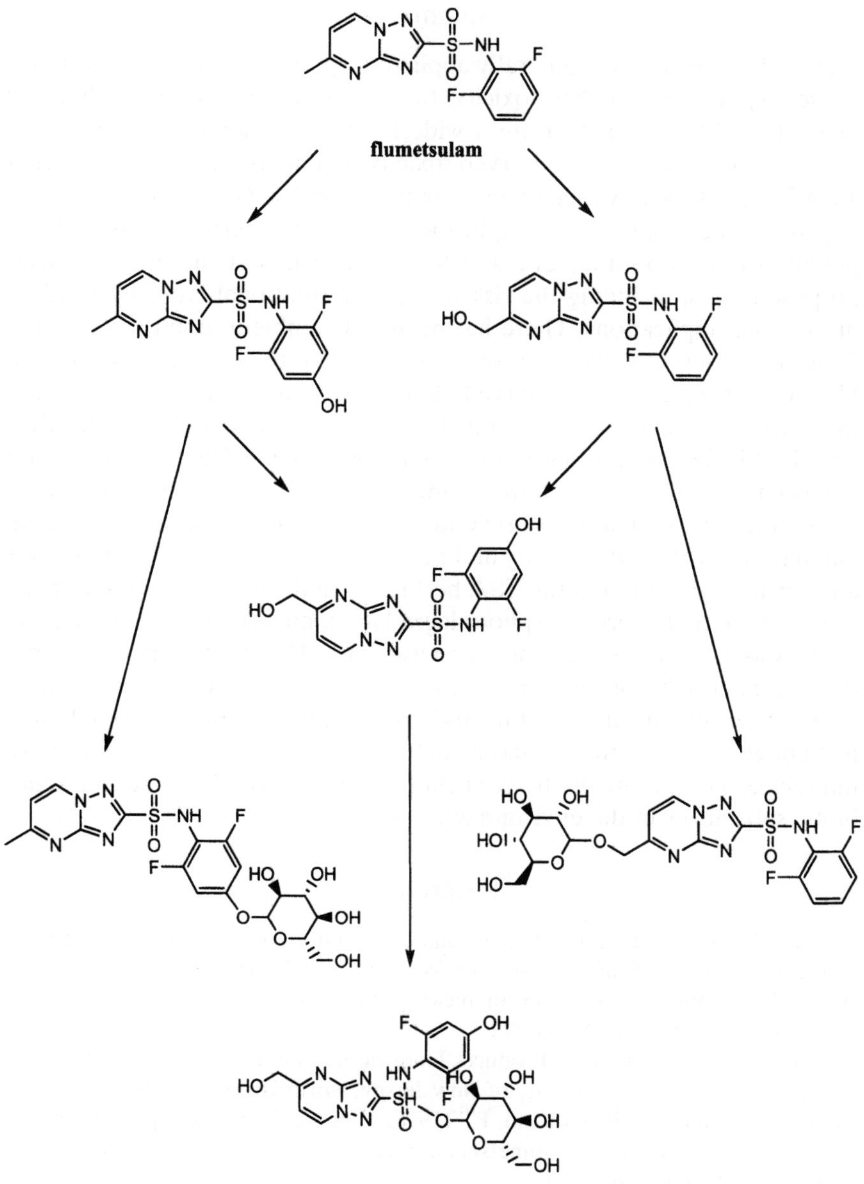

Fig. 7. Suggested metabolism of flumetsulam in tolerant corn, wheat, and barley seedlings by hydroxylation and glucose conjugation pathways (adapted from Frear et al. 1993).

recovered radioactivity appearing in the urine and feces. Only one metabolite, cloransulam, was identified, amounting to 9.5% of total radioactive residue in the liver (Lewer et al. 2000). Cloransulam is the corresponding acid formed by ester hydrolysis of cloransulam-methyl (see Fig. 6A).

Summary

The TSA herbicides are generally applied to agricultural fields at low rates. Possessing low mammalian toxicity, they pose little risk to either humans or wildlife. Plants demonstrate a wide range in sensitivity to these compounds. For example, the observed toxicity of flumetsulam to five different aquatic plant species varied by more than 10,000 fold (Peterson et al. 1994), beginning in the ppb range. Application of the TSA herbicides may pose a potential risk to both aquatic and terrestrial plants at the manufacturer-proposed rates, including the risk of injuring rotational crops by residues of previous applications. There has been little ambient monitoring for the TSA herbicides, but a study of selected low use rate herbicides in Midwest U.S. waters found that observed flumetsulam concentrations were not likely to be toxic to nontarget aquatic plants (Battaglin et al. 2000). Most TSA herbicides possess low mobility but moderate persistence in soils, with persistence increasing at lower temperatures. Environmental half-lives range from less than a day to several months. Penoxsulam, which is designated for use with rice, is highly mobile but not very persistent in both water and soils. Degradation of the TSA herbicides in the environment occurs by a combination of processes (photodegradation, microbial degradation, and hydrolysis). A major issue concerning use of the TSA herbicides is the rapid development of herbicide resistance in target species. There are major data gaps concerning the ambient fate and distribution of these compounds, and it is undetermined whether, and if so, to what extent, TSA herbicides have indirect ecological effects by contributing to the overall burden of xenobiotic substances in the environment.

References

Barnes JW, Oliver LR (2004) Cloransulam absorption, translocation, and efficacy on common broadleaf weed species. Weed Sci 52:634–641.

Barnwell P, Cobb AH (1994) Graminicide antagonisms by broadleaf weed herbicides. Weed Technol 18:763–772.

Baron J (2001) IR-4 New Products/Transitional Solution List—March, 2001. Rutgers, The State University of New Jersey, North Brunswick, NJ.

Baskaran S, Lauren DR, Holland, PT (1996) High-performance liquid chromatographic determination of flumetsulam, a newly developed sulfonamide in soil. J Chromatogr A 746:25–30.

Battaglin WA, Furlong ET, Burkhardt MR, Peter, CJ (2000) Occurrence of sulfonylurea, sulfonamide, imidazolinone, and other herbicides in rivers, reservoirs and ground water in the Midwestern United States, 1998. Sci Total Environ 248: 123–133.

Bernasconi P, Woodworth AR, Rosen BA, Subramanian MV, Siehl DL (1995) A naturally occurring point mutation confers broad range tolerance to herbicides that target acetolactate synthase. J Biol Chem 270(29):17381–17285.

Borges JH, García-Montelongo FJ, Cifuentes A, Rodríguez-Delgado MÁ (2005a) Analysis of triazolopyrimidine herbicides in soils using field-enhanced sample

injection-coelectroosmotic capillary electrophoresis combined with soli-phase extraction. J Chromatogr A 1100:236–242.
Borges JH, García-Montelongo FJ, Cifuentes A, Rodríguez-Delgado MÁ (2005b) Determination of herbicides in mineral and stagnant waters at ng/L levels using capillary electrophoresis and UV detection combined with solid-phase extraction and sample stacking. J Chromatogr A 1070:171–177.
Borges JH, Rodríguez-Delgado MÁ, García-Montelongo FJ, Cifuentes A (2005c) Analysis of pesticides in soy milk combining solid-phase extraction and capillary electrophoresis-mass spectrometry. J Sep Sci 28:948–956.
Boutsalis P, Powles SB (1995) Inheritance and mechanism of resistance to herbicides inhibiting acetolactate synthase in *Sonchus oleraceus* L. Theor Appl Genet 91:242–247.
Concha M, Shepler K, Merritt D (1994) Photodegradation of (^{14}C)-XDE-564 in aqueous buffered solution at pH 5 by natural sunlight. Dow Agro Sciences, Indianapolis, IN.
CambridgeSoft Corporation (2006) ChemFinder.com. http://chemfinder.cambridgesoft.com/.
Cupples AM, Sims GK, Hultgren RP, Hart SE (2000) Effect of soil conditions on the degradation of cloransulam-methyl. J Environ Qual 29(3):786–794.
deBoer GJ, Thornburgh S, Ehr RJ (2006) Uptake, translocation and metabolism of the herbicide florasulam in wheat and broadleaf weeds. Pestic Manag Sci 62:316–324.
Devine M, Eberlein CV (1997) Physiological, biochemical and molecular aspects of herbicide resistance based on altered target sites. In Roe RM, Burton JD, Kuhr RJ (eds) Herbicide Activity: Toxicology, Biochemistry and Molecular Biology. IOS Press, Amsterdam, pp 159–186.
Ferguson GM, Hamill AS, Tardif FJ (2001) ALS inhibitor resistance in populations of Powell amaranth and redroot pigweed. Weed Sci 49:448–453.
Foes MJ, Liu L, Tranel PJ, Wax LM, Stoller EW (1998) A biotype of common waterhemp (*Amaranthus rudis*) resistant to triazine and ALS herbicides. Weed Sci 46(5):514–520.
Frear DS, Swanson HR, Tanaka FS (1993) Metabolism of flumetsulam (DE-498) in wheat, corn, and barley. Pestic Biochem Physiol 45:178–192.
Gerwick BC, Kleschick WA (1991) DE-498: a new broad spectrum herbicide for soybeans and other crops. Weed Sci Abstr 31:28.
Hashem A, Bowran D, Piper T, Dhammu H (2001) Resistance of wild radish (*Raphanus raphanistrum*) to acetolactase synthase-inhibiting herbicides in the Western Australian wheat belt. Weed Technol 15:68–74.
Jabusch TW, Tjeerdema RS (2005) Partitioning of penoxsulam, a new sulfonamide herbicide. J Agric Food Chem 53(18):7179–7183.
Jabusch TW, Tjeerdema RS (2006a) Microbial degradation of penoxsulam in flooded rice soils. J Agric Food Chem 54:5962–5967.
Jabusch TW, Tjeerdema RS (2006b) Photodegradation of penoxsulam. J Agric Food Chem 54:5958–5961.
Jackson R, Ghosh D, Paterson G (2000) The soil degradation of the herbicide florasulam. J Agric Food Chem 56:1065–1072.
Kemp MS, Moss SR, Thomas TH (1990) Herbicide resistance in *Alopecurus myosuroides*. In: Green MR, LeBaron HM, Moberg WM (eds) Managing Resistance to Agrochemicals: From Fundamental Research to Practical Strategies. American Chemical Society, Washington, DC.

Kleschick WA, Gerwick BC (1989) The chemistry and biochemistry of triazolopyrimidinesulfonanilide herbicides, a new class of acetolactate synthase inhibitors. In: Copping L, Dodge AD, Dlaziel J (eds) Prospects for Amino Acid Biosynthesis Inhibitors in Crop Protection and Pharmaceutical Chemistry. Lavenham Press, Lavenham, UK, pp 133–145.
Kleschick WA, Gerwick BC, Carson CM, Monte WT, Snider SW (1992) DE-498, a new acetolactate synthase inhibiting herbicide with multicrop selectivity. J Agric Food Chem 40:1083–1085.
Krieger MS, Pillar F, Ostrander JA (2000a) Effect of temperature and moisture on the degradation and sorption of florasulam and 5-hydroxyflorasulam in soil. J Agric Food Chem 48:4757–4766.
Krieger MS, Wynn JL, Yoder RN (2000b) Extraction of cloransulam-methyl from soil with subcritical water and supercritical CO_2. J Chromatogr A 897:405–413.
Krieger MS, Yoder RN, Gibson R (2000c) Photolytic degradation of florasulam on soil and in water. J Agric Food Chem 48:3710–3717.
Krynitsky AJ (1997) Determination of sulfonylurea herbicides in water by capillary electrophoresis and by liquid chromatography/mass spectrometry. J AOAC Int 80(2):392–400.
Laganà A, Fago G, Marino A, Penazzi VM (2000) Liquid chromatography mass spectrometry tandem for multiresidue determination of selected post-emergence herbicides after soil column extraction. Anal Chim Acta 415:41–56.
Landstein D, Arad S, Barak Z, Chipman DM (1993) relationships among the herbicide and functional sites of acetohydroxy acid synthase from *Chlorella emersonii*. Planta (Berl) 191:1–6.
Lehmann RG, Miller JR, Fontaine DD, Laskowski DA, Hunter JH, Cordes RC (1992) Degradation of a sulfonamide herbicide as a function of soil sorption. Weed Res 32:197–205.
Lewer P, Finney-Brink KL, Duebelbeis DO (2000) Nature of the residue of [^{14}C]cloransulam-methyl in lactating goats. J Agric Food Chem 48:2532–2546.
Madafiglio GP, Medd RW, Cornisch PS, Van de Ven R (2000) Temperature-mediated responses of flumetsulam and metosulam on *Raphanum raphanistrum*. Weed Res 40:387–395.
Maycock R, Hastings M, Portwood D (1995) Ultra-trace analysis of metosulam, a new triazolopyrimidine herbicide. Int J Environ Anal Chem 58:1–4.
Mourad G, King J (1992) Effect of four classes of herbicides on growth and acetolactase-synthase activity in several variants of *Arabidopsis thaliana*. Planta (Berl) 188:491–497.
Murphy GP, Shaw DR (1997) Field mobility of flumetsulam in three Mississippi soils. Weed Sci 45(4):564–567.
O' Sullivan J, Thomas RJ, Bouw WJ (1999) Effect of flumetsulam and clopyralid soil residues on several vegetable crops and on sweet corn cultivars grown in rotation. Weed Technol 13:303–307.
PAN (2006) PesticideInfo Version 7.1. http://www.pesticideinfo.org/Index.html.
Parnell JS, Hall JC (1998) Development of an enzyme-linked immunosorbent assay for the detection of metosulam. J Agric Food Chem 46:152–156.
Patzoldt WL, Tranel PJ (2002) Molecular analysis of cloransulam resistance in a population of giant ragweed. Weed Sci 50:299–305.
Patzoldt WL, Tranel PJ, Alexander AL, Schmitzer PR (2001) A common ragweed population resistant to cloransulam-methyl. Weed Sci 49:485–490.

Peterson HG, Boutin C, Martin PA, Freemark KE, Ruecker NJ, Moody MJ (1994) Aquatic phyto-toxicity of 23 pesticides applied at expected environmental concentrations. Aquat Toxicol 28:275–292.

Renew JE, Huang C-H (2004) Simultaneous determination of fluoroquinolone, sulfonamide, and trimethoprim antibiotics in wastewater using tandem solid phase extraction and liquid chromatography–electrospray mass spectrometry. J Chromatogr A 1042:113–121.

Roberts DW, Knuteson JA, Jackson R (2003) The dissipation of penoxsulam in flooded rice fields. In: Pesticides in Air, Plant, Soil & Water Systems: XII Symposium Pesticide Chemistry, Piacenza, Italy, pp 349–357.

Rouchaud J, Neus O, Eelen H, Bulcke R (2002a) Dissipation and mobility of flumetsulam in the soil of corn crops. Meded Fac Landbouwwet Rijksuniv Gent 67(3):401–407.

Rouchaud J, Neus O, Eelen H, Bulcke R (2002b) Soil persistence and mobility in corn fields of flumetsulam applied at low doses. Bull Environ Contam Toxicol 69:785–791.

Rubin B (1996) Herbicide-resistant weeds—the inevitable phenomenon: mechanisms, distribution and significance. Z Pflanzenkr Pflanzenschutz XV:17–32.

Saari LL, Cotterman JC, Thill DC (1994) Resistance to acetolactase-inhibiting herbicides. In: Powles SB, Holtum JAM (eds) Plants: Biology and Biochemistry. Lewis, Boca Raton, FL, pp 88–139.

Schmitzer PR, Eilers RJ, Cséke C (1993) Lack of cross-resistance of imazaquin-resistant *Xanthium strumarium* acetolactase synthase to flumetsulam and chlorimuron. Plant Physiol 103:281–283.

Shackelford DD, Duebelbeis DO, Snell BE (1996) Determination of residues of cloransulam-methyl in soybeans and soybean forage, hay, and processed commodities by capillary gas chromatography with mass spectrometric detection. J Agric Food Chem 44:3570–3575.

Shaner DL, Singh BK (1997) Acetohydroxyacid synthase inhibitors. In: Roe RM, Burton JD, Kuhr RJ (eds) Herbicide Activity: Toxicology, Biochemistry and Molecular Biology. IOS Press, Amsterdam, pp 69–110.

Shaw DR, Murphy GP (1997) Field persistence of bioavailable flumetsulam. Weed Sci 45(4):568–572.

Sibony M, Rubin B (2003) Molecular basis for multiple resistance to acetolactate synthase-inhibiting herbicides and atrazine in *Amaranthus blithoides* (prostrate pigweed). Planta (Berl) 216:1022–1027.

Sprague CL, Stoller EW, Wax LM (1997) Response of an acetolactase synthase (ALS)-resistant biotype of *Amaranthus rudis* to selected ALS inhibiting and alternative herbicides. Weed Res 37:93–101.

Subramanian MV, Loney V, Pao L (1989) Mechanism of action of 1,2,4-triazolo[1,5-a]pyrimidine sulfonamide herbicides. In: Copping LG, Dalziel J, Dodge AD (eds) Prospects for Amino Acid Biosynthesis Inhibitors in Crop Protection and Pharmaceutical Chemistry. Society of Chemical Industry, Farnham, UK, pp 97–100.

Tjeerdema RS, Hill J, Fisher A, TenBrook P, Gunasekara A, Jabusch TW (2005) Project No. RP-5: The Environmental Fate of Pesticides Important to Rice Culture. Report to the California Rice Research Board. Department of Environmental Toxicology, College of Agricultural and Environmental Sciences, University of California, Davis, CA.

USEPA (1997) Pesticide Fact Sheet: Cloransulam. USEPA Office of Prevention, Pesticides and Toxic Substances, Washington, DC.

USEPA (2004a) Penoxsulam, 2-(2,2-difluoroethoxy)-*N*-(5,8-dimethoxy[1,2,4] triazolo[1,5-*c*]pyrimidin-2-yl)-6-(trifluoromethyl)benzenesulfonamide; pesticide tolerance. Fed Reg 69(185):57188–57197.

USEPA (2004b) Pesticide Fact Sheet: Penoxsulam. USEPA Office of Prevention, Pesticides and Toxic Substances, Washington, DC.

USEPA (2006) Pesticide products; registration applicants. Fed Reg 71(192): 58603–58604.

van Weesenbeeck IJ, Havens PL (1999) A groundwater exposure assessment of cloransulam-methyl in the U.S. soybean market. J Environ Qual 28(2): 513–522.

van Weesenbeeck IJ, Zabik JM, Wolt JD, Bormett GA, Roberts DW (1997) Field dissipation of cloransulam-methyl at four sites in the U.S. soybean market. J Agric Food Chem 45:3299–3307.

van Weesenbeeck IJ, Peacock AL, Havens PL (2001) Measurement and modeling of diclosulam runoff under the influence of simulated severe rainfall. J Environ Qual 30:553–560.

Ware GW, Whitacre DM (2004) The Pesticide Book. Meister Pro Information Resources, Willoughby, OH.

Wauchope RD, Yeh SJ, Linders BHJ, Kloskowski R, Tanaka K, Rubin B, Katayama A, Kördel W, Gerstl Z, Lane M, Unsworth JB (2002) Pesticide soil sorption parameters: theory, measurement, uses, limitations and reliability. Pestic Manag Sci 58:419–445.

Whaley CM, Wilson HP, Westwood JH (2006) ALS resistance in several smooth pigweed (*Amaranthus hybridus*) biotypes. Weed Sci 54:828–832.

Whitcomb CE (1999) An introduction to ALS-inhibiting herbicides. Toxicol Ind Health 15:232–240.

Wolt JD, Schwake JD, Batzer FR, Brown SM, McKendry LH, Miller JR, Roth GA, Stanga MA, Portwood D, Holbrook DL (1992) Anaerobic aquatic degradation of flumetsulam [*N*-2,6-difluorophenyl)-5-methyl[1,2,4]triazolo[1,5-*a*]pyrimidine-2-sulfonamide. J Agric Food Chem 40:2302–2308.

Wolt JD, Smith JK, Sims GK, Duebelbeis DO (1996) Products and kinetics of cloransulam-methyl aerobic soil metabolism. J Agric Food Chem 44:324–332.

Wood A (2006) Compendium of common pesticide names. http://www.alanwood.net/pesticides/index.html.

Yoder RN, Huskin MA, Kennard LM, Zabik JM (2000) Aerobic metabolism of diclosulam on U.S. and South American Soils. J Agric Food Chem 48:4335–4340.

Zabik JM, van Weesenbeeck IJ, Peacock AL, Kennard LM, Roberts DW (2001) Terrestrial field dissipation of diclosulam at four sites in the United States. J Agric Food Chem 49:3284–3290.

Zheng D, Patzoldt WL, Tranel PJ (2005) Association of the W574L ALS substitution with resistance to cloransulam and imazamox in common ragweed (*Ambrosia artemisiifolia*). Weed Sci 53:424–430.

Manuscript received April 12; accepted April 18, 2007.

Rev Environ Contam Toxicol 193:53–210

Parameters for Carbamate Pesticide QSAR and PBPK/PD Models for Human Risk Assessment

James B. Knaak, Curt C. Dary, Miles S. Okino, Fred W. Power, Xiaofei Zhang, Carol B. Thompson, R. Tornero-Velez, and Jerry N. Blancato

Contents

Communicated by James B. Knaak.

J.B. Knaak
Department of Pharmacology and Toxicology, School of Medicine and Biomedical Sciences, SUNY at Buffalo, 3435 Main Street, Buffalo, NY 14214, USA.

C.C. Dary · M.S. Okino · F.W. Power
U.S. Environmental Protection Agency, Human Exposure and Atmospheric Sciences Division, 944 E Harmon Ave, Las Vegas, NV 89193-3478, USA.

X. Zhang · C.B. Thompson
General Dynamics Information Technology, 181 N Arroyo Grande Blvd, Suite 105, Henderson, NV 89074-1624, USA.

R. Tornero-Velez · J.N. Blancato
U.S. Environmental Protection Agency, National Exposure Research Laboratory, Mail Code: D305-01 Research Triangle Park, NC 27711, USA.

I. Introduction

Our interest in providing parameters for the development of quantitative structure physiologically based pharmacokinetic/pharmacodynamic (QSPBPK/PD) models for assessing health risks to carbamates (USEPA 2005) comes from earlier work with organophosphorus (OP) insecticides (Knaak et al. 2004). Parameters specific to each carbamate are needed in the construction of PBPK/PD models along with their metabolic pathways. Parameters may be obtained by (1) development of QSAR models, (2) collecting pharmacokinetic data, and (3) determining pharmacokinetic parameters by fitting to experimental data. The biological parameters are given in Table 1 (Blancato et al. 2000).

The quantitative structure-activity relationship (QSAR) model approach requires (1) knowledge of the numerical values of the parameters of related chemicals and (2) specification of an equation for conducting the multiple regression analysis (Fouchecourt et al. 2001). TOPKAT (Accelrys Software Inc., San Diego, CA) is an excellent example of QSAR model development. The program consists of several toxicity QSAR models; one provides information on the acute oral toxicity of OP and carbamate insecticides. QSAR models predicting V_{max} and K_m values for human liver microsomes (HLMs) and CYP3A4 catalyzed hydroxylation reactions involving drug-based aromatic and alicyclic rings and aliphatic groups were recently developed by Enslein et al. (2007). Liu et al. (2005) developed a QSAR model for predicting tissue/blood partition coefficients, and Poulin and Theil (2002a)

Table 1. Biological Parameters Required for Carbamate Pesticide Physiologically Based Pharmacokinetic/Pharmacodynamic (PBPK/PD) Models.[a]

Biological process	Compartment	Parameter	Units
Absorption	GI tract		hr^{-1}
	Respiratory tract	Blood:air PC	Unitless
	Skin	Permeability constant	$cm\,hr^{-1}$
Distribution	Desired compartments	Blood:tissue PC	Unitless
Metabolism, P450 cytochromes, etc.	Liver	V^{b}_{max}	$\mu mol\,hr^{-1}\,kg^{-1}\,bw^{c}$
		K^{d}_{m}	$\mu mol\,kg^{-1}$
Excretion	Kidney	V^{b}_{max}	$\mu mol\,hr^{-1}\,kg^{-1}\,bw^{c}$
		K^{d}_{m}	$\mu mol\,kg^{-1}$
Toxicity	Blood, brain, liver	B esterase inhibition	$M^{-1}\,min^{-1}$
		Recovery	hr^{-1}
		Synthesis	hr^{-1}

[a]*Source*: Blancato et al. (2000).
[b]Maximum velocity.
[c]bw, body weight.
[d]Substrate concentration at half-maximum velocity.

used a mechanistic approach to predict tissue/blood partition coefficients. The Poulin and Theil (2002a) model was reexamined using log $D_{pH\,7.4}$ values and was used to estimate partition coefficients for the carbamates and their metabolites.

A search of the literature for metabolism/pharmacokinetic data turned up three carbamate reviews, written by Cool and Jankowski (1985), Kuhr and Dorough (1976), and Roberts and Hutson (1999). The reviews considered chemical degradation, degradation in soils, and metabolism in plants and animals, with metabolism in insects being covered by Kuhr and Dorough (1976). Data in these reviews and others provided information for metabolic flow charts or tables suitable for use in the development of 10 preliminary carbamate PBPK/PD models using the Advanced Continuous Simulation Language (ACSL) (The AEgis Technologies Group, Huntsville, AL).

A PBPK/PD model developed for isofenphos (CAS no. 25311-71-1) (Knaak et al. 2002) was used as a template for the development of models for aldicarb, carbaryl, carbofuran, formetanate, methiocarb, methomyl, oxamyl, pirimicarb, propoxur, and thiodicarb. Bimolecular rate constants, k_i ($pM^{-1}\,hr^{-1}$), were found describing the inhibition of the B-esterases [acetylcholinesterase (AChE), butyrylcholinesterase (BChE), and carboxylesterase (CaE)] by the carbamates or their toxic metabolites. The relationship between the inhibitory nature of the carbamates, structure of the target enzymes (e.g., B-esterases), location of the B-esterases in tissues, and the number of active sites/kg of tissue was reexamined using rate constants for substrate hydrolysis (e.g., k_{cat}). Metabolic rate constants, V_{max} and K_m, for the cytochrome P450 (CYP450) hydroxylation of ring compounds, their conjugation to glucuronic and sulfuric acid, and the hydrolysis of aryl and oxime carbamates by hydrolases were obtained, if available, from the literature.

No attempt was made in this review to obtain pharmacokinetic parameters by fitting the output of preliminary ACSL-based PBPK models to experimental data (e.g., absorption, distribution, metabolism, elimination, and toxicity data). This approach was considered to be outside the scope of the review. However, the metabolic pathways, molecular weights, tissue partition coefficients, bimolecular inhibition rates for AChE, ChE, CaE, the number of active sites in tissues, metabolic V_{max}, K_m values, and other parameters collected under approaches 1 and 2 were used by Zhang et al. (2006, 2007), Okino et al. (2005a,b), and Power et al. (2005) with experimental data and the Exposure Related Dose Estimating Model (ERDEM; Blancato et al., 2006) to model carbofuran, carbaryl, and aldicarb.

Under the Food Quality Protection Act (FQPA) of 1996, the U.S. Environmental Protection Agency (EPA) must consider available information concerning the cumulative effects on human health resulting from exposure to multiple chemicals that have a common mechanism of action, such as the *N*-methyl carbamates. ERDEM is capable of handling AMET (absorption, metabolism, elimination, and toxicity) of three or more carbamate pesticides depending upon the total number of metabolites being modeled. The upper limit of the current model (version 5.1) is 36 chemicals, which includes parent pesticides and or metabolites (Miles Okino; U.S. EPA, Las Vegas, NV).

II. Nature of Carbamate Insecticides

Carbamate insecticides are biologically active AChE inhibitors with the general formula $ROC(=O)NHCH_3$ for N-methylcarbamates and $ROC(=O)N(CH_3)_2$ for dimethylcarbamates. The structure of the R group, alkylphenyl, alkoxyphenyl, halogen phenyl, double ring and oxime, *N,N*-dimethyl, *N*-methyl, *N*-acyl, thiol (SC=O), thiono (OC=S), or dithio (SC=S), and its effects on AChE activity and toxicity, are briefly discussed in the following section, "Discovery". QSAR modeling played a very important role in relating the structure of carbamates to their cholinesterase (ChE) inhibitory properties and insect toxicity under the World Health Organization (WHO) Programme for the evaluation and testing of new insecticides (Wright 1971). This program led to worldwide synthesis, testing (adult mosquito, larval mosquito, and housefly), and the development of structurally related carbamates to control vector-borne diseases with the Departments of Entomology and Zoology, University of Illinois, Urbana-Champaign, IL, playing a key role.

A. Discovery

Physostigmine

The first known carbamate, physostigmine or eserine (CAS no. 57-47-6), was discovered in extracts of the calabar bean. The alkaloid was found to be miotic. Robertson (1863) suggested the drug for use in ophthalmology. "Alkaloids are compounds of nitrogen, capable of uniting with acids, like ammonia, and forming with them definite combinations, which are true salts" (de Coninck 1886). The work of Stedman and Barger (1925) led to the elucidation of the structure of physostigmine as the methylcarbamate of a substituted indole derivative. Physostigmine appears to be the only natural product within the *N*-methylcarbamate group and is unstable in water. The corresponding dimethylcarbamate (neostigmine, CAS no. 59-99-4) was prepared by Aeschlimann and Reinert (1931) and was found to be more stable and an effective miotic agent. Stemple and Aeschlimann (1956) reviewed the properties of these compounds. Somani and Khalique (1986) studied the distribution and pharmacokinetics of physostigmine in rats after intramuscular administration. The time course of BChE activity and plasma concentration showed that the maximum enzymatic inhibition (47%) occurred about the same time (7 min) as the peak plasma concentration ($583\,ng\,mL^{-1}$ at 5 min) with enzymatic activity recovering to 81% at 2 hr and 100% within 24 hr.

Dimethylcarbamates

Gysin (1954), at the Geigy Corporation, studied the insecticidal properties of the dimethylcarbamates of 5,5-dimethyldihydroresorcinol (Dimetan, CAS no. 122-15-6) and 1-iso-propyl-3-methyl pyrazole (Isolan, CAS no. 119-38-0). These two compounds, along with dimetilan (1-(dimethylcarbamoyl)-5-methyl-3-pyrazolyl dimethylcarbamate, CAS no. 644-64-4), underwent considerable evaluation as commercial insecticides. All these carbamates are potent anticholinesterase (antiChE) agents having I_{50} values for AChE in fruit flies of $5–6 \times 10^{-7}$ M for Dimetan to $7–10 \times 10^{-8}$ M for Isolan. The rate of hydrolysis of *p*-nitrophenyl methylcarbamate ($k_b = 5.4 \times 10^5\,M^{-1}\,min^{-1}$) is about 10^7 times greater than that of the corresponding dimethylcarbamate ($k_b = 3.4 \times 10^{-2}\,M^{-1}\,min^{-1}$). The dimethylcarbamates are more stable than the methyl carbamates; however, as antiChEs and insect toxicants, the methylcarbamates are far superior (Metcalf and Fukuto 1962).

Phenylmethylcarbamates

Metcalf and March (1950) were among the first to synthesize and study the activity of simple phenyl methylcarbamates. Substitution in the ortho or meta position with an alkyl group or halogen on the phenyl ring increased insecticidal activity. Maximum activity was found with m-tert-butylphenyl methylcarbamate. The mammalian/insect toxicity ratios ranged from >100 for butacarb (CAS no. 2655-19-8) to around 0.2 for aldicarb (CAS no. 116-06-3). Increasing the chain length of *N*-alkyl groups exponentially increased the recovery half-lives of inhibited insect AChE and decreased their inhibitory potential when going from *N*-methyl to *N*-ethyl groups (Yu et al. 1972a). The activity of alkoxyphenyl carbamates is governed by the same set of rules that apply to the alkylphenyl carbamates. AntiChE and toxicity for insects increase progressively with the size of the phenyl substituent, with maximum activity at the isopropyl and sec-butyl groups. The ortho positions are preferred over those of the meta positions. Goldblum et al. (1981) developed a QSAR for a set of 269 phenyl *N*-methylcarbamates inhibiting fly head AChE (pI_{50} values). The data were taken from the extensive studies of Metcalf and Fukuto (1962), Metcalf and Fukuto (1965a), Metcalf and Fukuto (1965b), Metcalf and Fukuto (1967), and Metcalf et al. (1965), which followed the earlier work of Metcalf and March (1950).

Halogen-substituted phenyl-, thio-, and nitrophenylcarbamates

The only halogen-substituted carbamate developed as a commercial insecticide was o-chlorophenyl methylcarbamate, which was marketed in both Japan and the United States (US). The electron-withdrawing properties of these carbamates produce a substantial degree of hydrolytic instability.

According to Metcalf (1971), none of the thiocarbamates approach the antiChE and insecticidal activity of the corresponding carbamate, indicating that these compounds are very different from the phosphorothiolates [e.g., Gutoxon (CAS no. 961-22-8), Amiton (CAS no. 78-53-5), etc.]. Replacing the carbamate linkage (OC (O) N) with thiol (SC=O), thiono (OC=S), or dithio (SC=S) reduced the antiChE activity of 2-isopropylphenyl *N,N*-dimethylcarbamate by 12 to 40 times (Petrovskii 1970).

Fifty-nine new *N*-methyl and *N,N*-dimethylcarbamates, meta-acylamidophenyl and meta-thioureidophenyl, were synthesized and tested as inhibitors of fly and bovine ChEs (Sacher and Olin 1972). Structure–activity correlation indicated that maximal antiChE activity was associated with a definite size of the alkyl substituent on the amide or thiourea grouping, with four to six carbons producing the most potent inhibitors. In spite of their good antiChE properties, most of these compounds were not highly effective as housefly insecticides.

A substantial number of nitrophenyl methylcarbamates and dimethylcarbamates were examined for pesticidal activity; none were outstanding either as an antiChE agent or as a toxicant (Fukuto et al. 1967). The *p*-nitrophenyl carbamates were not as toxic to insects as the *p*-nitrophenyl dialkylphosphates.

Multiple ring carbamates

The structure–activity relationships of multiring carbamates are of particular interest because of the significance of 1-naphthyl methylcarbamate (carbaryl) as an insecticide. Double-ring carbamates are merely extensions of doubly substituted phenyl methylcarbamates. Numerous other 2,3-ring additions exist, such as carbofuran, 2,3-dihydro-2,2-dimethyl-7-benzofuranyl methylcarbamate. This compound represents a spatial analogue of *o*-isopropoxyphenyl methylcarbamate (propoxur). Carbofuran is one of the most effective of all the carbamate insecticides and has one of the lowest LD_{50} values in the rat. The 3-OH and 3-C=O derivatives of carbofuran show greatly reduced affinity toward fly ChE, whereas the 3-$COCH_3$, 4-Cl, and 4-CH_3 derivatives are essentially unaffected (Metcalf et al. 1968). Substitution of 4- and 5-OH, on either the homonuclear ring or the heteronuclear ring of carbaryl, has little effect on the affinity for AChE, but it generally leads to a substantial reduction in toxicity for insects (Metcalf 1971).

N-*acylated carbamates*

Fraser et al. (1968) substituted various acyl groups for the proton on the *N*-methylcarbamates, producing *N*-acyl carbamates that retained high insecticidal activity and substantially reduced mammalian toxicity. Metabolism is believed to involve deacylation to the parent carbamate with a corresponding increase in antiChE activity (Fahmy et al. 1966). In general, the *N*-acyl derivatives are generally quite biologically unstable, although Sullivan et al. (1967) acetylated carbaryl as a means of increasing its stability during gas chromatography. Acetylation is believed to aid penetration to the site of action in the insect, where hydrolysis to the *N*-methylcarbamate releases the parent carbamate. No information was available concerning the metabolism of *N*-acyl-*N*-methylcarbamates in mammals. A reduction in toxicity may well be related to aromatic ring hydroxylation or hydrolysis of the carbamate over that of the acyl group.

Oxime carbamates

The structure–activity relationships of the *N*-methylcarbamoyl oximes were reviewed by Felton (1968), Fukuto et al. (1969), and Payne et al. (1966). The structural rigidity associated with the C=N bond is a unique feature of the oxime carbamates. These carbamates form pairs of syn (Z isomer) and anti (E isomer) isomers in which the oximino oxygen is either cis or trans to the aldehyde H. The syn isomer of methomyl has 100 times the antiChE activity and is 10 times more toxic to the housefly than the anti isomer. Aldicarb is believed to exist as the syn isomer (Weiden 1968), and it appears that this cis configuration allows greater access of the carbonyl carbon to the esteratic site and the methylthio group to the anionic site. Methomyl is applied as a foliar spray, whereas the more toxic and environmentally stable oxime carbamate, aldicarb, is applied to soil as a plant systemic.

B. Carbamates and Their Toxicity

Carbamate pesticides belong to four chemical classes: oxime-*N*-methylcarbamates, aryl *N*-methyl carbamates, *N*-phenyl carbamates, and methyl esters of substituted carbamic acids. Table 2 lists the common name, chemical class, nominal mass, and CAS number of the 10 carbamates of interest. Appendix A gives the chemical structures, physical parameters, and tissue:partition coefficients of parent carbamates and metabolites. Appendix B gives the metabolic pathways and preliminary metabolic rate constants (V_{max}, K_m) for the metabolism of parent carbamates and metabolites.

The toxicological data required by government agencies (e.g., USEPA and California Department of Pesticide Regulation) to register carbamate and other pesticides are extensive. A review of the registration requirements under the Federal Insecticide, Fungicide and Rodenticide Act (FIFRA) and California Department of Food and Agriculture (CDFA) regulations was made by Knaak et al. (1993). Acute toxicity studies provide information needed to run PBPK/PD models and carry out the required studies under FIFRA. Animal studies involving metabolism and ChE inhibition are considered under Special Toxicology in the FIFRA data requirements.

Acute oral toxicity

The acute oral toxicity of the 10 carbamates was taken from a number of sources, as indicated in Table 3. The most toxic carbamate is aldicarb, followed by carbofuran. Modelers should be aware of the LD_{50} of the carbamates and not exceed them in a modeling exercise. Enslein (1988) and Enslein et al. (1998) developed TOPKAT, a QSAR program for predicting

Table 2. General Information on the Carbamate Pesticides.

Common name	*N*-Methyl carbamate class	Molecular weight	CAS no.	Tables (in Appendices A and B)
Aldicarb	Oxime	190.26	116-06-3	14, 24
Carbaryl	Aryl	201.20	63-25-2	15, 25
Carbofuran	Aryl	221.25	1563-66-2	16, 26
Formetanate	Aryl	221.26	22259-30-9	17, 27
Methiocarb	Aryl	225.31	2032-65-7	18, 28
Methomyl	Oxime	162.21	16752-77-5	19, 29
Oxamyl	Oxime	219.26	23135-22-0	20, 30
Pirimicarb	Aryl	238.29	23103-98-2	21, 31
Propoxur	Aryl	209.24	114-26-1	22, 32
Thiodicarb	Oxime	354.47	59669-26-0	23, 33

Table 3. Acute Oral Toxicity (LD_{50}) of Carbamate Pesticides.

Carbamate	Rat, male $mg kg^{-1} bw$	Rat, female $mg kg^{-1} bw$	Mouse, male $mg kg^{-1} bw$	Mouse, female $mg kg^{-1} bw$
Aldicarb[a]	0.65[j]		0.3	
Carbaryl[b]	230[j]		128	
Carbofuran[a]	5.0 6.5[j]		2.0[k]	
Formetanate[c]	21[j]		18	
Methomyl[d]	17[j]	24	10	
Methiocarb[e]	14.0 10.0[j]	16.0		
Oxamyl[f]	3.1 2.5[j]	2.5	3.3	2.3
Pirimicarb[g]	33.1, 147 100[j]	68 to 221		107
Propoxur[h]	95 to 104 70[k]	86		
Thiodicarb[i]	50 to100 160[k]			

[a]Lewis (1996).
[b]Matsumura (1985).
[c]Worthing (1979).
[d]Tomlin (1994).
[e]USEPA (1994) RED, Methiocarb.
[f]Kennedy (1986).
[g]Gagua (1976); Hayes and Laws (1991).
[h]Verschueren (1983).
[i]WHO/FAO (2000).
[j]TOPKAT, Accelrys Software Inc., San Diego, CA.
[k]Fahmy et al. (1970).

the toxicity of chemicals including those of the carbamates (e.g., in Table 3). The acute oral LD_{50} values listed in Table 3 are in the TOPKAT training set. The LD_{50} model in TOPKAT consists of several class-specific models, with the class determined by structures, not use. There are 12–15 separate LD_{50} models. The training set has about 2,000 compounds distributed across the models. The toxicity data were obtained from RTECS (Registry of Toxic Effects of Chemical Substances, NIOSH, CDC, Atlanta, GA). The limitations of the models are primarily based on the quality of the data in RTECS. TOPKAT predicts the toxicity of compounds not in the training set and indicates when predicted values lie outside the training set.

Acute dermal toxicity

Myers and DePass (1993) reviewed acute toxicity testing by the dermal route. Regulatory guidelines recommend the rabbit for use in the acute

Table 4. Acute Dermal Toxicity (LD_{50}) of Carbamate Pesticides.

Carbamate	Rat, male, $mg\,kg^{-1}\,bw$	Rat, female, $mg\,kg^{-1}\,bw$	Rabbit, male, $mg\,kg^{-1}\,bw$
Aldicarb[a]	2.5		>5.0
Carbaryl[b]	4,000		2,000
Carbofuran[c]			10,200
Formetanate[d]			
Methomyl[e]	>1,000		>5,000
Methiocarb[f]	>5,000		
Oxamyl[g]	>1,200		740
Pirimicarb[h]		>500	>500
Propoxur[i]	>1,000		
Thiodicarb[j]	>2,000		

[a]Lewis (1996); Hansch and Leo (1979).
[b]Lewis (1996).
[c]Crop Protection Handbook (2007).
[d]No data.
[e]WHO (1996).
[f]Tomlin (1994).
[g]Kennedy (1986).
[h]Hayes and Laws (1991).
[i]Verschueren (1983).
[j]WHO/FAO (2000).

dermal toxicity test because of its sensitivity to dermally applied chemicals, large treatment area, docility, availability, and large database. The rat or guinea pig is often used for dermal studies when minimal chemical or funding is available. Toxicity is often sex dependent, with the LD_{50} values being greater for male than for female rats and rabbits (Myers and DePass 1993). Permeability constants are often two to three times greater for females than they are for males, the difference being due in part to the thickness of skin and the presence or absence of oils and waxes. The EPA requires animal studies under FIFRA guidelines (Knaak et al. 1993). The toxicity data in Table 4 were taken from a variety of sources. The Crop Protection Handbook (2007) probably contains the most complete list of carbamates used in agriculture along with their dermal toxicities. Aldicarb is the most dermally toxic carbamate of the carbamates considered here. Modelers should be aware of their dermal toxicity when building and running carbamate PBPK/PD models.

III. QSAR Models for Predicting Biological Parameters Used in PBPK/PD Models

Hansch et al. (2002) published an approach for organizing data on chemical-chemical and chemical-biological reactions. The importance of being able to relate structure to biological activity is obvious when one considers the large number of biological parameters (absorption/distribution, partition coefficients, metabolic rate constants, and biological response) that are needed in the construction of PBPK/PD models (see Table 1). The development of these models depends largely on the availability of biological data in the literature or on the development of good quality data in laboratory studies. The cost of the latter appears to be prohibitive in current drug or pesticide development programs. Sixty-one QSAR programs in the Hansch et al. (2002) database involving oxidoreductases of all types were found to focus on cytochrome P450 isozymes (CYPs) (e.g., algicides: Schmitt et al. 2000; antimalarials: Sinha et al. 1999).

Several structure–bioavailability relationship models have been proposed in the last few years (Clark and Pickett 2000; Ekins and Rose 2002; Klopman et al. 2000). The best known study is the "rule-of-five" developed by Lipinski et al. (2001) based on four molecular descriptors. The rule generates an alert about absorption problems if two of the following conditions are met: (1) molecular weight >500; (2) number of hydrogen-bonds acceptors >10; (3) number of hydrogen-bond donors >5; (4) calculated log P > 5.0; and (5) poor water solubility. Lipinski specifically states that the rule-of-five only holds for compounds that are not substrates for active transporters (Lipinski 2000; Lipinski et al. 2001). When the rule-of-five was developed, information about drug transporters was very limited (Wu and Benet 2005). Amidon et al. (1995) devised a Biopharmaceutical Classification System (BCS) that categorized drugs into four classes according to aqueous solubility and gastrointestinal permeability. A drug is considered to be "highly soluble" when the highest dose is soluble in 250 mL water over a pH range 1–7.5 at 37°C, whereas a drug is considered to be "highly permeable" when absorption of parent and metabolites is >90% of an administered dose. Wu and Benet (2005) suggested replacing permeability in the classification system with major routes of elimination. The change will facilitate predictions, expand the number of Class 1 drugs eligible for waiver of *in vivo* bioequivalence studies, and provide new insight into drug elimination. Class 1 drugs have high solubility, high membrane permeability, and are extensively metabolized (Wu and Benet 2005).

A. Human Oral Bioavailability

Yoshida and Topliss (2000) developed a QSAR model for predicting drug oral bioavailability using the quantitative structure–bioavailability relationship of 232 structurally diverse drugs. Oral bioavailability was assigned one of four ratings and was analyzed in relation to physicochemical and structural factors by the **or**dered **mu**lticategorical **c**lassification method using the **s**implex technique (ORMUCS) method. The bioavailability classifications are shown in Table 5.

Primary structural descriptors for drugs involved the distribution coefficient, log $D_{pH\ 6.5}$, where 6.5 is the pH of the small intestine. The observation that bioavailability generally followed acids > neutrals > bases led to the

Table 5. Bioavailability Classification.[a]

	Class 1	Class 2	Class 3	Class 4
Rating	1	2	3	4
Bioavailability (%)	≤20	20–49	50–79	>80

[a]Bioavailability Classification System (Yoshida and Topliss 2000).

formulation of a new descriptor, Δ log D (i.e., log $D_{pH\ 6.5}$ – log $D_{pH\ 7.4}$). The addition of 15 structural descriptors relating to well-known metabolic transformations yielded a satisfactory QSAR equation, $S(X) = \Sigma w_i s_i$ with S_i, the descriptor values, and W_i, the weighting factors. The optimum value for log $D_{pH\ 6.5}$ is –0.3, and a progressive negative impact on bioavailability occurs when values move away from this level. A log $D_{pH\ 6.5}$ value of 3.0 results in a reduction of the bioavailability score by 0.5 and a value of 5.0 by 1.3. This approach allowed the authors to obtain a correct classification rate of 71% for the training set and 60% for the 40 compounds included in the validation set. Pintore et al. (2003a) obtained the data set from Yoshida and Topliss (2000) and added 235 molecules. A large set of molecular descriptors was examined, and the most relevant parameters were selected using a procedure called genetic algorithm concepts and a stepwise technique (GA/SW). The aim of the project was to apply adaptive fuzzy partitioning (AFP) to two bioavailability data sets. The AFP model improved the power prediction about 10%–15% on the training and validation set molecules. Pintore et al. (2003a) did not strongly support the use of the physicochemical descriptors Δ log D and log P as they used topological and electrotopological descriptors instead. Neither article discussed the nature of the drug formulations used in the human bioavailability studies.

Carbamates are considered to be readily absorbed when dissolved in corn oil, PEG 400, propylene glycol, or other relatively nontoxic solvents. No information on the rate of absorption of these carbamates was found in the literature. Model optimization procedures were used in a carbofuran PBPK/PD models to simulate oral absorption (Zhang et al. 2006).

B. Toxicity Models

Venkatapathy et al. (2004) assessed the oral rat chronic lowest observed adverse effect level (LOAEL) model in TOPKAT (Enslein et al. 1998). The percentages of accurate predictions for nondatabase chemicals did not compare well with predictions for chemicals in TOPKAT's database. The National Center for Environmental Assessment (NCEA), EPA, Cincinnati, OH, used TOPKAT to screen/rank chemicals on the basis of their hazard potential. The modules that are used with the greatest frequency by NCEA are the rat chronic LOAEL, oral rat lethal dose (LD_{50}), rodent carcinogenicity, Ames mutagenicity, and developmental toxicity models. TOPKAT has a built-in quality control module that determines if a chemical's descriptors were within the model's prediction domain. TOPKAT also has a similarity search feature capable of determining whether a queried chemical was represented in TOPKAT's database.

In addition to TOPKAT, commercial QSAR software packages such as DEREK (LHASA, Leeds, UK), CASE and MULTICASE (Multicase, Cleveland, OH), COMPACT, HazardExpert (Compudrug, Budapest, Hungary), and ONCOLOGIC (Logichem, Boyertown, PA) are able to predict toxicity potentials for a wide variety of chemicals (Venkatapathy et al. 2004).

Pintore et al. (2003b) used AFP to predict the toxicity of pesticides. Data sets of 235 pesticide compounds, divided into three classes according to their toxicity toward rats, were used. Carbaryl, carbofuran, methiocarb, methomyl, oxamyl, pirimicarb, propoxur, and thiodicarb were included in the training sets.

Goldblum et al. (1981) used the QSAR Eq. 1 to describe the relationship between inhibition for 269 phenyl *N*-methylcarbamates inhibiting fly head AChE and the molecular descriptors MR (molar refractivity), E_s (Taft steric parameter), $\Im$ (an inductive parameter), σ (Hammett parameter, meta substituents), RGMR (MR of certain parts of ring substituents), CHG (indicator variable for charged substituents), and HB (hydrogen bonding) of the pesticides.

$$\begin{aligned} pI_{50} = {} & 0.557(\pm 0.08)MR_{3,4,5} + 1.558(\pm 0.020)MR_2 - 0.611(\pm 0.09)E_s \\ & - 0.940(\pm 0.10)\left(\sum \sigma^-_{0,p} + \sigma_m\right) + 1.431(\pm 0.31)CHG \\ & - 0.227(\pm 0.04)MR_2^2 - 5.236(\pm 1.27)\Im_{2,6}^2 + 3.465(\pm 0.09)\Im_{2,6} \\ & + 0.659(\pm 0.22)RGMR - 0.618(\pm 0.22)HB \\ & - 0.052(\pm 0.02)MR_3^2 - 0.563(\pm 0.29)E_{s_2} \times E_{s_6} + 3.458(\pm 0.21) \end{aligned} \quad (1)$$

where $n = 269$, $r^2 = 0.796$, $s = 0.485$, Ideal $\Im_{2,6} = 0.331(0.295 - 0.368)$, and Ideal $MR_2 = 3.34(3.16 - 3.79)$.

Using the best correlation equation, the authors were able to predict the concentration producing 50% inhibition for 269 different molecules within a factor of ±3 (i.e., antilog of 0.485) for a concentration range of 1 million. According to O'Brien (1960), it is more desirable to have k_i values (bimolecular rate of inhibition) for direct reference to the mechanism, since k_i values are directly related to the binding step:

$$EH^1 + CB \underset{k_{-1}}{\overset{k_{+1}}{\rightleftarrows}} EHCB \xrightarrow{k_2} EC \xrightarrow{k_3} EH^2 + C \quad (2)$$

where EH^1 = free enzyme, CB = carbamate inhibitor, EHCB = enzyme-carbamate complex, EC = carbamylated enzyme, EH^2 = regenerated enzyme, and C = carbamic acid.

$$K_d = k_{-1}/k_{+1} \quad \text{(equilibrium or affinity constant)} \quad (3)$$

$$k_i = k_2/K_d \quad \text{(bimolecular inhibition rate constant)} \tag{4}$$

Abd-Elroaf et al. (1977) and Nishioka et al. (1977) found k_2 to be more or less constant for a set of carbamates. The long reaction periods used in the I_{50} studies (Kolbezen et al. 1954) complicated the otherwise relatively straightforward relation of k_i to I_{50} (O'Brien 1960) because the decarbamylation rate constant, k_3, has a shorter half-life, $t_{1/2,}$ than the span of time used for the I_{50} measurements. The parameter k_i turns out to be insensitive to the concentration of the inhibitor used (Nishioka et al. 1976) in the kinetic runs, so it is of greater comparative value for results from different sources. Nevertheless, the authors believe that data with a large set of congeners can be of real value in enabling them to rationalize some of the salient events of the binding step and in exploring the characteristics of the binding site via physicochemical correlation studies.

C. Liver Microsomal P450 and CYP Hydroxylation Models

QSAR models were developed by Enslein et al. (2007) for predicting V_{max} and K_m phase I reactions involving the hydroxylation of (1) drug aromatic and (2) alicyclic rings and aliphatic groups by human liver microsomes (HLM) and by CYP3A4 isozyme. Additional drug-based QSAR-V_{max}, K_m models involving CYP1A2, 2D6, and 2C9, individually, are under development. Approximately 50% of the biotransformations involving drugs are hydroxylation reactions, the remaining transformations include N- and O-demethylations. Hydroxylated drugs are conjugated with glucuronic or sulfuric acid in phase II reactions. A number of carbamates are either ring-hydroxylated or their *N*-methyl groups are hydroxylated. Carbaryl (Tang et al. 2002) is converted to 4-OH, 5-OH, and hydroxymethyl derivatives, whereas carbofuran (Usmani et al. 2004a) is converted to 3-OH carbofuran. V_{max}, K_m data (e.g., V_{max}, nmol min^{-1}nmol^{-1} CYP3A4; K_m, µmol L^{-1}), from these two carbamates were used in the training sets for the CYP3A4 hydroxylation model along with data from 66 drugs. HLM QSAR – V_{max}, K_m models (e.g., V_{max}, µmol hr^{-1} kg^{-1} bw; K_m, µmol L^{-1}) were developed without carbamate data. The CYP3A4 models turned out to be stable and useful for predicting V_{max}, K_m values, while the HLM models were not stable and could not be used. The reason(s) for the instability appears to be associated with variation in CYP3A4 content in the microsomes.

D. Tissue:Blood Partition Coefficients

Zhang (2005) developed a nonlinear QSAR equation using (1) published tissue:blood partition coefficients (PC_ts) of 36 organic solvents in human fat, liver, brain, kidney, muscle, lung, and heart, and (2) PC_ts of 10 drugs for rabbit fat, brain, muscle, lung, and heart. No pesticidal carbamates were included in the training sets. The model takes into consideration the volume of the tissue composition (lipid, protein, and water), and the form(s) of the compound (cationic, anionic and neutral) at pH 7.4. The present research only included neutral and cationic forms. The QSAR equation took the following form:

$$\log PC_t = \log\left(\sum f_j 10^{\log P_{ij} + \log t/i}\right) \tag{5}$$

where subscript i = l, p, w (indicate the lipid, protein, and water in tissue, respectively); and subscript j = ui, +, – (indicate the neutral form, cationic form, and anionic form of a compound, respectively).

The author used the following molecular descriptors in the QSAR model: molecular polarizability (α), maximum positive charge (q^{+max}), sum of all positive partial atomic charges for all atoms in the molecule ($\Sigma Q+$), the sum of H-bond factor values for all acceptor substructures in the molecule (ΣCa), the sum of H-bond factor values for all donor atoms in a molecule (ΣCd) and the maximum H-bond acceptor descriptor in a molecule (Ca^{max}). All descriptors were calculated by the program package HYBOT/HYBOT-PLUS-98 (Raevsky 1997). Nonlinear regression analyses were performed using a standard regression program (GFA BASIC 4.38; GFA Software Technologies, Moenchengladbach, Germany). The descriptors were introduced into Equation 5 in a stepwise manner until the statistical results could not be further improved. Equation 5 states that the PC_t may be expressed as the weighted sum of three PC_ts (lipid:blood, protein:blood, and water:blood) of neutral and ionic forms. The equilibrium distribution of a chemical is essentially the sum of the distributions of the various chemical forms. Five parameters were used to fit neutral compound partitioning into tissues, and two parameters were used for ionic compounds. The partitioning of ionic compounds into lipids increases with increasing Ca^{max}. The q^{+max} value plays an important role in compounds partitioning into protein. The model equation is the expressive form of the Hansch equation (Hansch et al. 1962) in nonlinear spaces (multiphase system).

Liu et al. (2005) combined information on the tissue:blood partition coefficients of 35 organic compounds in seven different tissues with the lipid, protein, and water content of the seven tissues with their molecular descriptors to select the best descriptors for a QSAR model. The model represented one tissue composition descriptor, two electrostatic descriptors, one quantum chemical descriptor, and one geometric descriptor. The weight fraction of water (WF) in different tissues was the most important descriptor affecting a tissue:blood partition coefficient. A least squares-support vector machine (LS-SVM) was used to develop the nonlinear QSAR model. All calculations of the LS-SVM were performed using the MATLAB/C toolbox. No carbamates were present in the data set.

IV. Experimentally Derived Biological Parameters Used in PBPK/PD Models

A. Gastrointestinal Absorption

Bolus dose

Timchalk et al. (2002) used absorption rates of 0.01, 0.5, and 0.5 hr^{-1} for stomach to blood, intestine to blood, and stomach to intestine for chlorpyrifos (CAS no. 2921-88-2). A bolus dose (chemical in $mg\,kg^{-1}$ bw or $\mu mol\,kg^{-1}$ bw) is placed in the stomach at zero time. A number of first-order rate constants describe the absorption of the chemical from the stomach and intestine to blood. A first-order rate constant is also used to transfer chemical from stomach to intestine. A bioavailability factor of 0.8 (class 4) from Table 5 reduces the amount of the dose.

Rate of ingestion

In the case of a rate ingestion using the Exposure Related Dose Estimating Model (ERDEM) (USEPA 2004), a dose (concentration in dosing liquid, $mg\,mL^{-1}$ or $\mu mol\,mL^{-1}$) is administered at a uniform flow rate ($mL\,min^{-1}$) over a period of time (min) to give a dose in (mg or μmol) in the stomach. The dose (mg), as it arrives in the stomach, is then absorbed as indicated for a bolus dose. Total dose (mg) is calculated as follows: $mg\,mL^{-1} \times mL\,min^{-1} \times min$. In practice, if the time of administration is rapid, the dose acts like a single bolus dose.

B. Skin Absorption

History

Illnesses associated with the use of ChE-inhibiting pesticides (i.e., OP compounds and *N*-methylcarbamates) have generated considerable field research into the relationship between ChE inhibition and residues in air, on treated indoor surfaces, and on crop leaf surfaces (Honeycutt and Day 2001; Honeycutt et al. 1985; Siewierski 1984; Wang et al. 1989). This work was accompanied by the development of information pertaining to the science of percutaneous absorption (Bronaugh and Maibach 1985; Scott et al. 1990) and the linking of field exposures to percutaneous absorption and ChE inhibition (Knaak et al. 1989, 2002; Nigg and Knaak 2000; USEPA 1992). The information in these studies was used to develop code within ERDEM to predict the effects of dermal exposure to ChE-inhibiting pesticides on ChE activity in exposed individuals. Knaak et al. (2002) described the relationship between skin permeability, K_p ($cm\,hr^{-1}$), and partition coefficients of skin:air, skin:vehicle, and skin:blood. In the case of skin:air, a $K_{p\text{-vapor}}$ value ($cm\,hr^{-1}$) takes into consideration the partitioning of vapor between air and skin, leading to a concentration in the skin and a percutaneous absorption rate ($\mu g\,hr^{-1}\,cm^{-2}$). The permeability constants for pesticides in delivery solvents, $K_{p\text{-solvent}}$, are not determined in most studies. Skin absorption rates are reported in $\mu g\,hr^{-1}\,cm^{-2}$. The solvent, xylene, has been shown to enhance the penetration of methyl parathion over that of acetone in studies involving cadaver skin. Acetone has been used extensively for applying a pesticide on skin. The solvent rapidly evaporates, leaving the pesticide to be absorbed by skin or to evaporate into the air (Bronaugh and Maibach 1985).

Relationship between K_p (permeation constant), skin:air, and skin:vehicle partition coefficients

A direct relationship exists between the permeation constant, K_p ($cm\,hr^{-1}$), the diffusion coefficient D ($cm^2\,hr^{-1}$), the thickness of skin l (cm), the concentration C of the pesticide in skin ($g\,cm^{-3}$), and the skin:vehicle partition coefficient k_m (unitless) based on the following equation from Mattie et al. (1994).

$$\text{Flux} = Dk_mC/l = K_pC \qquad (6)$$

In the case of pesticides in vehicles such as water or air, skin:air and skin: water partition coefficients need to be determined and used in the equation to account for higher concentrations in skin caused by partitioning.

Skin:blood partition coefficients

Skin:blood partition coefficients are used in PBPK/PD models to transfer pesticides between skin and blood. Jepson et al. (1994) developed a laboratory procedure for measuring the partitioning of chemicals between skin and blood using low-binding Millipore cellulose filters, and Poulin and Theil (2002a) developed a mechanistic method for measuring tissue:blood partition coefficients. The procedures are discussed in detail in Section IV.C.

Dose–response, pharmacokinetic, and metabolism studies involving carbamate pesticides

Knaak et al. (1984) and Knaak and Wilson (1985) developed 7-d pharmacokinetic data on the percutaneous absorption of carbaryl ($43.5\,\mu g\,cm^{-2}$, $13.8\,cm^2$ skin) through the intact back skin of the rat using ^{14}C-naphthyl carbaryl. The skin permeability constant ranged from 4.0×10^{-3} to 7.1×10^{-3} ($cm\,hr^{-1}$) based, respectively, on the $t_{1/2}$ values for elimination from plasma or loss from skin. Skin absorption rates varied between 0.18 and $0.31\,\mu g\,cm^{-2}\,hr^{-1}$ (i.e., $K_p = 0.31/43.5\,\mu g\,cm^{-2}\,hr^{-1} = 7.1 \times 10^{-3}\,cm^{-2}\,hr^{-1}$). No information was obtained by Knaak et al. (1984) concerning the nature of metabolites in skin or eliminated in urine. The fate of ^{14}C naphthyl carbaryl in rat skin was studied by MacPherson et al. (1991) in the presence or absence of NADPH (nicotinamide adenine dinucleotide phosphate, reduced), UDPGA (uridine diphosphate glucuronic acid), or PAPS (3′-phosphoadenosine 5′-phosphosulfate). 1-Naphthol was formed by postmitochondrial fractions from the liver and skin with and without cofactors. In the presence of NADPH, hydroxycarbaryl was formed by the liver but not by skin postmitochondrial fractions. 1-Naphthyl glucuronide and sulfate were formed by postmitochondrial fractions from the liver and skin. No evidence of carbaryl metabolism was detected in studies involving intact skin.

In a pharmacokinetic percutaneous study with ^{14}C-labeled thiodicarb, thiodicarb was poorly absorbed from skin, with 4.4% found in urine, 2.5% in feces, 8.8% in the carcass, and 6.6% in respiratory air CO_2 and acetonitrile (Knaak and Wilson 1985). The remainder of the dose was found on the surface of the skin. Thiodicarb was absorbed during the first 24 hr at a rate of 0.27 μg cm^{-2} hr^{-1} ($K_p = 0.27/48 = 5.6 \times 10^{-3}$ cm hr^{-1}). The rate was similar to that of carbaryl. After 24 hr, the rate of loss from skin decreased to 0.042 μg cm^{-2} hr^{-1} ($K_p = 0.042/48 = 8.75 \times 10^{-4}$ cm hr^{-1}), and remained so until the study was terminated at 168 hr. The reason for the rate change appears to be associated with the physical properties of the applied thiodicarb, and its distribution across treated skin. Microscopic examination of treated skin revealed crystals of thiodicarb deposited on hair and on the surface of the skin suggesting that only a portion of the applied dose was readily available for absorption when using acetone as a vehicle. Vehicles such as alcohol have a tendency to increase percutaneous absorption.

Van de Sandt et al. (1993) topically applied ^{14}C propoxur at various concentrations to human, rabbit and porcine skin disks *in vitro*. Permeation rates in human skin were 9.2, 40.7, and 56.6 ng cm^{-2} hr^{-1}, respectively, for applied dosages of 25, 100, and 200 μg cm^{-2}. The experimental period was 6 hr. The metabolites in skin (1.1%–4.9%) were identified as IPP (2-isopropoxyphenol), IPP-glucuronide, IPP-sulfate, and an unknown metabolite. Human skin produced only sulfate conjugates, rabbit skin produced glucuronides, and porcine skin produced sulfates and glucuronides. Comparative *in vitro-in vivo* percutaneous absorption studies involving propoxur were carried out by Van de Sandt et al. (2000). For direct comparison, experimental conditions involving dose (150 μg propoxur cm^{-2}), vehicle (60% ethanol), and exposure time (4 hr) were standardized. Percutaneous penetration in human volunteers was measured by the analysis of 2-isopropoxyphenol in blood and urine. Maximal flux (μg cm^{-2} hr^{-1}) was calculated from the linear portion of the cumulative penetration curve. The permeability coefficient (K_p value in cm hr^{-1}) [H^3] H_2O was calculated according to ECETOC (1993), by dividing the maximal flux (μg cm^{-2} hr^{-1}) by the applied dose. In the *in vivo* rat studies, propoxur was applied to an area of 10 cm^2 (total dose, 1,500 μg), while in the human studies, an area of 100 cm^2 was utilized on the forearm (total dose, 15,000 μg). Maximal flux in the rat and human were 8.7 and 1.0 μg cm^{-2} hr^{-1}, respectively. K_p values were estimated to be $8.7/150 = 5.8 \times 10^{-2}$, and $1.0/150 = 6.6 \times 10^{-3}$ cm hr^{-1}, respectively, for the rat and human.

Additional pharmacokinetic studies involving pesticidal carbamates were not found in the published literature. The results of the studies involving carbaryl, thiodicarb, and propoxur suggest that carbamates are not as readily absorbed through skin as OP pesticides.

C. Tissue Partition Coefficients/Distribution

Partition coefficients from in vitro data

Tissue:blood partition coefficients are required for the development of PBPK models. Jepson et al. (1992, 1994), Knaak et al. (1995), and Kousba and Sultatos (2000) are examples of this work involving the use of nonvolatile chemicals. According to the procedure of Jepson et al. (1992), Knaak et al. (1995) used heparinized rat blood, homogenized rat tissues consisting of liver, kidney, muscle (thigh), fat (perirenal), and brain. The tissues (0.32 g) suspended in 20 mL 0.9% saline in triplicate vials (test and reference) were placed in a shaking water bath at 37°C to allow the contents to come to thermal equilibrium. Reference vials contained 20 mL 0.9% saline. Isofenphos (150 µg in 50 µL ethanol) was added and contents equilibrated for 24 hr in 20 mL 0.9% saline. A 2-mL aliquot was ultrafiltered, and 100-µL aliquots of the filtrate were analyzed for pesticide. Equation 7 was used to determine the amount of pesticide recovered in the reference saline after ultrafiltration.

$$R_f = C_{sal}/C_{fil} \tag{7}$$

where R_f is the recovered fraction, C_{fil} is the concentration of the pesticide in the ultrafiltrate from reference saline, and C_{sal} is the concentration of the pesticide in the reference saline before ultrafiltration.

The concentration in tissue or blood was calculated from the concentration found in the sample ultrafiltrate using Eqs. 8 and 9.

$$A_{ssal} = C_{ss}(V_{ss})/R_f \tag{8}$$

$$C_j = (A_{rsal} - A_{ssal})/V_j \tag{9}$$

where A_{ssal} = the amount of pesticide in the sample saline after filtration,
C_{ss} = the concentration of pesticide in the sample saline after filtration,
V_{ss} = the volume of saline in the test vial,
A_{rsal} = the amount of pesticide in the reference saline,
V_j = the volume of tissue, and
C_j = the concentration of pesticide in tissue j.
The tissue : saline and tissue : blood PC_ts are calculated using Eq. 10.

$$P_j = C_j(V_j)/A_{ssal} \tag{10}$$

where P_j is the partition coefficient for tissue j (blood, liver, etc.) and A_{ssal}, V_j, and C_j are described above.

Several difficulties are encountered in carrying out this procedure: (1) nonspecific binding sites on filters and filtration apparatus often retain pesticides; (2) filtration rates are slow even under pressure; and (3) enzyme inhibitors should be used to prevent metabolism. The PC_ts are then calculated by dividing the tissue : saline partition coefficients by the blood : saline partition coefficients. Isofenphos rat tissue partition coefficients were 48, 21, 17, 27, and 23, respectively, for fat, liver, brain, kidney and muscle (Knaak et al. 1995). The values obtained with the Poulin and Theil (2002a) mechanistic method described later in this review were 358, 11.6, 28.1, and 6.49, respectively, for fat, liver, brain, kidney, and muscle.

Jepson et al. (1994) modified their earlier 1992 method by changing tissue weights (1.0g fat, 1.5g muscle, 0.25g liver, and 0.5g blood) and adding centrifugation before ultrafiltration. Ultrafilration was carried out using a Millipore Low Binding Celluose Membrane Filter Unit with a 10,000 nominal molecular weight (Millipore Ultrafree-PF; Millipore, Bedford, MA). The membranes were attached to the bottom of a filter cup as part of the manufacturing process. The system with filter cups is shown in Fig. 1. Each cup contained a 10-mm. four-pointed Teflon stir bar (Starburst; Parkside Industries, Chicago, IL). The partition coefficients for parathion (CAS no. 56-38-2) were 103, 2.59, and 5.42, respectively, for fat, muscle, and liver. The values using the Poulen and Theil (2002a) mechanistic method were 295.9, 6.3, and 11.4, respectively, for fat, muscle, and liver.

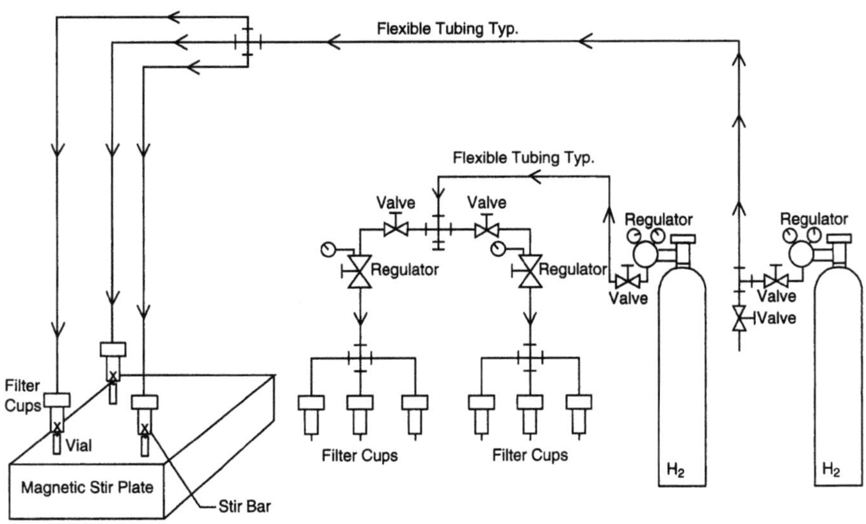

Fig. 1. Schematic of the system used to filter samples containing chemical equilibrated between tissue and saline. (Redrawn from Jepson et al. 1994 by permission of Oxford University Press.)

In the two experimental partition coefficient studies (Jepson et al. 1994; Knaak et al. 1995) involving OPs, the fat: blood PCs were significantly below those estimated by the Poulen and Theil (2000) method. An increase in the amount of fat (0.32 to 1.0 g) used by Jepson et al. (1994) in their parathion (log P, 3.84) studies had a positive effect on this PC over the value obtained by Knaak et al. (1995) for isofenphos (log P, 4.12). The differences between experimental PC values and those obtained using mechanistic models needs to be carefully examined before a recommendation can be made regarding the use of *in vitro* over mechanistic values.

Tissue: Blood Partition Coefficients from Mechanistic Models

QSAR models by Zhang (2005) and Liu et al. (2005) were described in Section III. D: Tissue: Blood Partition Coefficients. The models did not contain carbamates or their metabolites in the training set and may not be suitable for predicting tissue: blood partition coefficients for carbamates. The models look sound, and the procedures used in their development should be pursued in the development of QSAR models for carbamates.

Mechanistic approaches for prediciting partition coefficients from chemical structure information (log P) and tissue: plasma composition were developed and validated by Poulin and Krishnan (1996), Poulin and Theil (2000), Poulin et al. (2001), Poulin and Theil (2002a), and Poulin and Theil (2002b). Drug $P_{o:w}$ (*n*-octanol: buffer, nonionized species at pH 7.4), $D_{vo:w}$ and $D^*_{vo:w}$ (vegetable oil: buffer or olive oil: buffer, ionized and nonionized species at pH 7.4) were used in Eqs. 11 and 12 with tissue: plasma lipid and water, and unbound content in tissue: plasma to calculate $PC_{t:p\_nonadipose}$ and $PC_{t:p\_adipose}$ values (Poulin and Theil 2002a,b).

$$PC_{t:p\_nonadipose} = \frac{[P_{o:w}(V_{nlt} + 0.3V_{pht})] + [1(V_{wt} + 0.7_{pht})]}{[P_{o:w}(V_{nlp} + 0.3V_{php})] + [1(V_{wp} + 0.7V_{php})]} \times \frac{fu_p}{fu_t} \quad (11)$$

$$PC_{t:p\_adipose} = \frac{[D^*_{vo:w}/or/D_{vo:w}(V_{nlt} + 0.3V_{pht})] + [1(V_{wt} + 0.7V_{pht})]}{[D^*_{vo:w}/or/D_{vo:w}(V_{nlp} + 0.3V_{php})] + [1(V_{wp} + 0.7V_{php})]} \times \frac{fu_p}{1} \quad (12)$$

where V is the fractional tissue volume content of neutral lipids (nl), phospholipids (ph), water (w), tissue (t), and plasma (p), fu_t is the macromolecular unbound fraction in tissues, and fu_p is the macromolecular unbound fraction in plasma. The ratio fu_p/fu_t is equal to $C_{m\ tissue}/C_{m\ plasma}$ [(C = concentration; m = main binding macromolecules present in these two matrices, i.e., albumin, globulins and or lipoproteins (HDL, LDL)] with the ratio varying from 0.3 to 1 in mammalian tissues. Consequently, this contribution factor was not considered in determining PC_t when predicted values differed from experimental values by a factor less than 2 (Poulin and Theil 2000). Plasma protein binding was considered as a separate issue in a PBPK/PD model published by Timchalk et al. (2002) on chlorpyrifos.

In other terms, *in vivo* $PC_{t:p\_adipose}$ would only be dependent on the lipophilicity of nonionized species if the quantitative role of ionized species and plasma protein binding is not taken into account. The neutral nonpolar lipids of nonadipose tissues are mainly mixtures of triglycerides and cholesterol, and *n*-octanol mimics the lipophilicity of this mixture. The adipose tissue lipids are composed mostly of triglycerides, and olive oil is a good surrogate solvent. For cases involving only nonionized species in the aqueous phase, $D_{vo:w}$ was obtained using the relationship between $P_{o:w}$ and $D_{vo:w}$ in Eq. 13, as developed by Leo et al. (1971).

$$\log D_{vo:w} = 1.115 \log P_{o:w} - 1.35 \tag{13}$$

with $n = 104$, $r = 0.99$.

Fouchecourt et al. (2001) used the KowWin program (Meylan and Howard 1995) to obtain $\log P_{o:w}$ for *n*-hexane, and the Leo et al. (1971) equation for $\log K_{vo:w}$.

$$\log K_{vo:w} = 1.099 \log P_{o:w} - 1.31 \tag{14}$$

Equation 14 is comparable to Eq. 13. For the prediction of $P_{t:p\_adipose}$ with Eq. 12, data on $D^*_{vo:w}$ (nonionized and ionized species) are required instead of $D_{vo:w}$ (nonionized). $\log D^*_{vo:w}$ values were determined using Eq. 13 (i.e., $\log D_{ov:w}$) and the Henderson–Hasselbach equations for mono-, diprotic acids, bases, and zwitterionic chemicals as follows:

Monoprotic acid:

$$\log D^*_{vo:w} = \log D_{vo:w} - \log(1 + 10^{pH - pK_a1}) \tag{15}$$

Monoprotic base:

$$\log D^*_{vo:w} = \log D_{vo:w} - \log(1 + 10^{pK_a2 - pH}) \tag{16}$$

Diprotic acid:

$$\log D^*_{vo:w} = \log D_{vo:w} - \log(1 + 10^{pH - pK_a1 + pH - pK_a2}) \tag{17}$$

Diprotic base:

$$\log D^*_{vo:w} = \log D_{vo:w} - \log(1 + 10^{pK_a1 - pH + pK_a2 - pH}) \tag{18}$$

Zwitterionic:

$$\log D^*_{vo:w} = \log D_{vo:w} - \log(1 + 10^{-pK_a2 + pH + pK_a1 - pH})$$
$$pK_a1 = \text{acid},\ pK_a2 = \text{base} \tag{19}$$

Note that Eqs. 15–19 require information on the pK_a of the acid. Poulin and Theil (2002a) assumed that $D^*_{vo:w}$ and $D_{vo:w}$ differ by a factor corresponding to the ionized species in the aqueous phase. According to Lewis (2000), $\log D_{7.4}$ involving octanol:water is a combination of lipophicity (the $\log P$ term) and ionization (the pK_a term) with the relationship being $\log D^{7.4}_{o:w} = \log P_{o:w} - \log(1 - 10^{pH - pK_a})$ for monoprotic acids. This equation is the same as Eq. 15 except for the fact that distribution takes place between vegetable oil:water in Eq. 15 and not octanol:water as described by Lewis (2000).

The relationship between $\log P_{o:w}$ and $\log D_{vo:w}$ for adipose tissue : plasma and adipose tissue : blood is shown in Figs. 2 and 3. Data from Parham et al. (1997) for polychlorinated biphenyls (PCBs) are included in the adipose : plasma figure, and data from Jepson et al. (1994) are included in the adipose : blood figure. Tissue : blood PC_ts are routinely used in PBPK models. The use of $\log D_{vo:w}$ reduces the partition coefficients by approximately a factor of 10; however, this factor is not greatly reduced for chemicals with $\log P_{o:w}$ greater than 3. All the PCBs had $\log P_{o:w}$ partition coefficients greater than 4.0; this is also the case with most of the OPs, but not with the carbamates.

The $\log D_{vo:w}$ model appears unbiased when $\log P = 0$ to 2 and biased when $\log P > 3$. The experimental parathion log P values of Jepson et al. (1994) for parathion are shown along with values predicted by the model.

Selection of $\log P_{o:w}$ and $\log D_{o:w}$ Values for Use in Mechanistic Models

The key to using the mechanistic approach is the selection of a log P/log D model that adequately predicts values for the 10 carbamates and their metabolites (ionized and nonionized species). Log P models include Clog P (Leo 1993), KLog P (Klopman et al. 1994), HLog P (Viswanadhan et al. 2000), ALog P (Wildman and Crippen 1999), XLog P (Wang et al. 2000), ACDLog P (Petrauskas and Kolovanov 2000), SMILELOG (Convard et al. 1994), CHEMICALC (Suzuki and Kudo 1990), and LOGKOW

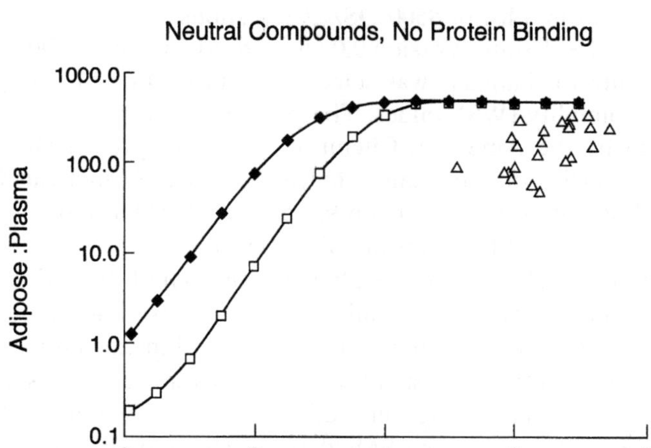

Fig. 2. Adipose : plasma partition coefficients vs. $\log P_{o:w}$ (and $\log D_{vo:w}$). ◆ Log $P_{o:w}$ (Knaak et al. 2005); □ $\log D_{vo:w}$ (Leo et al. 1971); Δ PCB data (Parham et al. 1997).

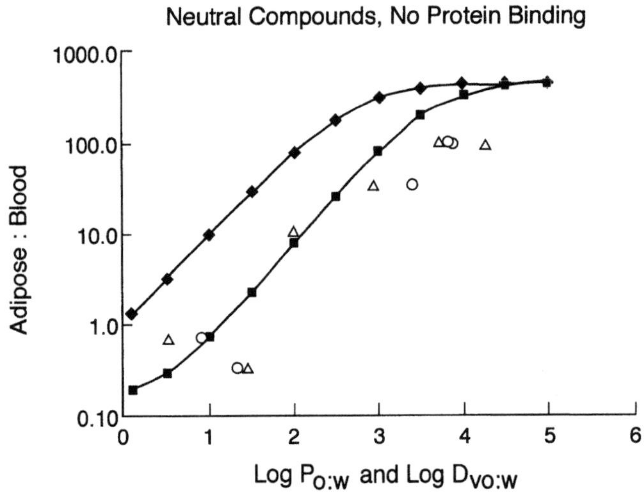

Fig. 3. Adipose:blood partition coefficients vs. $\log P_{o:w}$ (and $\log D_{vo:w}$). ◆ $\log P_{o:w}$ (Knaak et al. 2005); ■ $\log D_{vo:w}$ (Leo et al. 1971); ○ experimental $\log P_{o:w}$ (Jepson et al. 1994); Δ predicted $\log P_{o:w}$ (Jepson et al. 1994). Points at the bottom are for paraoxon; points at the top are for parathion.

(Meylan and Howard 2000). Hansch et al. (2002) currently have a database containing almost 30,000 experimentally measured octanol/water log P and log D values for constructing QSAR equations. The importance of log P values cannot be overestimated, as Hansch et al. (2002) lists 4614 QSARs with log P terms in a list of 8500 QSAR equations.

ACD LogD Sol Suite, Version 9.0 (Advanced Chemistry Development, Toronto, Ontario, Canada) was selected for predicting pK_a, log P, log D, and water solubility (WS) values. The program is commercially available and is used by the American Chemical Society to predict chemical and physical properties of chemicals listed in SciFinder (American Chemical Society, Washington, DC). Version 9.0 predicts log D at any pH and provides a printout of log D values in units of 0.1 pH.

As part of the selection process, the WS for each of the carbamates and their metabolites were individually plotted against $\log D_{pH7.4}$ values to determine whether a linear, but inverse, relationship existed based on the knowledge that metabolism significantly increases the water solubility of administered pesticides, drugs, and other chemicals. Water solubility was calculated using the equation of Meylan and Howard (1994a,b) where

$$\text{Log S(mol/L)} = 0.796 - 0.854 \log K_{o:w} - 0.00728\text{MW} + \text{corrections} \quad (20)$$

ACD log D values were used in the equation in place of $\log K_{o:w}$ to calculate water solubilities. This procedure yielded linear plots with r^2 values greater

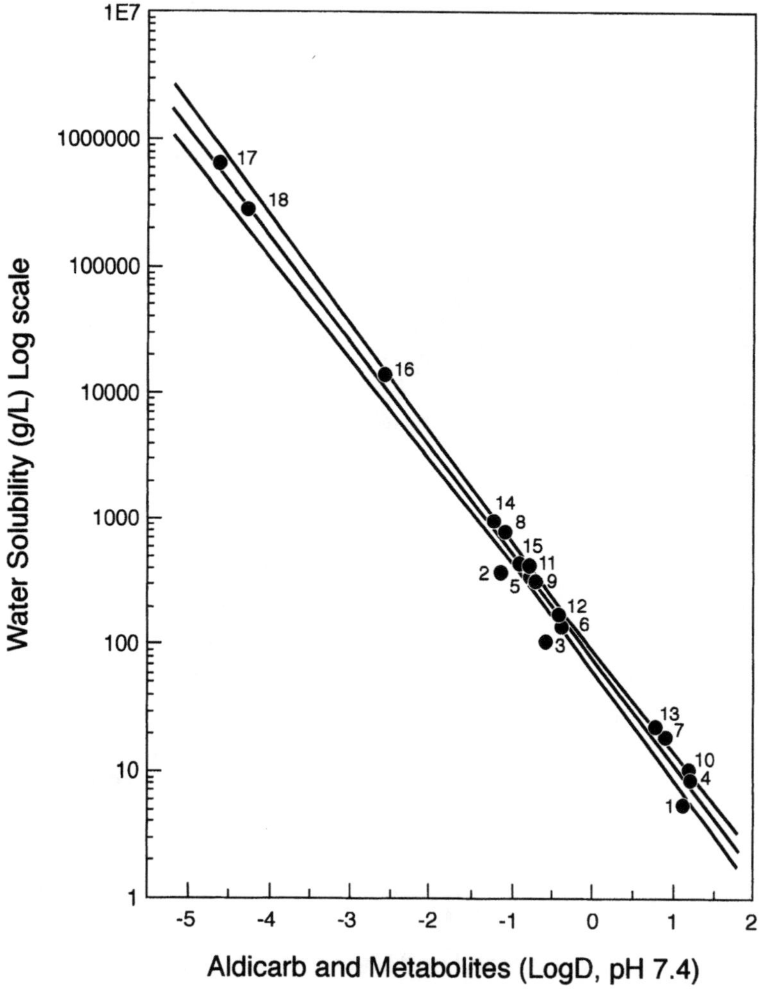

Fig. 4. Water solubility ($g\,L^{-1}$, Eq. 20) of aldicarb and metabolites as a function of ACD log $D_{pH7.4}$. Aldicarb and metabolites are identified in Table 14 (Appendix A). Outer bounds represent the 95% confidence interval.

than 0.9 for all 10 carbamates and their metabolites. Figures 4 and 5 show the plots for aldicarb and carbaryl. Water solubilities >10,000 ($g\,L^{-1}$) were calculated by the Meylan and Howard (1994a,b) equation for conjugated carbamate metabolites but were not provided by the ACD log D suite.

These results were not obtained using $\log K_{o:w}$ values (Discovery Studio Accord for Excel; Accelrys, San Diego, CA) for carbamate metabolites because $\log K_{o:w}$ alone did not adequately predict values for water-soluble ionized metabolites.

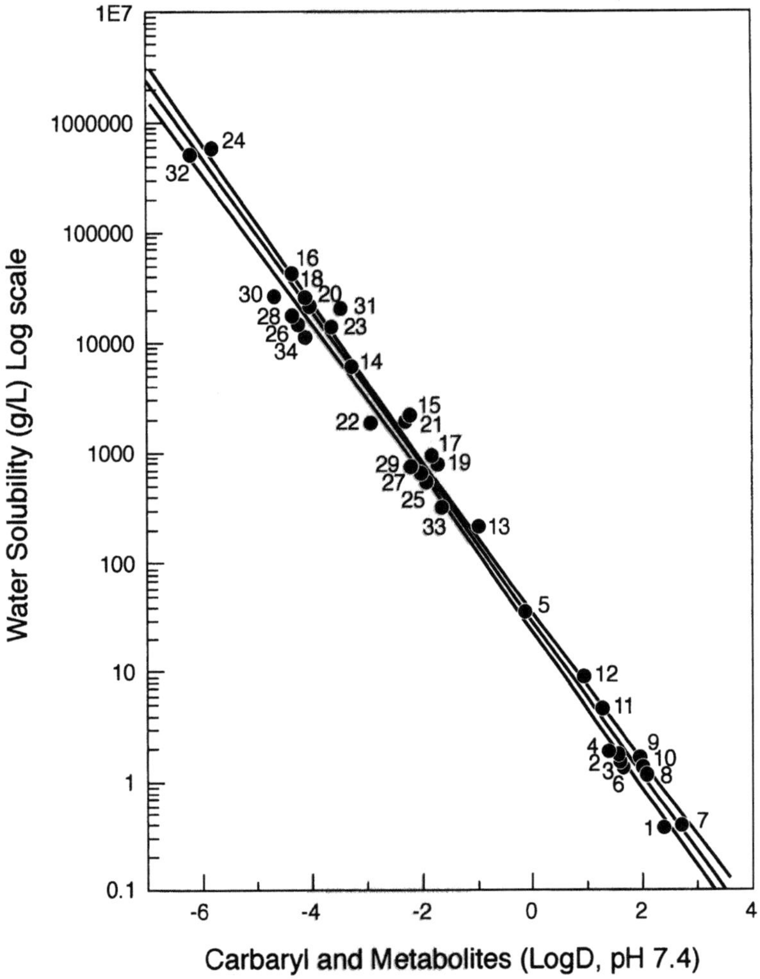

Fig. 5. Water solubility ($g L^{-1}$, Eq. 20) of carbaryl and metabolites as a function of ACD Log $D_{pH7.4}$. Carbaryl and metabolites are identified in Table 15 (Appendix A). Outer bounds represent the 95% confidence interval.

Partition Coefficients for Carbamates and Metabolites

On the basis of the foregoing findings involving water solubility for aldicarb acid, $\log D_{pH7.4}$ (−2.56), determined by the ACD LogD Suite was used in conjunction with an Excel spreadsheet (Table 6) to calculate $\log D_{vo:w}$ (−4.20) from $\log D_{pH7.4}$ (−2.56) (Eq. 13) with $\log D^*_{vo:w} = \log D_{vo:w}$ (−4.20) (Eq. 15, no pK_a correction factor required) to obtain the adipose tissue:blood partition coefficients for aldicarb acid (see Table 14, Appendix A) using Eq 22. In the case of nonadipose tissue, Eq. 21 was used. When $\log P$ (0.87) was used in Eq. 13, a value of −0.38 was obtained for $\log D_{vo:w}$

and a value of –4.10 with Eq. 15 for $\log D^{*}_{vo:w}$ (pH, 7.4 and pK_a, 3.68). The results for $\log D^{*}_{vo:w}$ compared favorably with the value obtained using $\log D_{pH7.4}$.

In the case involving a glucuronide such as 4-OH carbaryl glucuronide, $\log D^{*}_{vo:w}$ (pH, 7.4 and pK_a, 2.75) was –6.63 using $\log P$ (–0.57). When $\log D_{pH7.4}$ (–4.28) was used the value for $\log D^{*}_{vo:w}$ was –6.12. In the case of a sulfate, such as 4-OH carbaryl sulfate, $\log D^{*}_{vo:w}$ (pH, 7.4; pK_{a1}, –4.58 and pK_{a2}, 11.64) was –6.17 using $\log P$ (1.57). When $\log D_{pH7.4}$ (–1.93) was used the value was –3.50 for $\log D^{*}_{vo:w}$. The reason for the difference is related to the manner in which ACD LogD Suite computes $\log D_{pH7.4}$ versus the Henderson–Hasselbach equation (Eqs. 15 and 17) for a mono- or diprotic acid. On the basis of the differences observed we decided to use $\log D_{pH7.4}$ values for determining partition coefficients for all carbamate metabolites.

The tissue:plasma and tissue:blood partition coefficients were determined using Eqs. 21 and 22 (see Table 6). Tissue:blood partition coefficients are depicted below.

$$PC_{t:b\_nonadipose} = \frac{[D_{o:w}/or/P_{o:w}(V_{nlt} + 0.3V_{phlt})] + [1(V_{wt} + 0.7V_{phlt})]}{[D_{o:w}/or/P_{o:w}(V_{nlb} + 0.3V_{phlp})] + [1(V_{wb} + 0.7V_{phlb})]} \times \frac{fu_b}{fu_t} \tag{21}$$

$$PC_{t:b\_adipose} = \frac{[D^{*}_{vo:w}/or/D_{vo:w}(V_{nlt} + 0.3V_{pht})] + [1(V_{wt} + 0.7V_{pht})]}{[D^{*}_{vo:w}/or/P_{vo:w}(V_{nlb} + 0.3V_{phb})] + [1(V_{wb} + 0.7V_{phb})]} \times \frac{fu_b}{1} \tag{22}$$

where $PC_{t:b\_nonadipose}$ and $PC_{t:b\_adipose}$ = partition coefficient between tissue and blood, pH 7.4, distribution coefficient (D), partition coefficient (P), octanol:water (o:w), vegetable oil:water (vo:w), V is the fractional tissue volume content of neutral lipids (nl), phospholipids (ph), water (w), tissue (t), and blood (b), fu_t is the macromolecular unbound fraction in tissues, and fu_b is the macromolecular unbound fraction in blood.

The partition coefficients for the 10 carbamates and their metabolites in adipose tissue, brain, rapidly perfused tissue (e.g., heart), kidney, liver, slowly perfused (e.g., muscle), and skin are shown in Tables 14–23 (Appendix A).

Graphs of $\log D_{pH\,7.4}$ vs. Adipose Tissue:Blood and Liver:Blood Partition Coefficients

The plots of $\log D_{o:w}$ (pH 7.4) vs. $PC_{t:b\_liver}$ and $\log D^{*}_{vo:w}$ (pH 7.4) vs. $PC_{t:b\_adipose}$ for aldicarb and carbaryl are shown in Figs. 6 and 7.

In Figs. 6 and 7, the water-soluble metabolites, with $\log D_{pH7.4} < 0$, all have $PC_{t:b\_liver}$ *and* $PC_{t:b\_adipose}$ values ≤1.0 at the bottom of each plot, while the parent carbamates and their neutral metabolites, with $\log D_{pH7.4} > 0$, have $PC_{t:b}$ values ranging from 1.0 to 40.0 extending up and on the right side of each plot.

Table 6. Example Spreadsheet Used in Calculating Partition Coefficients (Eqs. 21, and 22) for aldicarb acid in Table 14 (Appendix A).

Input parameters	
log $D_{pH7.4\ o:w}$	–2.6
pH^a	7.4
pK_a^a	3.68
log $D^*_{vo:w}$	–4.2
log $D_{vo:w}$	–4.2
fu_p	1
fu_t	1
fu_p/fu_t	1

				Lipids						
				Neutral	Phospho					
Rat tissue	Eq. no.[b]	Tissue $(V_t)^c$	Water $(_{vw})^c$	$(V_{nl})^c$	$(V_{pl})^c$	$K_{o:w}{}^d$	Numerator[e]	$P_{t:b}{}^f$	$P_{t:p}{}^g$	$V_t{}^*P_{t:p}{}^h$
Adipose	22	0.0761	0.12	0.853	0.002	6E-05	0.1215	0.14	0.13	0.01
Adipose	21	0.0761	0.12	0.853	0.002	0.0028				
Bone	21	0.041476	0.446	0.0273	0.0027	0.0028	0.4480	0.53	0.47	0.02
Brain	21	0.0057	0.788	0.0392	0.0533	0.0028	0.8255	0.98	0.86	0.00
Gut	21	0.027	0.749	0.0292	0.0138	0.0028	0.7588	0.90	0.79	0.02
Heart	21	0.0033	0.779	0.014	0.0118	0.0028	0.7873	0.94	0.82	0.00
Kidney	21	0.0073	0.771	0.0123	0.0284	0.0028	0.7909	0.94	0.82	0.01
Liver	21	0.0366	0.705	0.0138	0.0303	0.0028	0.7263	0.86	0.76	0.03
Lung	21	0.005	0.79	0.0219	0.014	0.0028	0.7999	0.95	0.83	0.00
Muscle	21	0.404	0.756	0.01	0.009	0.0028	0.7623	0.91	0.79	0.32
Skin	21	0.19	0.651	0.0239	0.018	0.0028	0.6637	0.79	0.69	0.13
Spleen	21	0.002	0.771	0.0077	0.0136	0.0028	0.7806	0.93	0.81	0.00
Plasma	21	0.0449	0.96	0.00147	0.00083	0.0028	0.9606			0.631[i]
Plasma	22	0.0449	0.96	0.00147	0.00083	6E-05	0.9606[h]			
Whole blood	21	0.0816	0.84	0.0013	0.002	0.0028	0.8414[g]			
Whole blood	22	0.0816	0.84	0.0013	0.002	6E-05	0.8414[f]			
Erythrocytes	21	0.0367	NU[j]	NU	NU					

				Lipids						
				Neutral	Phospho					
Human tissue	Eq. no.[b]	Tissue (V_t)[c]	Water (v_w)[c]	(V_{nl})[c]	(V_{pl})[c]	$K_{o:w}$[d]	Numerator[e]	$P_{t:b}$[k]	$P_{t:p}$[l]	$V_t*P_{t:p}$[m]
Adipose	22	0.11957	0.18	0.79	0.002	6E-05	0.1814	0.22	0.19	0.02
Adipose	21	0.11957	0.18	0.79	0.002	0.0028				
Bone	21	0.085629	0.439	0.074	0.0011	0.0028	0.4400	0.54	0.46	0.04
Brain	21	0.02	0.77	0.051	0.0565	0.0028	0.8097	0.99	0.86	0.02
Gut	21	0.0171	0.718	0.0487	0.0163	0.0028	0.7296	0.89	0.77	0.01
Heart	21	0.0047	0.758	0.0115	0.0166	0.0028	0.7697	0.94	0.81	0.00
Kidney	21	0.0044	0.783	0.0207	0.01622	0.0028	0.7944	0.97	0.84	0.00
Liver	21	0.026	0.751	0.0348	0.0252	0.0028	0.7688	0.94	0.81	0.02
Lung	21	0.0076	0.811	0.003	0.009	0.0028	0.8173	1.00	0.86	0.01
Muscle	21	0.4	0.76	0.0238	0.0072	0.0028	0.7651	0.93	0.81	0.32
Skin	21	0.0371	0.718	0.0284	0.0111	0.0028	0.7259	0.88	0.77	0.03
Spleen	21	0.0026	0.788	0.0201	0.0198	0.0028	0.8019	0.98	0.85	0.00
Plasma	21	0.0424	0.945	0.0035	0.00225	0.0028	0.9466			0.525[i]
Plasma	22	0.0424	0.945	0.0035	0.00225	6E-05	0.9466[m]			
Whole blood	21	0.0771	0.82	0.0032	0.002	0.0028	0.8214[l]			
Whole blood	22	0.0771	0.82	0.0032	0.002	6E-05	0.8214[k]			
Erythrocytes	21	0.0347	NU	NU	NU					
All tissue less plasma		0.956								

[a]Values for monoprotic acids. Used in Eq. 15 to calculate $D^*_{vo:w}$.
[b]Equation number used to calculate partition coeofficients: $P_{t:b}$ and $P_{t:p}$.
[c]Values used in Eqs. 21 and 22 as presented in Poulin and Thiel (2002a). Values for Eq 6 in Poulin and Thiel (2002a) are: E:P (erythrocyte:plasma *in vivo*) = 1, B:P (blood:plasma *in vitro*) = 1, and Ht (hematocrit in blood) = 0.45.
[d]$K_{o:w}$ as calculated in Eq. 14 .
[e]Numerator value from Eq. 21 or 22 depending on tissue.
[f]Denominator value used in Eq. 21 or 22 for $P_{t:b}$ in the rat.
[g]Denominator value used in Eq. 21 or 22 for $P_{t:p}$ in the rat.
[h]Denominator value used in Eq. 21 or 22 for $Vt*P_{t:p}$ in the rat.
[i]Volume distribution at steady state (Vdss; $L kg^{-1}$).
[j]NU, not used in Poulin and Thiel (2002a).
[k]Denominator value used in Eq. 21 or 22 for $P_{t:b}$ in the human.
[l]Denominator value used in Eq. 21 or 22 for $P_{t:p}$ in the human.
[m]Denominator value used in Eq. 21 or 22 for $Vt*P_{t:p}$ in the human.

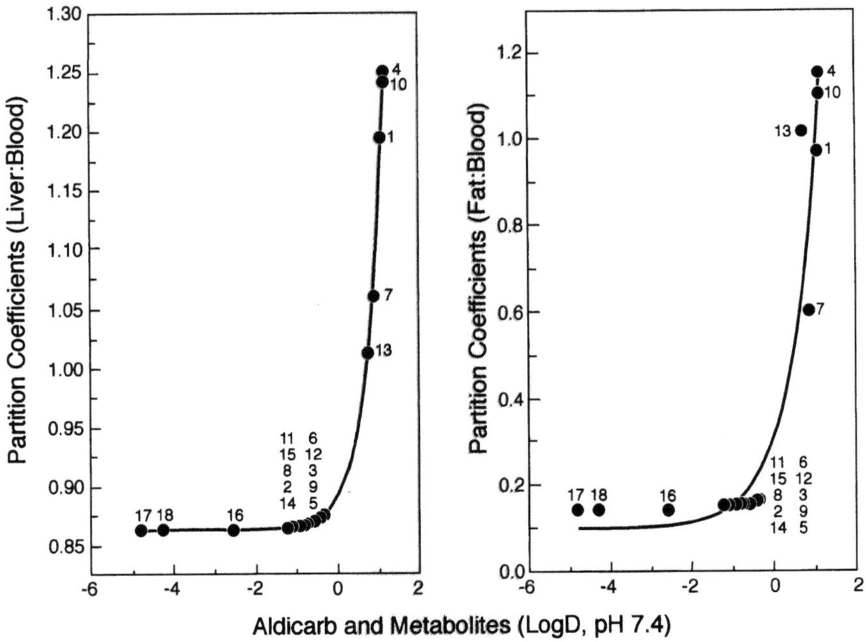

Fig. 6. Liver : blood and fat : blood partition coefficients of aldicarb and metabolites as a function of ACD $\log D_{pH7.4}$. Eq. 21 involving ACD $\log D_{pH7.4}$ was used to calculate partition coefficients for liver. Eq. 22 involving the conversion of ACD $\log D_{pH7.4}$ to $\log D^*_{vo:w}$ using Eq. 13 was used to calculate partition coefficients for fat. The Henderson–Hasselbach equations were not used. Aldicarb and its metabolites are identified in Table 14 (Appendix A).

D. Glucuronidation and Transcellular Transport

In recent years investigators have revealed that metabolism can be altered by changes occuring in drug transport. According to Benet et al. (2003), the incorporation of efflux and uptake processes lead to better predictions of drug clearances from *in vitro* systems. Of interest is the interaction between P-glycoprotein and CYP3A4 (efflux-metabolism alliance); OATPs (organic anion-transporting peptides) and CYP3A4 (uptake transporter-metabolism); and MRP2 (multidrug-resistant protein) and UGTs (UDP-glucuronosyltransferases).

Hepatic transporters are categorized into two groups: efflux transporters that are localized on the canalicular (apical) membrane, and the uptake transporters that are found on the sinusoidal (basolateral) membrane (Lam and Benet 2004) of hepatocytes. According to Chang and Benet (2005), the V_{max} and K_m constants for glucuronidation by human liver microsomes are 20.2 nmol min^{-1} mg^{-1} and 216 µM, respectively, with the exit of 1-naphthyl glucuronic acid being a slower or faster (not determined) process. Endogenous transporters in conjunction with partition coefficients are believed to

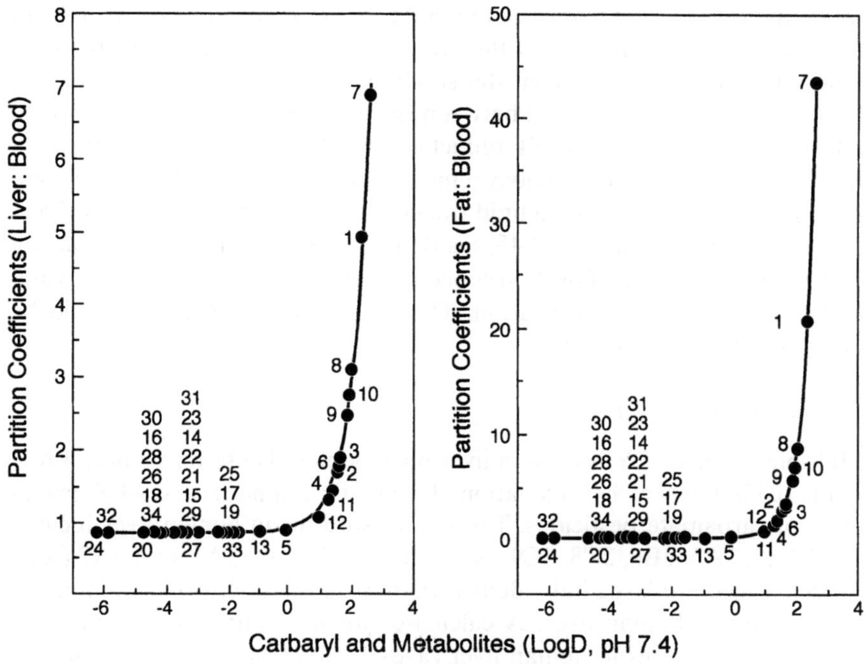

Fig. 7. Liver : blood and fat : blood partition coefficients of carbaryl and metabolites as a function of ACD $\log D_{pH7.4}$. Equation 21 involving $\log D_{pH7.4}$ was used to calculate partition coefficients for liver. Eq. 22 involving the conversion of ACD $\log D_{pH7.4}$ to log D*vo:w using Eq. 13 was used to calculate partition coefficients for fat. The Henderson–Hasselbach equations were not used. Carbaryl and its metabolites are identified in Table 15 (Appendix A).

be active in determining the distribution of formed metabolites. The importance of the activities of transporters and metabolizing enzymes in the absorption, distribution, metabolism, and elimination (ADME) of carbamate pesticides has yet to be established. Inhibitors of these systems may be used to study their effect on ADME (Lau et al. 2003).

E. Metabolic Enzymes, P450s, and CaEs

Cytochrome P450s

CYP450 was first shown to be a hemoprotein by Omura and Sato (1964). It has a noncovalently bound iron protoporphyrin IX prosthetic group, similar to the heme in *b*-type cytochromes, hemoglobin, and myoglobin. The heme prosthetic group may be removed from the protein by acidic acetone, alkali, or pyridine. The binding of carbon monoxide provides a means for their analysis and quantification. The enzymes of the CYP-dependent monooxygenase system are embedded in the endoplasmic reticulum of liver hepatocytes and the cells of other organs. The endoplasmic reticulum is harvested from a cell homogenate by centrifugation at 16,000 g to remove large particles and then at 105,000 g to pellet the microsomes (endoplasmic reticulum) from the supernatant (Lu and West 1980). The P450 enzymes are distributed between the two surfaces of the microsomal bilayer with the NADPH-CYP reductase and P450 facing the cytoplasm of the cell. Therefore, the monooxygenase system is composed of three components: reductase protein, a lipid fraction, and CYP protein. The purification of rat and human liver P450s is described by Guengerich and Martin (1998). An estimation of protein concentrations provides a means for evaluating the progress of purification. The content of CYP heme should be about 18 nmol mg^{-1} protein.

Multiple Forms of CYPs

Multiple forms of CYPs exist in liver microsomes. These forms play a role in the oxidation (i.e., hydroxylation of aromatic, aliphatic, and alkyl groupings) of carbamate pesticides. The major isoforms in human liver include CYP 1A2, 2A6, 2B6, 2C8, 2C9, 2C19, 2D6, 2E1, 3A4, 3A5, and 3A7 (fetal livers). There are large individual variations in the microsomal content of these forms in human liver. A calculator for determining the content of these P450 isozymes in human liver (ages 1–18 yr) was recently presented by Foxenberg et al. (2006, 2007). The recombinant human CYPs are available from a Baculovirus-insect cell expression system along with the coexpression of NADPH-CYP reductase (Supersomes; Gentest, Woburn, MA). Yields range from 50 to 250 pmol P450 mg^{-1} microsomal protein as determined by CO difference spectral analysis (Hood et al. 1998).

In Vitro Metabolism of Aldicarb, Carbaryl, and Carbofuran by Human and Rat Liver Microsomes and Individual CYPs

Aldicarb was converted by rat liver microsomes to aldicarb sulfoxide (Pelekis and Krishnan 1997) with V_{max} and K_m values of 5.41 nmol min^{-1} mg^{-1} microsomal protein and 184 μM, respectively. Negligible quantities of aldicarb sulfone were found after incubation. The results imply that under in vivo conditions aldicarb sulphoxidation is not likely to be saturated at lethal dose levels. CYP and the flavin monooxygenase systems (FMO) are involved in this process (Perkins et al. 1999). Selective *in vitro* inhibition of flavin-containing and CYP monooxygenases confirmed that the former enzymes mainly catalyze sulfoxide production whereas the latter catalyze sulfone formation (Montesissa et al. 1994).

Three major metabolites of carbaryl (i.e., 5-hydroxy-, 4-hydroxy-, and methylolcarbaryl) were formed by human liver microsomes or individual CYP isoforms (Tang et al. 2002). CYP1A1 and 1A2 formed 5-hydroxylcarbaryl while CYP3A4 and 1A1 were the most active in forming 4-hydroxycarbaryl. Activity assays were performed by the incubation of carbaryl (up to 500 μM) with microsomes (1 mg protein) or CYPs (18–50 pmol P450) in the presence of an NADPH-generating system, 100 mM Tris (pH 7.4) buffer containing 5 mM $MgCl_2$ for 15 min. In the case of the CYPs the buffers used were recommended by Gentest. Analysis was performed using HPLC and analytical standards.

Kinetic assays used carbaryl concentrations ranging from 10 to 1,000 µM under the conditions described above. V_{max} and K_m values for HLMs and individual CYPs are given in Appendix B, Tables 34 and 35 for carbaryl. 1-Naphthol was not a major product of CYP-catalyzed reactions in the presence or absence of NADPH. Conditions did not appear to be favorable for the hydrolysis of carbaryl by CaEs. Carbaryl is believed to be both an inhibitor and a substrate for these enzymes. In Table 34, the in vitro carbaryl HLM V_{max} values (µmol min^{-1} mg^{-1} protein) may be expressed in vivo values (µmol hr^{-1} kg^{-1} bw). This calculation is accomplished by multiplying the *in vitro* values by 60 min hr^{-1}; 30 mg microsomal protein/g of liver; and by 27 g liver/kg bw to give the in vivo value. In Table 35, the in vitro CYP values for carbaryl are expressed in nmol min^{-1} $nmol^{-1}$ P450. These values may be converted to *in vivo* values (nmol hr^{-1} kg^{-1} bw) by multiplying the values by CYP content in nmol mg^{-1} microsomal protein, 60 min hr^{-1}; 30 mg microsomal protein/g liver; and by 27 g liver/kg of bw to give the *in vivo* value.

The effects of carbaryl were investigated in conjunction with microsomal hepatic lipid peroxidation, and NADPH-dependent reductase activities. Carbaryl did not affect lipid peroxidation under *in vivo* conditions. Moreover, following administration of the compound, the activities of NADPH-cytochrome reductase as well as NADPH-neotetrazolium reductase was not influenced by carbaryl (Beraud et al. 1989).

In a paper by Cho et al. (2006), CYP3A4 converted 1-naphthol to 1,4-dihydroxydihydronaphthalene (4-OH naphthol) with V_{max}, K_m values of (1.58×10^{-3} µmol min^{-1} mg^{-1} protein), 77 µmol hr^{-1} kg^{-1} bw, and 80 µM, respectively. A number of other metabolites were found but not identified. These metabolites may be 2-OH naphthol, 5-OH napththol, 3,4-dihydroxydihydrol naphthol, and 5,6-dihydroxydihydrol naphthol in Appendix B under carbaryl. On the basis of protein content in the liver CYP3A4, 1A2, and 2C19 were considered the most important isoforms for 1-naphthol metabolism.

Usmani et al. (2004a) studied the *in vitro* metabolism of carbofuran by human liver microsomes (HLMs), mouse liver microsomes (MLMs), and rat liver microsomes (RLMs), and CYPs to 3-OH carbofuran (and two minor metabolites). K_m values were 1.974, 0.207, and 0.551 mM for HLMs, RLMs, and MLMs, respectively. The V_{max} values were 3.30 and 5.50 nmol min^{-1} mg^{-1} protein for HLMs and RLMs. The intrinsic clearance rates for RLMs and MLMs were 14 fold greater than HLMs. CYP3A4 is the major isoform responsible for the oxidation of carbofuran in humans, with CYP1A2 and 2C19 possessing some but less activity. The activity of HLMs from 17 single donors varied by more than 5 fold, indicating that the specific content of at least CYP3A4 was not constant in these microsomes. The HLM and individual CYP V_{max} and K_m values for carbofuran are summarized in Tables 34 and 35, respectively.

Pretreatment of rats with tetraisopropylpyrophosphoramide (iso-OMPA) (CAS no. 513-00-8) 1 hr before carbofuran potentiated the toxicity of this *N*-methylcarbamate insecticide threefold. CaE activity in a variety of organs including brain, muscle, liver, and plasma was significantly reduced, while acetylcholinesterase (AChE) activity was unchanged. Significant inhibition of AChE was observed after the combination of iso-OMPA and carbofuran (Gupta and Dettbarn 1993).

Buronfosse et al. (1995) showed the sulfoxidation of methiocarb is catalyzed by FMO and CYP with 50% being carried out by each enzyme system. Stereoselective sulfoxidation occurs only for FMO with an enantiomeric excess of 88% in favor of the (A)-enantiomer. Lung and kidney microsomes have high flavin-containing monooxygenase levels. FMO is important relative to CYP in these tissues. Thioether-containing organophosphates are effective substrates for the flavin-containing monooxygenase in mouse liver microsomes, with K_m values between 3.5 and 36 nM. According to Tynes and Hodgson (1985), thioether-containing carbamates are less effective substrates in other animals having K_m values near 280 nM (Tynes and Hodgson 1985). The sulfoxidation of thioethers by HLMs was found to be predominantly metabolized by P450 isozymes (85%–90%) compared with FMO isoforms. The highest rates were obtained with CYP1A1, 1A2, 3A4, 2B6, 2C9*1, 2C19, 2D6*1, and FMO1 (Usmani et al. 2004b).

Very little if any work has been done on the *in vitro* metabolism of oxamyl in humans. *In vitro* rat liver studies were carried out by Harvey and Han (1978). NADPH fortified microsomal plus soluble rat liver fractions failed to yield N-demethylated oxamyl oxime or oxamyl acid. The data supports the N-demethylation of oxamyl as the pathway to des *N*-methyl oxime or des *N*-methyl oxamyl acid.

Hydrolysis of Carbamates by CaEs

Mammalian CaEs are located in the endoplasmic reticulum, cytosol, and mitochondria of cells. Hosokawa et al. (1990) compared the liver microsomal CaE activity in nine animal species and humans. A review of mammalian carboxylesterases by Satoh and Hosokawa (1998) lists (in table 1) the catalytic activity of highly purified CaE isozymes from mammalian liver microsomes. Microsomal CaEs hydrolyze *p*-nitrophenylacetate, malathion, butanilicaine (CAS no. 3785-21-5), and isocarboxazid (CAS no. 59-64-0) with RLMs having the highest specific activity and humans the lowest. Activity was reported in $\mu mol\,min^{-1}\,mg^{-1}$ of microsomal protein.

CaEs are believed to be involved in the hydrolysis of the carbamic ester bond of pesticidal carbamates (Gupta and Dettbarn 1993). 1-Naphthol ([S], 125 μM carbaryl) was generated at the rate of $0.034\,nmol\,min^{-1}\,mg^{-1}$ protein with pooled HLM in the presence and absence of an NADPH-regenerating system (Tang et al. 2002). NADPH was used to determined whether an oxidative enzyme was involved in the hydrolysis. This rate is equivalent to 1.65 μmol hr^{-1} kg^{-1} body weight. 1-Naphthol was also formed by cytosolic enzymes from carbaryl (125 μM) at a rate of 0.008 nmol min^{-1} mg^{-1} protein. These rates are low when compared to rates (specific activity, 9.1 μmol min^{-1} mg^{-1} protein; K_m, 12.1 μM) obtained using *p*-nitrophenyl acetate as substrate (Stok et al. 2004).

Ross et al. (2006) determined the activity of recombinant CE (i.e., CaE) and liver microsomes by measuring the production of *p*-nitrophenol liberated from *p*-nitrophenyl valerate, acetate, or butyrate ([S], 500 μM) at 400 nm on a spectrophotometer. An extinction coefficient of 13 cm^{-1} mM^{-1} was used to convert the slopes of each activity curve to specific activities. Specific activities for CaEs in rat and pooled HLMs were 9.33 and 2.21 μmol min^{-1} mg^{-1} protein, respectively, for *p*-nitrophenyl acetate. The kinetic parameters (V_{max}, nmol min^{-1} mg^{-1}; K_m, μM) for the hydrolysis of *trans*-permethrin by RLMs and pooled HLMs were, respectively, 1.34, 19.16 and 1.12, 20.66. CaE activity was 39 times greater toward *trans*-permethrin than toward carbaryl. The hydrolysis of carbaryl may have been reduced by the inhibitory effects of carbaryl on CaE (Stok et al. 2004).

Rabbit serum albumin catalyzes the hydrolysis of carbaryl (k_{cat}, 7.1×10^{-5} sec^{-1} or 0.256 hr^{-1}; K_m, 240 μM) and *p*-nitrophenyl butyrate, with butyrate ester being a competitive inhibitor of the hydrolysis of carbaryl (Sogorb et al. 2002). The inhibition of carbaryl hydrolysis by sulfhydryl-blocking agents suggests that a cysteine residue plays an important role in the active center. Human, chicken, and bovine serum albumins are also able to hydrolyze carbaryl.

F. Models for *In Vivo* Metabolism of Carbamates

The metabolic pathways for the 10 carbamates of interest are given in Tables 24–33 (see Appendix B). The tables were developed from pathway data in published literature and from physiological models developed from the pathways. A physiological model for aldicarb is presented in Appendix C. Similar physiological models were developed for each of the remaining 9 carbamates but are not included in Appendix C. ACSL programs were written for each physiological model. The models may be used to develop graphic models using ACSL's Graphic Modeler. ERDEM is an example of a graphic model. The metabolic pathway data in Appendix B may be used in command files for programming ERDEM. The following paragraphs briefly describe the metabolic information found in the literature for the 10 carbamates of interest.

Aldicarb

The metabolic pathway is described in Table 24 (Appendix B). This carbamate pesticide is readily biotransformed by animals and humans (Andrawes et al. 1967; Baron 1994; Knaak et al. 1966; Marshall and Dorough 1977,

1979; Tobia et al. 2004). Urinary metabolic products of aldicarb include the more toxic aldicarb sulfoxide (~35%), the less toxic sulfone (~1.0%), oxime sulfoxide (~35.5%), oxime sulfone (~2.0%), and neutrals or polar metabolites (~30.5%) as the major products (Knaak et al. 1966). The neutrals are most likely nitrile sulfoxide and sulfone and alcohol sulfoxide and sulfone. The structures of the metabolites are listed in Table 14 (see Appendix A). Andrawes et al. (1967) reported that 80% of an oral dose to rats was eliminated in urine and 4% in feces. Approximately 50% of the urinary metabolites were water soluble and could not be extracted into organic solvents. The hydrolysis products, aldicarb oxime, oxime sulfoxide and oxime sulfone, have no significant toxicity to animals or insects. The nitriles come from the degradation of aldicarb, aldicarb sulfoxide and aldicarb sulfone and are the primary stable products of aldicarb degradation (Cobb et al. 2001). Eighteen analytical standards were prepared by Durden et al. (1970) for studies involving the metabolism of aldicarb in plants and animals.

Carbaryl

The metabolic pathway is described in Table 25 (Appendix B). The principal metabolic pathways of carbaryl are ring hydroxylation and hydrolysis (Dorough and Casida 1964; Knaak et al. 1965; Knaak and Sullivan 1967; Knaak et al. 1968). As a result, numerous metabolites (aglycones) are formed and subjected to conjugation with the formation and urinary excretion of water-soluble sulfates and glucuronides (Cardona and Dorough 1973; Knaak et al. 1967). Hydrolysis results in the formation of 1-naphthol, carbon dioxide, and methyl-amine. Hydroxylation produces 4-hydroxycarbaryl, 5-hydroxycarbaryl, *N*-hydroxy-methylcarbaryl, 5,6-dihydro-5,6-dihydroxycarbaryl, and 1,4-naphthalendiol (Leeling and Casida 1966; Sullivan et al. 1972). The principal metabolites in humans are the glucuronides of 1-naphthol and 4-hydroxycarbaryl and the sulfate of 1-naphthol (Knaak et al. 1968). Chen and Dorough (1979) found no evidence for the conjugation of naphthyl-^{14}C-carbaryl with glutathione or mercapturic acid in their studies. The hydrolytic pathway appears to be a major pathway in humans with ring hydroxylation of either 1-naphthol or carbaryl as secondary pathways (Knaak et al. 1968). Under normal exposure conditions, the accumulation of carbaryl in animals is unlikely.

Enterohepatic cycling of carbaryl metabolites in rats has been reported by Marshall and Dorough (1979) and Struble et al. (1983). According to Marshall and Dorough (1979), 37.5% of an oral dose ($0.01\,mg\,kg^{-1}$ bw) was excreted in bile after 3 hr, principally as glucuronides. The glucuronides are absorbed, metabolized, and recycled, resulting in 1.4% of the dose ultimately being eliminated in feces. Metabolites identified in bile were 5,6-dihydro-5,6-dihydroxycarbaryl glucuronide (12%–18% of biliary ^{14}C), and the conjugated isomer of hydroxymethylcarbaryl (2% of biliary ^{14}C). The percentage of biliary ^{14}C decreased with higher dosages, with doses of 1.5–30.0 mg kg^{-1} bw resulting in three times more biliary ^{14}C than a 300 mg kg^{-1} bw dose (Struble et al. 1983).

Carbaryl, 1-naphthol, and their hydroxylated metabolites were separated on thin-layer glass plates or silica gel columns and identified by cochromatography with known standards. The conjugates, glucuronides and sulfates, were separated by weak ion-exchange chromatography (i.e., DEAE cellulose) and identified by cochromatography with known standards (Knaak et al. 1965). Separation of closely related conjugates (i.e., glucuronides of 4-OH and 5-OH carbaryl) was difficult to achieve. In many cases it was necessary to hydrolyze the conjugates using glucuronidase or sulfatases and to identify the aglycones using silica gel column or thin layer chromatography. Knaak et al. (1967) used GC to separate several glucuronides on a SE-30 column as the acetylated or silylated derivatives of their methyl esters. In practice, methylation was carried out after acetylation, whereas in the case of the silylated derivatives, methylation was performed first followed by silylation.

Carbofuran

The metabolic pathway for carbofuran is described in Table 26 (Appendix B). The information on the pathway was extracted from Dorough (1968), Ferguson et al. (1984), Fukuto (1972), Ivie and Dorough (1968), Knaak et al. (1970), Knaak (1971), Marshall and Dorough (1979), Metcalf et al. (1968), Roberts and Hutson (1999), and Usmani et al. (2004a).

The fate of ring-C^{14} and carbonyl-C^{14} carbofuran was studied by Knaak et al. (1970) in the dairy cow with 83% and 12% of the dose, respectively, being eliminated in urine. Less than 4% of the administered ring-labeled carbofuran was eliminated in feces. The major metabolites were the sulfates of carbofuran phenol (45.4%) and 3-keto-7-phenol (10.4%) and the glucuronides of carbofuran phenol (14.4%), 3,7-diol (15.9%), and 3-hydroxy carbofuran (4.2%). 3-Hydroxycarbofuran-carbonyl-C^{14} is hydrolyzed in the cow to almost the same extent as carbonyl-C^{14}-carbofuran.

Ivie and Dorough (1968) separated the urinary metabolites of ring-C^{14} carbofuran from the cow into organo extractables and water solubles, with 90%–97% of the label being present as water solubles. Carbofuran phenol (34.36%) was the major hydrolysis product recovered in the water phase of a 12-hr urine after acid hydrolysis along with 3-keto-7-phenol and the 3,7-diol. Small quantities of 3-OH carbofuran (9.14%) and 3-OH-*N*-hydroxymethyl carbofuran (2.16%) were also reported. Acid hydrolysis failed to completely release the aglycones of carbofuran from their conjugates, leaving up to 57% in the water phase. The differences reported between the two cow studies appear to result from (1) the administration of carbofuran orally by bolling gun, (2) the administration via a rumen fistula, and (3) the analytical procedures used to determine the urinary metabolites.

In rats, oxidative and hydrolytic metabolites accounted for about 70% and 21%, respectively, of the administered radioactivity (Roberts and Hutson 1999). Oxidative metabolism was the predominant pathway in mice (Dorough 1968). Ferguson et al. (1984) presented tissue-time course data for the disappearance of carbofuran and the appearance of 3-OH carbofuran in the rat after the oral and intravenous administration of carbonyl-C^{14}-labeled carbofuran. Organoextractable metabolites were analyzed using thin-layer chromatography (TLC), while water-soluble conjugates were first hydrolyzed and then analyzed by TLC. No information was obtained by Ferguson et al. (1984) regarding the fate of the hydrolyzed products (i.e., carbofuran phenol, 3,7-diol, and 3-keto-7-phenol).

According to Marshall and Dorough (1979), biliary elimination of glucuronide metabolites amounted to 2.6% of dose 3 hr after dosing and 28.5% after 48 hr. The glucuronides, principally 3-OH carbofuran glucuronide, were absorbed from the gastrointestinal tract, metabolized, and eliminated in urine or absorbed back into bile to be recycled. The overall result was the elimination of 4% of the dose in feces and the remainder in urine.

Formetanate

The metabolic pathway for formetanate is described in Table 27 (Appendix B). Information on the pathway was taken from Sen-Gupta and Knowles (1970), Fukuto (1972), and Roberts and Hutson (1999).

In rats treated with formetanate, 60%–80% of the urinary excretory products were water-soluble conjugates. The major pathway was the formation of the formamino derivative, 3-formaminophenyl *N*-methylcarbamate, followed by the enzymatic removal of the *N*-methylcarbamoyl group to give formaminophenol. Deformylation of formaminophenol (Ahmad and Knowles 1971) resulted in 3-aminophenol whereas subsequent N-acetylation produced 3-acetamidophenol.

Major aglycones of sulfuric and glucuronic acids were 3-acetamidophenol, 3-formaminophenol, and 3-aminophenol. Water-soluble carbamates include 3-formaminophenyl-*N*-methylcarbamate and demethylformetanate. Sen-Gupta and Knowles (1970) found 3-aminophenyl *N*-methylcarbamate and formetanate phenol in plants but did not find them in rat urine. An oral dose of ^{14}C-formetanate was rapidly absorbed and excreted by rat, 85% in urine and 8% in feces. After 72 hr, 2% of the ^{14}C was still retained.

Methiocarb

The metabolic pathway of methiocarb is described in Table 28 (Appendix B). The information was taken from Kuhr and Dorough (1976), Menzie (1974), Oonnithan and Casida (1968), Roberts and Hutson (1999), Strother (1972), and Van Hoof and Heyndrickx (1975).

When rat and human liver enzymes were incubated with [^{14}C-carbonyl]-, [^{14}C-4-methyl]-, or [^{14}C-*N*-methyl] methiocarb, about a dozen metabolites were observed. The major metabolite was methiocarb sulfoxide, with the human liver producing slightly less than the rat (13% vs. 16%), according to Strother (1972). Hydrolysis yielded the phenolic moiety of methiocarb, oxidation of methiocarb sulfoxide yielded the sulfone, while hydroxylation produced *N*-hydroxymethyl methiocarb. Products were eliminated in free and conjugated forms. Several metabolic products from human liver studies were not observed in rat studies (Strother 1972). Rats given a single oral dose of 5.0mg methiocarb eliminated up to 2.3% of the dose in the urine as unchanged methiocarb and 3.3% as its phenolic metabolites, mostly within 48hr of administration (Hayes and Laws 1991).

Methomyl

The metabolic pathway of methomyl is described in Table 29 (Appendix B). Information on the metabolic pathway was taken from Harvey et al. (1973), Huhtanen and Dorough (1976), and Reiser et al. (1997). In the study by Harvey et al. (1973), radiolabeled methomyl (*S*-methyl [1-^{14}C] *N*-[(methylcarbamoyl)-oxy] thioacetimidate) administered orally to rats is metabolized and rapidly eliminated within 24hr in the ratio of one part [^{14}C]carbon dioxide, two parts [1-^{14}C]acetonitrile, and one part urinary metabolites. The identities of the urinary metabolites were not established by Harvey et al. (1973); however the absence of methomyl, *S*-methyl *N*-hydroxythioacetimidate, the *S*-oxide of methomyl, and the *S,S*-dioxide of methomyl and conjugates was demonstrated. Methomyl can exist in two geometric configurations (Huhtanen and Dorough 1976), Z and E isomers. In rats, carbonyl- or oximino-labeled *syn*-methomyl (Z isomer) was metabolized to carbon dioxide and acetonitrile at a 2:1 ratio. In contrast, the anti isomer (E isomer) was metabolized predominantly to acetonitrile. It was concluded on the basis of this and other evidence that a Beckman rearrangement of the syn- and anti oximes occurs before formation of carbon dioxide and acetonitrile.

The methomyl PBPK/PD model developed in this review, but not presented, contains a lung compartment for removing ^{14}C-carbon dioxide and acetonitrile formed during the metabolism of methomyl in the liver. The Z isomer of methomyl is partially converted to the E isomer during metabolism. The Z isomer is metabolized to carbon dioxide, while the E isomer goes to acetonitrile. AChE inhibition is believed to be largely caused by the Z isomer, although the E isomer may also produce inhibition.

The acetonitrile formed by hydrolysis of methomyl is conjugated with glutathione. The glutathione conjugate is hydrolyzed to form the glutamylcysteine, cysteinylglycine, and the cysteine conjugates of acetonitrile. The cysteine conjugate goes on to form an *N*-sulfate–cysteine conjugate (Reiser et al. 1997).

The main metabolites of methomyl are acetonitrile per se (5%–34% of dose), respiratory carbon dioxide (19%–34% of dose), and unknown urinary metabolites in a ratio of 2:1:1. One of the urinary metabolites (1%–4% of dose) was identified as the *N*-sulfate–cysteine conjugate. A large number of the urinary metabolites are unknown products.

Oxamyl

The metabolic pathway for oxamyl is described in Table 30 (Appendix B). This information was extracted from Chang and Knowles (1979) and Harvey and Han (1978). Oxamyl is degraded by three pathways, hydrolysis to the oxamyl oxime (oximino metabolite), des *N*-methyl oxamyl, or enzymatic conversion via oxamyl nitrile (DMFC) to oxamyl acid (*N,N*-dimethyloxamic acid). The oxime and acid are both conjugated with glucuronic acid along with des *N*-oxamyl oxime and des N-oxamyl acid (N-methyloxamic acid) from des N-methyl oxamyl before elimination in urine and feces (>70%) (Harvey and Han 1978). No oxamyl or other organo-soluble metabolite was detected in urine, feces, or tissues. ^{14}C from oxamyl oxime is extensively converted to $^{14}CO_2$ or incorporated into amino acids accounting for most (>50%) of the radioactivity retained in the tissues. Similar products were found in the urine and feces of the mouse from intraperitoneally (IP)-administered oxamyl-^{14}C.

In the intact rat administered ^{14}C-oxamyl (2.5–4.6 mg kg^{-1} bw), most of the dose (68%–72%) was recovered in urine with lesser amounts in feces after 72 hr. Conjugates of the oximino compound, the acid, and their monomethyl derivatives constituted >70% of the metabolites excreted in the urine and feces. Less than 0.3% of the oxamyl was exhaled as carbon dioxide (Harvey and Han 1978). Mice treated IP with ^{14}C-oxamyl eliminated 96.4% of the dose by 96 hr, 88.7% in urine and 7.7% in feces (Chang and Knowles 1979).

Extensive degradation/metabolism of (1-^{14}C) oxamyl was observed in the goat. Radioactive thiocyanate was the major metabolite identified in milk as well as in the methanol/water extracts for all tissue samples. Oxamyl-derived residues in the urine derived from oxamyl nitrile have been identified as thiocyanate, des *N*-oxamyl acid, oxamide, and des *N*-methyloxamide (Li et al. 1997). Thiocyanate is generated from cyanide through the reaction with thiosulfate catalyzed by rhodanese (Solomonson 1981). The metabolic pathway in the ruminant (goat) appears to be significantly different from the pathways determined in monogastric animals such as the rat and mouse.

Pirimicarb

The metabolic pathway of this dimethylcarbamate is described in Table 31 (Appendix B). Information on the pathway was taken from Baron (1991). Pirimicarb's major urinary metabolites in the rat, dog, and cow are similar, resulting from oxidative and hydrolytic mechanisms, and consisting primarily of hydroxypyrimidines with modifications of the alkyl constituents of the heterocyclic moiety. Of the administered dose, 2-dimethylamino-5, 6-dimethyl-4-hydroxypyrimidine (DDHP) accounted for 10%–16.3%, 2-methylamino-5,6-dimethyl-4-hydroxypyrimidine (MDHP) for 20.5%–41%, 2-amino-5,6-dimethyl-4-hydroxypyrimidine (ADHP) for 12.9%–21%, and 2-dimethylamino-6-hydroxymethyl-5-methyl-4-hydroxypyrimidine (DHHP) for 1.8%–5.7%. The major metabolites were eliminated unconjugated. The glucuronide of pirimicarb phenol and hydroxymethyl pirimicarb are included in Table 31 (Appendix B) as possible metabolites. The glucuronide is more water soluble than the phenol at pH 7.4.

DDHP, MDHP, and ADHP were detected in urine samples of seven workers who had applied pirimicarb (Hardt et al. 1999; Hardt and Angerer 1999). Concentrations of MDHP and ADHP were much higher than that of DDHP, indicating a considerable demethylation capacity in humans. No metabolites were found in urine specimens of controls. The pyrimidines investigated represent sensitive and specific parameters for biological monitoring of exposure to pirimicarb (Hardt et al. 1999).

In an earlier metabolism study involving [1,3-^{15}N-2-2-^{13}C]-labeled pirimicarb mixed with 2-C^{14}-labeled pirimicarb, 10 metabolites were found in the excreta of rats (Hendley and Lam 1981). The metabolites were DHHP, MDHP, ADHP, 2-[(methoxymethyl) methylamino)-4-hydroxy-5,6-dimethylpyrimidine (CAS no. 78195-32-1), 2-[(methoxymethyl) amino]-4-hydroxy-5,6-dimethylpyrimidine (CAS no. 78195-33-2), and 2-[(hydroxymethyl)methylamino]-4-hydroxy-5,6-dimethylpyrimidine (CAS no. 78195-31-0). The last three metabolites were formed by hydroxylation of a methylamino group, methylation of the hydroxy group, and demethylation of an *N*-methyl group, but were not found in the human studies (Hardt et al. 1999). Their log $D_{pH\ 7.4}$ values, 0.15, –0.06, and –0.49, respectively, suggest that these metabolites are more water soluble than pirimicarb phenol, and if present would appear in urine unconjugated.

Propoxur

The metabolic pathway for propoxur is described in Table 32 (Appendix B). Information on the pathway was obtained from the in vitro work of Oonnithan and Casida (1968) and the review of Knaak (1971). RLMs, in the presence of NADPH, converted carbonyl-^{14}C, *N*-methyl-^{14}C, and ring-^{14}C-propoxur to seven metabolites possessing the intact carbamate and isopropoxy group, 2-isopropoxy-4-hydroxyphenyl methylcarbamate, 2-isopropoxy-5-hydroxyphenyl methylcarbamate, 2-isopropoxyphenyl *N*-hydroxymethylcarbamate, and unidentified products. The one metabolite with the phenyl O—C (O)—N—C linkage was 2-hydroxyphenyl methylcarbamate, whereas the three phenols with the isopropoxy group (C—(C) C—O) were 2-isopropoxyphenol, 2-isopropoxy-4-OH phenol, and 2-isopropoxy-5-OH phenol. In the presence of UDP-glucuronic acid, these products would be conjugated with glucuronic acid. Sulfates appeared in the urine of rats after the 2-wk daily oral administration of 30–50 mg kg^{-1} bw of propoxur (Foss and Krechniak 1980; Krechniak and Foss 1982, 1983). At least one of the unidentified metabolites and 2-isopropoxy-5-hydroxyphenyl methylcarbamate are more potent ChE inhibitors than propoxur (Oonnithan and Casida 1968).

Propoxur is biotransformed *in vivo* by depropylation to 2-hydroxyphenol-*N*-methyl-carbamate, and by hydrolysis to the phenol. Ring hydroxylation and isopropoxy hydroxylation followed by conjugation account for the glucuronides detected in urine (ACGIH 1991). The *in vivo* fate of individual 7.5 µmol kg^{-1} bw dosages (IP) of carbonyl-^{14}C and N-methyl-^{14}C propoxur in the rat were studied by Krishna and Casida (1966). ^{14}C-CO_2 was recovered in exhaled air from the carbonyl-(31.2%) and methyl-(21.7%) labeled propoxur after 48 hr, with the remaining percentage being recovered in urine (>60%), feces (~2%), and body tissues (~2%–5%).

According to Hayes and Laws (1991), the relationship of urinary excretion of 2-isopropoxyphenol to propoxur concentration was nearly linear following inhalation exposure. This metabolite was found in the urine 3 d after termination of exposure to 78 mg m^{-3}, 2 d after exposure to 9 mg m^{-3}, and 24 hr after exposure to 0.4 mg m^{-3}.

Thiodicarb

The metabolic pathway is described in Table 33 (Appendix B). Information on the pathway was obtained from WHO/FAO (2000), and the metabolic work on methomyl in Table 29 (Appendix B) (Harvey et al. 1973; Huhtanen and Dorough 1976; Reiser et al. 1997).

Thiodicarb (acetamide-^{14}C) is rapidly absorbed from the gastrointestinal tracts of rats, metabolized to unstable intermediates (i.e., methomyl, methomyl oxime, etc.), and finally to acetonitrile, CO_2, and low molecular weight metabolites. The initial step in the process involves hydrolysis to methomyl. The nature of the isomer, either (E) or (Z), was not determined. In the rat, an oral dose of 2.0 mg kg^{-1} bw thiodicarb (acetamide-^{14}C) reached a peak in plasma after 10 hr and was eliminated in urine (30%) and exhaled air (40%). The urine contained water-soluble materials while the exhaled air contained radiolabeled carbon dioxide and acetonitrile. The overall NOAEL for erythrocytic and splenic effects was 3 mg kg^{-1} bw/d in a 79-wk study (WHO/FAO 2000). Brain ChE depression (>60%) occurred at 5 mg kg^{-1} bw, with pin-point pupils and reduced body temperature.

G. Response: *In Vivo* antiChE activity

The development of antiChE time course data for carbamates has relied heavily upon the administration of the carbamate (e.g., oral, dermal, inhalation) followed by the determination of ChE activity in tissues such as the blood and brain at selected time intervals using spectrophotometric assays such as the Ellman assay (Ellman et al. 1961). Under normal Ellman assay conditions, significant spontaneous reactivation occurs with carbaryl (Nostrandt et al. 1993). Assay modifications that include (1) preincubation of concentrated tissue with concentrated chromogen (i.e., DTNB), (2) dilution to final reaction volume immediately before measurement, and (3) measurement of ChE over a short period of time (5–10 min) gave results comparable to a radiometric method. Significant spontaneous reactivation may still occur with the modified method if the assay is carried out for >10 min (Nostrandt et al. 1993).

Aldicarb

The major metabolite of aldicarb, aldicarb sulfoxide, has been shown to have 76 times greater antiChE activity than the parent compound (DHEW/NCI 1979). Rats receiving an oral dose (0.33 mg kg^{-1} bw) of aldicarb recovered from AChE inhibition 2 hr earlier than rats receiving aldicarb sulfoxide (Knaak et al. 1966). AChE measurements were made using a pH Stat. Cambon et al. (1979) studied the effect of aldicarb on the activity of AChE in tissues from pregnant rats and fetuses. Dosages of 0.001, 0.01, and 0.10 mg kg^{-1} bw were administered to pregnant rats (oral LD_{50} = 1 mg kg^{-1} bw; Kuhr and Dorough 1976) on day 18 of gestation. Tissue samples taken from animals at the two highest dosages 1 and 5 hr post administration showed a significant decrease in AChE activity. The effect persisted in maternal and fetal blood beyond 24 hr. The lowest dose did not change AChE activity after 1 hr in maternal tissues except for the liver, while the AChE activity in fetal blood, brain, and liver were significantly inhibited. AChE determinations were made using the method of Ellman et al. (1961).

Carbaryl

Short-term studies in animal species confirm that carbaryl can cause toxicity from ChE inhibition. Wide variations in the dosage required to induce toxicity in either different species or in one species by different routes of administration can in part be explained by differences in drug disposition. Limited long-term exposure studies in rats and dogs have not demonstrated unexpected adverse effects. However, long-term exposure in pigs results in a progressive neuromyopathy that is associated with structural damage and is not acutely reversible with atropine (Branch and Jacqz 1986a,b).

Motor activity and neuromotor function were examined in adult Sprague-Dawley rats exposed to carbaryl, and behavioral effects were compared with the time course of ChE inhibition. Rats received an IP injection of 0, 4, 8, 16, or 28 mg kg^{-1} bw carbaryl in corn oil 20 min before testing. Dosages of 8, 16, and 28 mg kg^{-1} bw decreased rat maze activity whereas 16 and 28 mg kg^{-1} bw reduced open field activity. Maximum effects of carbaryl on blood and brain ChE and motor activity were seen within 15 min. Maze activity had returned to control levels within 30–60 min, whereas ChE levels remained depressed for 240 min (Ruppert et al. 1983).

Carbofuran

Cambon et al. (1979) studied the effect of carbofuran on the activity of AChE in tissues from pregnant rats and fetuses. Dosages of 0.05, 0.25, and 2.50 mg kg^{-1} bw were administered to pregnant rats (oral LD_{50} = 11 mg kg^{-1} bw; Kuhr and Dorough 1976) on day 18 of gestation. The largest dose, 2.5 mg kg^{-1} bw, resulted in signs of AChE inhibition within 5 min. Eight of 32 rats treated at this level died within 30 min. AChE activity in maternal brain was 27% of control activity whereas activity in fetal brain was 80%. AChE activity was significantly reduced in maternal and fetal tissues taken from surviving dams 1 hr after treatment at the high dose. Activity was considered to be normal after 24 hr. AChE activity in tissues taken from animals at the lower dosages (0.25 and 0.05 mg kg^{-1} bw) 1 hr after administration was less than control values in certain tissues. Decreases were observed in blood from dams and fetuses and maternal liver 1 hr after the administration of 0.05 mg kg^{-1} bw. AChE determinations were made using the method of Ellman et al. (1961).

Formetanate

Phenyl *N*-methylcarbamates exert their insecticidal activity by inhibiting enzyme ChE and do not require activation by metabolism before exhibiting their enzyme-inhibiting properties (Parke 1968). Moser and MacPhail (1987) studied the effects of ChE inhibition on operant behavior in rats. A dose of 0.5 mg kg^{-1} bw formetanate produced a pronounced suppression of response rates in trained rats.

Methiocarb

The antiChE insecticide, methiocarb, inhibits serine hydrolases by carbamylating a serine residue at the catalytic site. This insecticide is viewed as a potential inhibitor of serine hydrolase-dependent immune functions, including interleukin 2 (IL-2) signaling (Casale et al. 1993).

Methomyl

The inhibition of ChE by methomyl is quickly reversed in dogs given 10 mg kg^{-1} bw methomyl without atropine pretreatment. Signs of intoxication disappeared in 2 hr, and blood ChE levels returned to normal in less than 4 hr (ACGIH 2001).

Oxamyl

For subchronic or prechronic exposure in female rats given oxamyl at dietary levels of 100 ppm or higher for 29 d, AChE activity was decreased in the plasma from day 7 on and in the brain at sacrifice. There were no clear effects on erythrocyte AChE levels. In males, AChE activity was marginally depressed in plasma and erythrocytes. AChE activity was not affected by oxamyl at 50 ppm (2.5 $mg\,kg^{-1}\,bw\,d^{-1}$) (Hayes and Laws 1991). According to Kennedy (1986), a 4.86 $mg\,kg^{-1}$ bw single bolus dose of oxamyl dropped blood ChE activity in the rat from 4.3 ± 1.0 to 2.6 ± 0.8 μmol substrate hydrolyzed min^{-1} at 5 min after treatment. Activity fell to 1.8 ± 0.7 μmol substrate hydrolyzed min^{-1} at 4 hr after treatment and returned to normal (5.7 ± 0.8) 24 hr post dosing.

Pirimicarb

Cambon et al. (1979) studied the effect of pirimicarb on the activity of AChE in tissues from pregnant rats and fetuses. Dosages of 2.0, 20, and 50–150 $mg\,kg^{-1}$ bw were administered to pregnant rats (oral LD_{50} = 145 $mg\,kg^{-1}$ bw; Kuhr and Dorough 1976) on day 18 of gestation. All the rats dosed with 50 $mg\,kg^{-1}$ bw died soon after administration. AChE activity in the maternal brain was 20% of control activity while activity in the fetal brain was 65%. There was significant inhibition of AChE activity in all tissues taken from animals at the mid-dose (20 $mg\,kg^{-1}$ bw) 1 hr after administration. The effects were still apparent in blood and liver from dams and fetuses after 5 hr. The lowest dose (2 $mg\,kg^{-1}$ bw) produced a significant decrease in AChE activity in several tissues after 1 and 5 hr with no apparent effect after 24 hr. AChE determinations were made using the method of Ellman et al. (1961).

Propoxur

A nontoxic dose (5 $mg\,kg^{-1}$ bw) of propoxur is potentiated by IP pretreatment of rats with 1.0 $mg\,kg^{-1}$ bw iso-OMPA (Gupta and Kadel 1990). CaE activity was markedly reduced, indicating that propoxur toxicity is enhanced by the inhibition of this hydrolase. Carbamylation of AChE produces an accumulation of ACh and the picture of muscarinic and nicotinic poisoning. Spontaneous hydrolysis of the carbamate–ChE complex occurs in vivo, leading to the disappearance of clinical effects within 24 hr. Penetration of the blood–brain barrier by the carbamates is insignificant; for this reason, few central nervous system (CNS) symptoms occur (Ellenhorn and Barceloux 1988).

Thiodicarb

Thiodicarb is hydrolyzed to methomyl in animals and humans. The antiChE properties of thiodicarb are related to the rate that methomyl is formed in metabolizing tissues and made available to tissue ChEs. In a rat dose–response study, a dermal dose of 33.3 $mg\,kg^{-1}$ bw (322 $\mu g\,cm^{-2}$, 25 cm^2 treated skin) resulted in 50% red cell ChE inhibition in 24 hr (Knaak and Wilson

1985). Carbaryl at dermal dose levels of 417 mg kg^{-1} bw (4,000 μg cm^{-2}, 25 cm^2 treated) produced no red cell ChE inhibition after 24 hr. An automated Ellman procedure using a Technicon AutoAnalyzer II Continuous-Flow Analytical system was used to analyze blood for AChE inhibition (Knaak et al. 1978, 1980). No information is available concerning AChE reactivation using the automated Ellman procedure.

V. Target Enzymes, the ChEs and CaEs

AChE terminates nerve impulses by catalyzing the hydrolysis of the neurotransmitter ACh. Carbamate insecticides control insects by the direct inhibition of AChE in their nervous systems, which eventually leads to respiratory failure and death. A somewhat similar set of events occurs in humans exposed to toxic amounts of these insecticides, except the environmental dose received by exposed individuals is usually lower per kg bw, and recovery generally takes place before respiratory failure or other problems occur.

To model (i.e., PBPK/PD model) these events in animals and humans, the following information should be available: (1) inhibited enzymes (i.e., AChE and BChE), (2) preferred substrate, (3) location of enzymes in tissues, (4) specific content (μmol kg^{-1} tissue), and (4) their bimolecular inhibition rates.

Nigg and Knaak (2000) reviewed the status of the ChEs as part of a review on blood ChEs as human biomarkers of OP pesticide exposure. Carbamate insecticides have been implicated in cases of food (Hunter et al. 1997; Wilson et al. 1989) and applicator poisoning (Knaak, personal communication). The food poisoning resulted from the application of aldicarb (unregistered use) on watermelons in California, while the applicator poisoning involved clothing contaminated with both methomyl and parathion in Imperial County, CA, during the 1970s. The rapid recovery of ChE inhibition by carbamate insecticides has prevented efforts to clearly relate reported illnesses to carbamate exposures. Blood samples must be collected as soon as the illnesses are reported and appropriately analyzed for ChE inhibition (Wilson et al. 1997, 2002). Knowledge of the location, function and structure of the ChEs is needed for understanding their physiological importance and for properly modeling inhibition and recovery.

The CaEs in liver and other tissues are also inhibited by carbamate insecticides. The B-esterases (CaEs) hydrolyze carboxylesters of natural products and drugs, and function as scavengers in the removal of carbamates and OPs. These enzymes are also believed to hydrolyze carbamates, rendering them useless as insecticides.

A. Structure, Multiple Forms, and Distribution

AChE

The structure of AChE from the eel *Torpedo californica* was reported by Sussman et al. (1991). The monomer is an α/β protein that contains 537 amino acids and structurally is a 12-stranded mixed β-sheet surrounded by 14 α-helices. Amino acid sequences are known only for *Torpedo* AChE, *Drosophila* AChE, 85% of fetal bovine AChE, and human BChE. The postulated "anionic site" of AChE that binds the quaternary ammonium ion of ACh is located in a gorge represented by 14 aromatic residues. Charges in the anionic site and active site are believed to stabilize the Ch group. The esteratic or active site serine is at amino acid 200 for red blood cell (RBC) AChE, and at amino acid 198 for plasma BChE (Lockridge et al. 1987; Sutton et al. 1991). The binding sites are represented in Fig. 8 by Sussman et al. (1991). The esteratic site is embedded in a gorge about 20 Å long that reaches halfway into the three-dimensional structure of the protein and is responsible for the hydrolysis of ACh.

AChE exists as mono-, di-, and tetramers of catalytic subunits (G_1, G_2, G_4) with each unit containing active sites. The most complex form, A_{12}, has 12 subunits. The forms are either hydrophilic (soluble) or tightly bound to a phospholipid membrane. The molecular differences in AChE and BChE were reviewed by Chatonnet and Lockridge (1989) and are given in Table 7. Asymmetric or immobilized forms of AChE are found only in peripheral nerves and muscles of vertebrates. Membrane-bound G_4 AChE is found in mammalian brain, while membrane-bound G_2 AChE is found in erythrocytes.

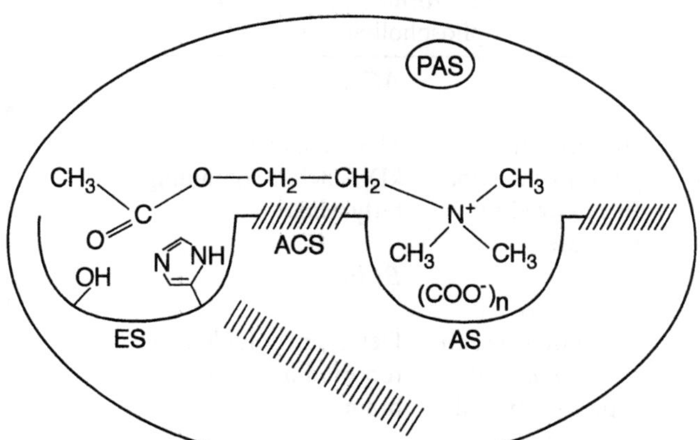

Fig. 8. Schematic representation of the binding sites of AChE based upon previous kinetic, spectroscopic, and chemical modification studies: *ES*, esteratic site; *AS*, anionic substrate-binding site; *ACS*, active site-selective aromatic cation-binding site; *PAS*, peripheral anionic binding site or sites. The *hatched areas* represent putative hydrophobic binding regions. The ACh molecule is shown spanning the esteratic and anionic sites of the catalytic center. Imidazole and hydroxyl side chains of His and Ser are shown with the esteratic site. Within the anionic site, (COO-)n represents six to nine putative negative charges. (Redrawn from Sussman JL et al. 1991 with permission from AAAS).

Table 7. Molecular Isoforms of AChE and BChE.

Hydrophilic, water-soluble forms secreted into body fluids			
		AChE	
G_1[a]	G_2[a]	G_4[a]	A_{12}[b]
Degradation product of G4	Degradation product of G4	Secreted by adrenal gland; peripheral nerve cells, found in plasma, spinal fluid	
		BChE	
G_1	G_2	G_4	A_{12}
Degradation product of G4	Degradation product of G4	Human plasma BChE, 95% of activity in plasma	
Immobilized forms, asymmetrical			
		AChE	
G_1	G_2	G_4	A_{12}
			Torpedo californica muscle of primitive vertebrate
		BChE	
G_1	G_2	G_4	A_{12}
			Muscle of mammals and birds
Amphiphilic globular forms (membrane bound anchored to phospholipids bilayers)			
		AChE	
G_1	G_2	G_4	A_{12}
	Erythrocyte, glycolipid anchor-PI[c], cleaved by protease	Mammalian brain, 20-kDa anchor containing fatty acids	
		BChE	
G_1	G_2	G_4	A_{12}
	Heart of *Torpedo marmorata* and superior cervical ganglion of rat	Detergent-soluble form reported in mammalian brain	

Source: Chatonnet and Lockridge (1989).

[a]Globular forms: G1, G2, and G4 contain one, two, or four subunits. G4 tetramer is an association of two dimmers linked by disulfide bonds; dimmers and monomers appear to be degradation products of tetramers.

[b]A12 contains three tetramers.

[c]PI, phosphatidylinositol.

Genomic DNA analysis and genetic lineage studies from different species suggest that all molecular forms of AChE in higher vertebrates are a product of a single gene (Maulet et al. 1990; Randall et al. 1987; Rotundo 1988; Sikorav et al. 1987), while structural diversity (i.e., G_1, G_2, G_4, A_{12}) is a product of posttranscriptional and translational events.

BChE

BChE is present in regions of the brain in positions not related to AChE, such as capillary endothelial cells, in glial cells, and in neurons. Human BChE is synthesized in the brain. Plasma BChE is a 24% carbohydrate sialoglycoprotein synthesized in the liver (Khoury et al. 1987). In plasma, the products of the BChE gene (BCHE) exist in multiple molecular forms. Three main forms can be recognized by gel electrophoresis: they are designated in order of decreasing mobility as G_1, G_2, and G_4. The G_4 form represents about 95% of the activity. The forms represent globular (G) monomeric, dimeric, and tetrameric forms of the enzyme. G_4 is a highly stable enzyme (Masson and Goasdoue 1986). Earlier studies show that a part of G_4 can dissociate spontaneously into G_1 and G_2 (La Motta et al. 1965; Masson 1979).

Studies involving genomic blots and restriction maps indicate that there is only one human BChE gene (Arpagaus et al. 1990; Lockridge and La Du 1991) located on the q arm of chromosome 3 (Sparkes et al. 1984; Zakut et al. 1989). The enzyme shows a complex genetic and molecular polymorphism that appears to be mainly the result of posttranslational modifications of the gene products. Genomic blots provide strong evidence for one gene in the monkey, dog, rat, mouse, guinea pig, cow, sheep, pig, rabbit, and chicken. The complete amino acid sequence of the usual form of BChE (Lockridge et al. 1987) was used to find the cDNA clone (McTiernan et al. 1987), while cDNA was used to find the gene (Arpagaus et al. 1990).

No genetic variants are known for human RBC AChE. About 10 genetic variants are known for plasma BChE (Lockridge 1990). The human BChE gene contains four exons. Exons 2, 3, and 4 contain sequences for the globular, tetrameric BChE. Exon 2 encodes 83% of the mature protein, from the N-terminal of the protein, Glu 1 to Gly 478. Exon 2 encodes the active site serine, Ser 198, the putative anionic site component, Asp 70, the potential active site histidine, His 438, seven of the total of nine carbohydrate chains per subunit, and two disulfide loops, Cys 65–92 and Cys 252–263.

Five genetic variants of human BCHE have been mapped to exon 2. Variants H, J, and K have 90%, 66%, and 33% lower activity, respectively. Atypical is the dibucaine resistant variant, whereas Silent-1 has no activity. The atypical variant has a reduced K_m for all positively charged substrates and inhibitors but a normal k_{cat}. Atypical BChE has a glycine in position 70 in place of aspartic acid.

The J variant has been mapped to exon 3. Antibody assays indicate that the amount of BChE protein is reduced. In the J variant, the charged amino acid, Glu 497, is replaced by the hydrophobic residue valine.

The K variant has been mapped to exon 4. Antibody assays (Rubinstein 1978) indicate a reduced amount of serum protein. Glycine is substituted for Ala at position 539. The K variant is the most common genetic variant, occurring in homozygous form in 1.7% of the Caucasian population and in heterozygous form in 22.6% of the population. The form is not associated with any abnormal drug response.

CaE

CaEs are members of the α/β hydrolase family of enzymes, which contain a catalytic triad consisting of a nucleophile (Ser-221), a base (His-467), and an orientating acid (Glu-353) and are localized in the endoplasmic reticulum of many tissues (Hosokawa et al. 1987, 1990; Satoh and Hosokawa 1998). Mammalian liver CaEs belong to a family of proteins encoded by multiple genes. The isozymes were initially classified by their substrate specificities and pI (isoionic point). The isozymes are now classified into four main CES groups (carboxylesterase, EC 3.1.1.1) and several subgroups. Two major human liver isozymes, hCE-1 and hCE-2, belong to classes CES1 and CES2. By Northern blot, a single band of approximately 2.1 kilobases (kb) was seen for hCE-1 (Riddles et al. 1991), and three bands of approximately 2, 3 and 4.2-kb were seen with hCE-2 (Schwer et al. 1997). The intensities of the 2.1-kb band were liver ≫ heart > stomach > testis = kidney = spleen > colon > other tissues. In the case of hCE-2, the 2-kb band was located in liver > colon > small intestine > heart, the 3-kb band in liver > small intestine > colon > heart, and the 4.2-kb band in brain, testis, and kidney only. Analysis of substrate structure versus efficiency for the ester or carbamate substrates reveals that the two CaEs recognize different structural features of the substrate (i.e., acid, alcohol, etc.). The catalytic mechanism involves the formation of an acyl-enzyme on an active serine. Knowledge of the substrate structure–activity relationships and the tissue distribution of CE (CaE = CE) are critical for predicting the metabolism and pharmacokinetics of pesticides in humans.

B. AChE, BChE, and CaE Substrate Selectivities (Activity in Crude Tissue Preparations, Concentrations in Tissues, and Effects of Tetramer on Active Sites and Purification)

Mutation studies indicate that BChE catalyses ACh hydrolysis as efficiently as AChE, when six of the active site gorge aromatic residues, with human AChE numbers (HuAChE) 72, 124, 286, 295, 297, and 337, are replaced by aliphatic amino acids (Cygler et al. 1993; Harel et al. 1992). On the other hand, AChE and BChE exhibit distinct substrate and inhibitor selectivities, and some of the differences are associated with the nature of the amino acids at each of these positions. Early work indicated that the main functional difference between the AChE and BChE active sites is related to the structure of the acyl pocket, where residues corresponding to Phe295 (288) and Phe297 (290) are replaced by Leu and Val, respectively. BChE is more reactive than AChE toward bulky substrates such as butyrylcholine (BCh) (where $k_{cat}/K_m = 22 \times 10^8$ vs. $0.3 \times 10^8 M^{-1} min^{-1}$) or OP inhibitors such as diisopropyl phosphofluoridate (DFP, CAS no. 55-91-4) (where $k_i = 1.0 \times 10^5$ vs. $1.66 \times 10^7 M^{-1} min^{-1}$) and paraoxon (CAS no. 311-45-5) (Kaplan et al. 2001). The primary biological role of human CaE (hCE1) is xenobiotic metabolism. The enzyme is capable of cleaving ester, amide, and thioester linkages in a wide variety of compounds. Single site mutations in the active site of hCE1 convert the enzyme into an efficient OP insecticide hydrolase that is able to detoxify nerve agents (Redinbo et al. 2003).

Determination of AChE, BChE and CaE in crude tissue preparations

Maxwell et al. (1987) determined the esteratic binding sites of AChE, BChE, and CaE in tissues of control rats for purposes of determining the relationship between inhibition and detoxification. The methods that were used to obtained information on enzyme activity and turnover rates are important because the results (esteratic sites) were used in PBPK/PD models developed by Gearhart et al. (1990, 1994) and Timchalk et al. (2002) to predict inhibition. The methods used by Maxwell et al. (1987) to obtain enzyme activity are briefly described below.

Rats were exsanguinated and perfused via cardiac puncture with 150 mL saline before tissue harvesting and homogenization. Brain tissue was homogenized in 9 vol ice-cold saline containing 1% (v/v) Triton X-100 using a Potter-Elvehjem tissue homogenizer. Spleen, lung, kidney, and liver were homogenized in the same manner except that 4 vol saline with 1% (v/v) Triton X-100 was used. The diaphragm, heart, intestine, and skeletal muscle were homogenized in 4 vol saline by three 10-sec pulses of a Polytron homogenizer (Brinkman Instruments, Westbury, NY) at a speed setting of eight. Triton X-100 was added equal to 1% of the final volume and gently rotated for 10 min. All homogenates were centrifuged at 15,000 *g* for 10 min in a refrigerated centrifuge (4°–6°C), and the supernatants were stored frozen at –20°C before ChE analysis.

Assays of ChE were performed (Groff et al. 1976) using acetylthiocholine (ATCh) as substrate. Estimates of the total ChE and AChE was accomplished by ChE determinations in the absence and presence of iso-OMPA, a specific inhibitor of BChE (Austin and Berry 1953). Inhibition of BChE was performed by the method of Michalek et al. (1983) using 10 μM iso-OMPA for all tissues except plasma, where 100 μM iso-OMPA was used (Grubic et al. 1981).

CaE activity was measured using an automatic pH stat (Radiometer America, Cleveland, OH) with 0.2% tributyrin in saline as substrate at pH

Table 8. Enzyme Activities in Control Rat Tissues.[c]

	Enzyme activity (μmol min^{-1} g^{-1} tissue)			
	Cholinesterase		Detoxifying enzymes	
Tissue	Total ChE[a]	AChE[a]	BChE[b]	CaE[a]
Brain	8.12 ± 1.42	7.34 ± 1.25	0.78	1.0 ± 0.2
Lung	1.82 ± 0.25	0.38 ± 0.06	1.44	23.4 ± 1.9
Spleen	1.86 ± 0.18	1.00 ± 0.15	0.86	5.4 ± 0.3
Muscle	1.66 ± 0.09	1.29 ± 0.14	0.37	4.1 ± 0.5
Diaphragm	1.73 ± 0.16	1.29 ± 0.18	0.44	5.3 ± 0.4
Intestine	6.76 ± 0.80	0.71 ± 0.3	6.05	385.7 ± 59.6
Kidney	0.26 ± 0.04	0.09 ± 0.03	0.17	29.8 ± 6.3
Heart	5.65 ± 0.59	0.84 ± 0.07	4.81	3.6 ± 0.5
Liver	0.67 ± 0.06	0.17 ± 0.04	0.5	32.4 ± 4.2
Plasma	0.48 ± 0.07	0.22 ± 0.02	0.26	7.6 ± 1.7

[a]Total ChE, AChE and CaE activities are expressed as x ± standard deviation, $n = 9$.
[b]BChE = total ChE − AChE.
[c]Adapted from Maxwell et al. (1987).

8.0. The reaction was started by the addition of 2–200 μL homogenate. CaE activity was determined by titrating liberated acid with 0.01 N NaOH. The enzyme activities (μmol min^{-1} g^{-1} tissue) in control rat tissues as determined by Maxwell et al. (1987) are given in Table 8. BChE activity was determined by subtracting AChE activity from total ChE activity.

Active sites, a measure of B-esterase concentrations in tissue

The number of active sites of AChE, BChE, and CaE (i.e., B-esterase concentrations in tissues) were used in PBPK/PD models by Gearhart et al. (1990, 1994), Knaak et al. (2002), and Timchalk et al. (2002) to describe inhibition by DFP/paraoxon, paraoxon/isofenphos-oxon, and chlorpyrifos-oxon, respectively. The numbers of active sites (enzyme concentration g^{-1} tissue) were estimated by Maxwell et al. (1987) in plasma, liver, brain, and other regions by dividing esterase activity (μmol min^{-1} g^{-1} tissue) in control rat tissues (see Table 8) by their turnover rates (substrate hydrolyzed/min/active site). The turnover rates (expressed in hr^{-1}) are shown as AChE, 1.81×10^5 from Wang and Murphy (1982); BChE, 6.10×10^4 from Main et al. (1972); and CaE, 1.81×10^3 from Ikeda et al. (1977) in Tables 9, 10, and 11, respectively.

In practice, Knaak et al. (2002) used the esterase binding sites from Maxwell et al. (1987) in μmol B-esterase L^{-1} tissue, bimolecular inhibition rate constant (μM^{-1} hr^{-1}), concentration of carbamate in tissue (μmol L^{-1}), and volume of tissue (L) in their PBPK/PD model to determine inhibition levels, as shown in Eq. 23 for the brain.

Table 9. Estimated AChE Available for Inhibition/Detoxification of Carbamate Pesticides.[a]

	Enzyme activities		Esterase binding sites based on turnover rates					
Tissue	$\mu mol\, min^{-1} g^{-1}$	$\mu mol\, hr^{-1} kg^{-1}$	$\mu mol\, kg^{-1}$[c]	$\mu mol\, kg^{-1}$[d]	$\mu mol\, kg^{-1}$[e]	$\mu mol\, kg^{-1}$[f]	Mean $\mu mol\, kg^{-1}$	SD[b] $\mu mol\, kg^{-1}$
Brain	7.34	4.40×10^5	3.76×10^{-2}	2.86×10^{-2}	4.89×10^{-2}	5.65×10^{-2}	4.29×10^{-2}	1.23×10^{-2}
Lung	0.38	2.28×10^4	1.95×10^{-3}	1.48×10^{-3}	2.53×10^{-3}	2.92×10^{-3}	2.22×10^{-3}	6.36×10^{-4}
Spleen	1.0	6.00×10^4	5.13×10^{-3}	3.89×10^{-3}	6.67×10^{-3}	7.69×10^{-3}	5.85×10^{-3}	1.67×10^{-3}
Muscle	1.29	7.74×10^4	6.62×10^{-3}	5.02×10^{-3}	8.60×10^{-3}	9.92×10^{-3}	7.54×10^{-3}	2.16×10^{-3}
Diaphragm	1.29	7.74×10^4	6.62×10^{-3}	5.02×10^{-3}	8.60×10^{-3}	9.92×10^{-3}	7.54×10^{-3}	2.16×10^{-3}
Intestine	0.71	4.26×10^4	3.64×10^{-3}	2.76×10^{-3}	4.73×10^{-3}	5.46×10^{-3}	4.15×10^{-3}	1.19×10^{-3}
Kidney	0.09	5.40×10^3	4.62×10^{-4}	3.50×10^{-4}	6.00×10^{-4}	6.92×10^{-4}	5.26×10^{-4}	1.51×10^{-4}
Heart	0.84	5.04×10^4	4.31×10^{-3}	3.27×10^{-3}	5.60×10^{-3}	6.46×10^{-3}	4.91×10^{-3}	1.41×10^{-3}
Liver	0.17	1.02×10^4	8.72×10^{-4}	6.62×10^{-4}	1.13×10^{-3}	1.31×10^{-4}	9.94×10^{-4}	2.85×10^{-4}
Plasma	0.22	1.32×10^4	1.13×10^{-3}	8.57×10^{-4}	1.47×10^{-3}	1.69×10^{-3}	1.29×10^{-3}	3.68×10^{-4}
Blood		1.32×10^4	1.128×10^{-3}					

[a]Turnover rates (substrate hydrolyzed hr^{-1} active $site^{-1}$) used for calculations.
Enzyme activities ($\mu mol\, min^{-1} g^{-1}$ tissue) for calculations of binding sites were taken from Table 8.
Esterase binding sites = (enzyme activity)/turnover rate.
[b]Coefficient of variation = 28.64%.
[c]Turnover: $1.17 \times 10^7 hr^{-1}$; Wang and Murphy (1982).
[d]Turnover: $1.54 \times 10^7 hr^{-1}$; Torpedo, Selwood et al. (1993).
[e]Turnover: $9.00 \times 10^6 hr^{-1}$; Torpedo/recombinant, Radic et al. (1992).
[f]Turnover: $7.80 \times 10^6 hr^{-1}$; Mouse/wild type, Radic et al. (1992).

Table 10. Estimated BChE Available for Inhibition/Detoxification of Carbamate Pesticides.[a]

	Enzyme activity		Esterase binding sites based on turnover rates		
Tissue	$\mu mol\ min^{-1} g^{-1}$	$\mu mol\ hr^{-1} kg^{-1}$	$\mu mol\ kg^{-1}$ [b]	$\mu mol\ kg^{-1}$ [c]	$\mu mol\ kg^{-1}$ [d]
Brain	0.78	4.68×10^4	1.28×10^{-2}	5.94×10^{-3}	7.09×10^{-3}
Lung	1.44	8.64×10^4	2.36×10^{-2}	1.10×10^{-2}	1.31×10^{-2}
Spleen	0.86	5.16×10^4	1.41×10^{-2}	6.54×10^{-3}	7.82×10^{-3}
Muscle	0.37	2.22×10^4	6.07×10^{-3}	2.82×10^{-3}	3.36×10^{-3}
Diaphragm	0.44	2.64×10^4	7.21×10^{-3}	3.35×10^{-3}	4.00×10^{-3}
Intestine	6.05	3.63×10^5	9.92×10^{-2}	4.60×10^{-2}	5.50×10^{-2}
Kidney	0.17	1.02×10^4	2.79×10^{-3}	1.29×10^{-3}	1.55×10^{-3}
Heart	4.81	2.89×10^5	7.89×10^{-2}	3.66×10^{-2}	4.37×10^{-2}
Liver	0.5	3.00×10^4	8.20×10^{-3}	3.81×10^{-3}	4.55×10^{-3}
Plasma	0.26	1.56×10^4	4.26×10^{-3}	1.98×10^{-3}	2.36×10^{-3}

[a]See Table 9 for calculation of enzyme activity and esterase binding sites based on turnover rates.
[b]Turnover: $3.66 \times 10^6\ hr^{-1}$; Main et al. (1972).
[c]Turnover: $7.88 \times 10^6\ hr^{-1}$; *Torpedo*, Golicnik et al. (2002) .
[d]Turnover: $6.60 \times 10^6\ hr^{-1}$; Kaplan et al. (2001).

Table 11. Estimated CaE Available for Inhibition/Detoxification of Carbamate Pesticides.[a]

	Enzyme activity		Esterase binding sites based on turnover rates	
Tissue	$\mu mol\ min^{-1} g^{-1}$	$\mu mol\ hr^{-1} kg^{-1}$	$\mu mol\ kg^{-1}$ [b]	$\mu mol\ kg^{-1}$ [c]
Brain	1.0	6.00×10^4	5.52×10^{-1}	1.68×10^0
Lung	23.4	1.40×10^6	1.29×10^{-1}	$3.93 \times 10^{+1}$
Spleen	5.4	3.24×10^5	2.98×10^0	9.09×10^0
Muscle	4.1	2.46×10^5	2.27×10^0	6.90×10^0
Diaphragm	5.3	3.18×10^5	2.93×10^0	8.92×10^0
Intestine	385.7	2.31×10^7	2.13×10^2	$6.48 \times 10^{+2}$
Kidney	29.8	1.79×10^6	$1.65 \times 10^{+1}$	$5.02 \times 10^{+1}$
Heart	3.6	2.16×10^5	1.99×10^0	6.06×10^0
Liver	32.4	1.94×10^6	$1.79 \times 10^{+1}$	$5.44 \times 10^{+1}$
Plasma	7.6	4.56×10^5	4.20×10^0	$1.29 \times 10^{+1}$

[a]See Table 9 for calculation of enzyme activity and esterase binding sites based on turnover rates.
[b]Turnover: $1.09 \times 10^5\ hr^{-1}$; Ikeda et al. (1977).
[c]Turnover: $3.56 \times 10^4\ hr^{-1}$; Stok et al. (2004).

Equation for the inhibition of blood AChE by carbamates:

$$dA_{iAChEB}/dt = (K_{iAChEB}C_{AChEB}C_{carbamate}V_B)\ (\mu mol\, hr^{-1}) \quad (23)$$

where V_B = volume of blood (L),
K_{iAChEB} = AChE bimolecular inhibition rate constant ($\mu M^{-1} h^{-1}$),
C_{AChEB} = concentration of free AChE in blood ($\mu mol\, L^{-1}$), and
$C_{carbamate}$ = concentration of carbamate in blood ($\mu mol\, L^{-1}$).

In all three PBPK/PD models [Gearhart et al. (1990, 1994), Knaak et al. (2002), and Timchalk et al. (2002)], the active site data from the rat were used in human models. The PD portion of these models is highly dependent on these values. New estimates are needed to validate OP and carbamate pesticide PBPK/PD models. Additional turnover numbers, k_{cat}, (Radic et al. 1992; Pyror et al. 1992; Selwood et al. 1993; Radic et al. 1993; Kaplan et al. 2001) for the hydrolysis of thiocholinesters by AChE and BChE were found and used to obtain new estimates of µmol active sites per kg tissue. One k_{cat} value was found for CaE (Stok et al. 2004). The results are shown in Tables 9 and 10.

A 28.6% coefficient of variation for AChE was determined for the various tissues based on the new turnover numbers in Table 9. This variation suggests that the output (inhibition) from PBPK/PD models may vary by at least this percentage. The source of enzymes and methods for determining turnover numbers was found to vary between studies.

In a study by Pryor et al. (1992), the active sites of electric eel (EE) AChE were measured by titration with umbelliferyl diethyl phosphate (UDP) (CAS no. 299-45-6). The release of umbelliferone (CAS no. 93-35-6) from the reaction with EE-AChE was monitored by fluorescence spectroscopy. The time course of fluorescence emission increased following the addition of AChE to the titration solution. The data (fluorescence vs. time in sec) fit the following first-order equation:

$$F = (F_0 - F_{inf})e^{-kt} + F_{inf} \quad (24)$$

where F, F_0, and F_{inf} are the emission intensities of time *t*, 0, and infinity, respectively; k is the first-order rate constant ($3.1 \times 10^7 hr^{-1}$). The enzyme concentration was determined from the value of $F_{inf} - F_b$ (F_b, background fluorescence) using the calibration plot. Division of k by the concentration of UDP gave a second-order bimolecular rate constant of $2.77 \times 10^{11} M^{-1} hr^{-1}$. This rate constant is similar to the rate constant ($1.71 \times 10^{11} M^{-1} hr^{-1}$) determined by Kaplan et al. (2001) for human AChE and ATCh in which the hydrolytic parameters (K_m, k_{cat}, and k_{cat}/K_m) of human recombinant AChE and BChE against ATCh and BTCh (acetyl and butyrylthiocholine) were assayed according to Ellman et al. (1961) and monitored using a Thermomax microplate reader (Molecular Devices, Sunnyvale, CA). Enzyme concentration was determined by ELISA (enzyme-linked immunosorbent assay) (Shafferman et al. 1992) and by active site titration (Velan et al. 1991) using the PsCs-soman stereomer (CAS no. 24753-16-0). Human AChE displayed a 50-fold difference in reactivity for ATCh over BTCh with k_{cat} being $2.4 \times 10^7 hr^{-1}$ and $4.8 \times 10^5 hr^{-1}$, respectively. Human BChE was more reactive against BTCh than AChE ($k_{cat} = 6.6 \times 10^6 hr^{-1}$ and $4.8 \times 10^5 hr^{-1}$, respectively).

Effects of tetramer and molecular forms on active site measurements

The tetramer is the most important form for AChE in physiological conditions (i.e., neuromuscular junction and the nervous system). A study by Zhang et al. (2005) involving diffusion of ACh to the active sites of the tetramer show that the reaction rates for three mouse AChE tetramers differ for individual active sites in the tetramer. Depending upon salt concentration, the rates for active sites in the tetramer are equivalent to 67%–75% of the rate for the monomer. Measurements made on AChE activity in tissues may underestimate the amount of available enzyme for carrying out ACh hydrolysis. In an earlier study by Gordon et al. (1978), the turnover numbers of different forms (18 S, 14 S, and 8 S asymmetrical forms and the 11 S form) from electric eel were measured using an active phosphorylating inhibitor as a titrating agent. The 8 S and 11 S forms were approximately 35% more active than the 18 S and 14 S forms.

Purification of AChE, BChE, and CaE

Purified/recombinant enzymes are needed for studying the kinetics of their catalytic reactions. Rosenberry and Scoggin (1984) purified human erythrocyte AChE on an acridinium resin affinity column. Five mg of enzyme were obtained from 10 L of outdated erythrocytes. The purified enzyme had a specific activity of 5000–5800 units mg^{-1} protein (unit = $1.0 \mu mol\, min^{-1}$; $5.0–5.8\, mmol\, min^{-1} mg^{-1}$ protein) and was free of polypeptide contaminants by gel electrophoresis. In detergents, the isolated enzyme corresponded to a disulfide-linked dimer (G2) that was converted to 75-kDa subunit monomers (G1) by reduction with dithiothreitol. G1 contained 1.7 mol free sulfhydryl groups mol^{-1} subunit. This study provided strong evidence that erythrocyte AChE is an amphipathic protein.

Affinity chromatography, of mutant mouse AChE in serum free media, with trimethyl (*m*-aminophenyl) ammonium linked through a long tether arm to Sepharose CL-4B resin (Sigma, St. Louis, MO) permitted a one-step purification of AChE, both mutant and wild-type in amounts between 5 and 25 mg, as previously described. Purity was assessed by SDS-PAGE (sodium dodecyl sulfate-polyarcylamide gel electrophoresis) and by comparisons of specific activity with absorbance at 280 nm to measure protein concentration ($\varepsilon_{280} = 1.14 \times 10^5 M^{-1} cm^{-1}$) (Boyd et al. 2004). The catalytic activity of each unlabeled and labeled mutant was measured with the Ellman assay. K_m and the dissociation constant of a ternary complex resulting in substrate inhibition or activation, K_{ss}, were kinetically evaluated.

The x-intercept of a plot of the residual catalytic activity versus the concentration of the irreversible inhibitor 7-(methylethoxyphosphinyloxy)-1-methylquinolinium iodide (MEPQ) (CAS no. 95230-44-7) yielded the enzyme concentration, and, in turn, k_{cat}. No data were supplied for k_{cat}, K_m, and K_{ss}.

Two types of CaE (esterase I and II) were purified from rat liver microsomes after 90% of the triacylglycerol lipase activity was released by heparin treatment. Purification of the residual fraction involved diethylaminoethyl (DEAE)-cellulose chromatography, hydroxyapatite chromatography, isoelectric focusing, and Sephadex G-200 chromatography. The final preparation of esterases I and II, which were purified 70- and 140 fold, respectively, gave single protein bands on polyacrylamide gel and sodium dodecyl sulfate-gel electrophoresis. The molecular weights of esterases I and II were calculated to be about 70,000 and 160,000 by gel filtration on Sephadex G-200. Esterase I differed immunologically from esterase II, and it was found to constitute about 30% of the total esterase activity in the microsomal fraction; esterase II constituted 50%–60% (Ikeda et al. 1977).

Sanghani et al. (2004) purified human isozymes hCE1 and hCE2 from class 1 and 2 CaEs, respectively, using a modification of the method of Humerickhouse et al. (2000). Briefly, 85 g human liver was first homogenized and centrifuged at 35,000 *g* for 40 min. The supernatant was added to DE52 resin, the resin washed, and the CaEs eluted from the resin and further purified on a 50-mL concanavalin A column. Human isozymes hCE1 and hCE2 were separated on a preparative nondenaturing gel electrophoresis column in a Bio-Rad Prep Cell model 491 (Bio-Rad, Hercules, CA). Purified hCE1 and hCE2 exhibited specific activities of about 7 and 140 units mg^{-1}, respectively, using methylumbelliferyl acetate as substrate. A unit is defined as 1.0 μmol substrate hydrolyzed $min^{-1} mg^{-1}$ enzyme.

VI. Acylation and Decarbamylation of ChEs and CaEs

A. Hydrolysis of Substrates

The hydrolysis of ACh and BCh by ChE is extremely rapid, making it difficult to measure the rate of hydrolysis. To handle this problem, activity measurements are made under steady-state conditions or after the partial inhibition of ChE by an irreversible inhibitor. In most preparations the actual molar concentrations of enzyme are unknown and the amount of enzyme can be expressed only in terms of its specific activity (i.e., units mg^{-1} protein; units μmol^{-1} enzyme; V_{max} mg^{-1} protein). Turnover number or rate, k_p, k_2, or k_{cat}, is defined as follows:

$$k_{cat}(min^{-1}) = \frac{V_{max}}{[E]_t} = \frac{\text{pmoles(substrate} \rightarrow \text{product)}\,min^{-1}\,kg^{-1}}{\mu\text{moles(enzyme)}kg^{-1}} \quad (25)$$

where total enzyme is given as

$$[E]_t = [E] + [ES] \tag{26}$$

The reciprocal of k_{cat} gives the time required for a single catalytic cycle (Segel 1993). The following section is a review of the kinetics of the reaction.

Acylation/hydrolysis, AChE, BChE, and CaE

AChE (EH) combines with acetylcholine (AB) resulting in a Michaelis–Menten complex (EHAB) that is converted to an acylated enzyme (EA) and a leaving group (BH) (Aldridge and Reiner 1972). The acylated enzyme decomposes into an enzyme product as indicated below:

$$EH + AB \underset{k_{-1}}{\overset{k_{+1}}{\rightleftarrows}} EHAB \xrightarrow{k_{+2}} EA + BH \underset{+H_2O}{\xrightarrow{k_{+3}}} EH + AOH \tag{27}$$

where k_{+1}, k_{-1}, k_{+2} and k_{+3} are the velocity (sec^{-1}) of the equilibrium reactions. BH and AOH are the products of substrate hydrolysis, where BH is Ch and AOH is acetic acid, respectively. The EA formed from ACh hydrolyzes very quickly ($t_{1/2} \leq 2.3 \times 10^{-6}$ min) (Wilson 1951a,b; Cohen et al. 1955). These rates are initially measured under steady state conditions when the concentrations of all species (EH, EHAB, and EA) are constant. The steady-state kinetic parameters (Pryor et al. 1992) are given by the following equations:

$$V_{max}/K_m = (k_{cat}[E]_T)/K_m = \frac{k_{+1}k_{+2}[E]_T}{k_{-1} + k_{+2}} \tag{28}$$

$$V_{max} = k_{cat}[E]_T = \frac{k_{+2}k_{+3}[E]_T}{k_{+2} + k_{+3}} \tag{29}$$

where $[E]_T$ is the total enzyme.

The second-order rate constant, k_{cat}/K_m, for ACh and ATCh hydrolysis is $\sim 10^8\,M^{-1}sec^{-1}$ ($3.6 \times 10^{11}\,M^{-1}hr^{-1}$) and is diffusion controlled. The turnover number, k_{cat}, is $\sim 3.6 \times 10^7\,hr^{-1}$. Pryor et al. (1992) determined the kinetic parameters V_{max} and K_m by nonlinear least-squares fitting of V_i and [S] to the Michaelis–Menten equation, where V_i and [S] are the initial velocity and substrate concentration, respectively:

$$V_i = \frac{V[S]}{K + [S]} \tag{30}$$

Values of k_{cat} and k_{cat}/K_m were calculated by dividing V_{max} and V_{max}/K_m, respectively, by the enzyme concentration (mass of 70,000 per active site). Initial velocities were measured by linear least-squares fitting of time courses for ≤5% of total substrate turnover using the assay of Ellman et al. (1961).

Kinetic activity of AChE, BChE, and CaE

Pryor et al. (1992) determined the kinetic constants (k_{cat}, k_{cat}/K_m) for the hydrolysis of TCh esters by *Electrophorus electricus* AChE (Sigma, St. Louis, MO), purified human red blood cell AChE (Rosenberry and Scoggin 1984), fetal bovine serum AChE (De La Hoz et al. 1986), and *Torpedo californica* AChE (Sussman et al. 1988).

In a study by Radic et al. (1993), mouse wild-type and mutant AChE and BChE activities were measured in 0.1 M Na_3PO_4 (pH 7.0) at 22°C by the assay of Ellman et al. (1961) according to Eq. 31:

$$\begin{array}{ccccc} S+E & \rightleftharpoons & ES & \xrightarrow{k_{cat}} & E+P \\ + & & + & & \\ S & & S & & \\ K_{ss}\updownarrow & & K_{ss}\updownarrow & & \\ S+SE & \rightleftharpoons & SES & \xrightarrow{bk_{cat}} & SE+P \end{array} \qquad (31)$$

where S combines at two different sites on the enzyme, forming two binary complexes, ES and SE, of which only ES results in substrate hydrolysis. Values for K_m, K_{ss}, and b were calculated using nonlinear computer fitting according to the following equation:

$$v = \left(\frac{1+b[S]/K_{SS}}{1+[S]/K_{SS}}\right)\left(\frac{V_{max}}{1+K_m/[S]}\right) \qquad (32)$$

where K_{ss} is a substrate inhibition constant, and b reflects the efficiency of hydrolysis of the ternary complex, SES, relative to ES (Aldridge and Reiner 1972; Webb and Johnson 1969). Active sites were quantitated by titration with 7-(methylethoxyphosphinyloxy)-1-methylquinolinium iodide (MEPQ), a high-affinity phosphorylating agent (Levy and Ashani 1986; Radic et al. 1992). Inhibition by reversible inhibitors was measured at four to five substrate and three inhibitor concentrations. Reciprocal and regression plots were used to obtain kinetic parameters. Inhibitor and substrate binding to AChE and BChE was modeled by an energy minimization docking program (Discover-Insight 2.2.0; Biosym Technologies, San Diego, CA) on a Silicon Graphics Indigo Elan (Radic et al. 1992).

Stok et al. (2004) determined the kinetic parameters for the hydrolysis of p-nitrophenyl acetate by rat CaE 6.1. The values of k_{cat}, the turnover number, and k_{cat}/K_m, the second-order rate constant, were determinded to be $3.5 \times 10^4\,hr^{-1}$ and $3.09 \times 10^9\,M^{-1}\,hr^{-1}$, respectively. These values are 1000 and 116 times slower than those reported by Pryor et al. (1992) for AChE and ACh, respectively. The hydrolytic action of AChE on ACh is considered to be the most rapid reaction recorded. Carbaryl is an inhibitor of CaE 6.1, with K_d, k_2 and k_3 values of 35 μM, 108 hr^{-1}, and 39.6 hr^{-1}, respectively.

Complexities of AChE and BChE kinetic activity

The kinetics of BChE catalyzed reactions display a number of complexities that could reflect microheterogeneity: (1) substrate hydrolysis deviates from Michaelis–Menten kinetics (Christian and Beasley 1968; Eriksson and Augustinsson 1979; Kalow 1964); (2) irreversible inhibition does not follow first-order kinetics (Main 1969b); and (3) kinetics presents a nonlinear temperature dependence with a break at 18°–21°C in Arrhenius plots (Ferro and Masson 1987). In addition, where there are four active sites per tetramer (Eriksson and Augustinsson 1979; Lockridge and La Du 1978), lower active site numbers have been reported, e.g., two (Muensch et al. 1976).

B. Carbamylation (Inhibition) and Decarbamylation of ChEs and CaEs

Rate of inhibition characterized by second-order rate constant. The inhibition of AChE by carbamate insecticides occurs by the following reactions:

$$\mathrm{EH + AB} \underset{k_{-1}}{\overset{k_{+1}}{\rightleftharpoons}} \mathrm{EHAB} \xrightarrow{k_{+2}} \mathrm{EA + BH} \xrightarrow[+H_2O]{k_{+3}} \mathrm{EH + AOH} \quad (\mathrm{EH + AB} \xrightarrow{k_{-i}} \mathrm{EA + BH}) \tag{33}$$

where EH and AB are the enzyme and carbamate, respectively, EHAB is the enzyme–inhibitor complex, and $K_a = k_{-1}/k_1$, which is now called the dissociation constant, K_d, is the binding, affinity constant. Main (1969a) described a procedure for measuring the binding constant K_a (M), carbamylation constant k_2 (hr^{-1}), and the overall bimolecular rate constant k_i ($M^{-1} hr^{-1}$). The bimolecular rate constant was determined by plotting the reciprocal of the inhibition rate against the reciprocal of the inhibitor concentration. The slopes give k_i, and the intercepts of the extrapolated lines on the ordinate and abscissa give ($-K_a^{-1}$) and (k_2^{-1}).

Saxena et al. (1997) and Ordentlich et al. (1993) measured the inhibition bimolecular rate constants (k_i) for OPs under pseudo-first-order conditions from the plots of slopes of ln E (enzyme) versus time at different inhibitor concentrations. Rate constants under second-order conditions were determined from nonlinear regression of plots of ln $[E/(I_0 - (E_0 - E))]$ vs. time (I_0, initial inhibitor concentration).

Inhibition of more than one form of enzyme

Main (1969b) observed that the plots of the log velocity against time for serum ChE and erythrocyte AChE were not linear with the OP tetram (CAS no. 78-53-5). Similar curving was observed with neostigmine under conditions of negligible regeneration. Curving was caused by the inhibition of two or more forms of ChE. The use of a purified preparation of eel AChE and neostigmine resulted in little or no curving until 99% of the enzyme had been inhibited.

Inhibitory activity as function of carbamate structure

The application of the Main (1964) equation to inhibition of AChE by carbamates has been studied in detail by Hastings et al. (1970) and O'Brien (1968). Fukuto et al. (1967) assumed that these compounds behave as competitive substrates for ChE with high affinities for the enzyme and low turnover numbers. The K_m for the hydrolysis of ACh chloride by bovine erythrocytes was reported to be 0.29×10^{-3} M (Hetnarski and O'Brien 1975b). This constant is equivalent to the binding of a carbamate to the enzyme surface as measured by the dissociation constant, K_d. According to Hetnarski and O'Brien (1973), variations exist in the inhibitory activity of aryl methylcarbamates and aryl dimethylcarbamates toward AChE with the latter compounds inhibiting only by complex formation, and not by complex formation followed by carbamylation, as is the case with aryl methylcarbamates. Most carbamates, however, give rise to a carbamylated enzyme. Rapid decarbamylation results in recovered enzyme when the enzyme is removed from the surplus inhibitor. Hetnarski and O'Brien (1973) considered recovery "pseudoreversibility" in the sense that the enzyme recovers, and the carbamate is hydrolyzed in the course of the reaction. Potency is normally a function of both affinity and innate carbamylation activity. The validity of the K_d and k_2 estimations depend upon inhibition following first-order kinetics when inhibitor concentrations are constant (Iverson and Main 1969).

CaE inhibition by carbamates

The protective effect of CaE against AChE inhibition by carbamates has not been well characterized as it has been for the OPs. In the case of OPs, CaE performs the role of a high-affinity/low-capacity detoxification system with k_i values $> 10^6 M^{-1} min^{-1}$. CaEs appear to be less effective in the case of carbamates as Stok et al. (2004) determined the k_i for carbaryl to be $5.14 \times 10^4 M^{-1} min^{-1}$.

Maxwell (1992) measured the activity of lung, plasma, and liver CaE against 1-naphthyl acetate using the spectrophotometric method of Ecobichon (1970) in the presence of 10^{-5} M physostigmine to prevent hydrolysis by BChE. Physostigmine was not required for measurement of purified plasma CaE activity. The ratio of the inhibition rate constants (k_i ratio: plasma CaE/brain AChE) for paraoxon (5/1) and a number of OPs of military significance were plotted against their protective ratios (LD_{50} of controls/CaE inhibited animals; ratio of 2 for paraoxon). Small k_i ratios of less than 1.0 gave protective ratios of 4 or greater. The larger number of binding sites in liver suggests that CaE detoxication may be more important in this tissue than in plasma or the lung.

Decarbamylation of Inhibited Molecular Forms, First-Order Rate Constant

Spontaneous decarbamylation of both plasma and erythrocyte ChE inhibited by carbamates is an established feature. It is commonly accepted that the binding of carbamates to ChE is reversible and that some hydrolysis of the carbamate occurs on the enzyme (Aldridge and Reiner 1972; Rotenberg and Almog 1995). Reactivation of aliquots of carbamylated AChE was followed after extensive dilution of the inhibited enzyme using the method of Ellman et al. (1961) (Dawson 1994; Dawson and Poretski 1985).

C. Bimolecular Rate Inhibition Constants, k_i, for the Ten Carbamates

Kinetic constants for the inhibition of AChE and BChE are shown in Tables 12 and 13 for the 10 carbamates and their active metabolites (Dawson 1995; Herzsprung et al. 1992; Hetnarski and O'Brien 1975a; Main 1979; Vilarem et al. 1989; Yu et al. 1972b). Noteworthy is the large amount of work carried out by Herzsprung et al. (1992). k_2 is assumed to be equivalent for all the carbamates (Goldblum et al. 1981). K_d was calculated using k_2 and k_i.

Herzsprung et al. (1992) determined the k_i for the inhibition of AChE from electric eel, AChE from bovine erythrocytes, and BChE from human and horse serum by aldicarb, aldicarb sulfoxide and sulfone, carbofuran, 3-hydroxycarbofuran, carbaryl, methomyl, oxamyl, pirimicarb, propoxur and 2-hydroxypropoxur using the colorimetric method of Ellman et al. (1961). According to Herzsprung et al. (1992), the reaction of the carbamates with ChE may be considered a pseudo-first-order reaction when the concentration of enzyme is small compared to the concentration of inhibitor:

$$\begin{gathered} EH + IB \xrightarrow{k_i} EI + BH \\ [EH_t] = [EH_0] * \exp(-k_i^* [IB] * t) \end{gathered} \tag{34}$$

where [IB] = concentration of carbamate in the incubation mixture [mol L^{-1}], [EH] = concentration of the active enzyme, EI = enzyme inhibitor complex, BH = leaving group, and t = inhibition time.

The bimolecular rate constant k_i is given by

$$k_i = \frac{\ln 2}{t_{1/2} * [IB]} \tag{35}$$

The reaction half-time, $t_{1/2}$, and k_i were determined from a plot of log AC_t vs. t. Dawson (1994) calculated the second-order rate constant for carbamylation (k_i), and the first-order rate constant for decarbamylation (k_3) for carbaryl from the data described by Aldridge and Reiner (1972). Bovine erythrocyte AChE (soluble preparation), human erythrocyte AChE (membrane-bound), and rat brain AChE (solubilized with Triton X-100) were used as sources of AChE by Dawson (1994). The kinetic data presented in Tables 12 and 13 along with other data are briefly discussed next for each carbamate.

Table 12. Kinetic Constants for the Inhibition of AChE by Carbamate Pesticides and for Their Reactivation.

Carbamate/ Metabolite	Dissociation constant, K_d, (M)	Carbamylation constant k_2, (min^{-1})	Decarbamylation constant k_3, (min^{-1})	Bimolecular rate constant k_i, ($M^{-1} min^{-1}$)
Aldicarb[a]	1.0×10^{-2}	146	0.0109[g]	1.46×10^4
Aldicarb[b]	1.23×10^{-4}	1.72		1.39×10^4
Aldicarb[c]				5.0×10^4
Aldicarb sulfoxide[b]	—	—	0.010	5.6×10^5
Aldicarb sulfoxide[c]				1.3×10^6
Aldicarb sulfone[b]	—	—	0.010	6.0×10^3
Aldicarb sulfone[c]				1.5×10^4
Carbaryl[a]	5.0×10^{-2}	20.0	0.0103[g]	4.0×10^2
Carbaryl[d]	1.3×10^{-5}	1.83		1.4×10^5
Carbaryl[e]		0.71		2.5×10^4
Carbaryl[c]				3.3×10^4
Carbofuran[f]	2.1×10^{-4}	21.6	0.010	1.029×10^5
Carbofuran[c]				1.7×10^6
3-OH Carbofuran[c]				3.5×10^5
Formetanate[g]			0.010	
Methiocarb[a]	6.7×10^{-6}	1.2	0.010	1.79×10^5
Methiocarb[c]				1.0×10^5
Methiocarb sulfoxide[c]				0.7×10^5
Methiocarb sulfone[c]				0.7×10^4
Methomyl[b] (Z isomer)	5.40×10^{-6}	0.62		1.14×10^5
Methomyl[h]	3.305×10^{-5}	2.3	0.0113[g]	6.96×10^4
				1.5×10^5
Oxamyl[b]	9.71×10^{-6}	1.28	0.010	1.31×10^5
Oxamyl[c]				3.5×10^5
Pirimicarb[g]			0.010	
Propoxur[a]	1.0×10^{-5}	1.1	0.010	1.1×10^5
Propoxur[d]	2×10^{-5}	1.47		7.35×10^5
Propoxur[c]				1.3×10^5
2-OH Propoxur[c]			0.010	0.8×10^2
Thiodicarb[g]			0.010	

[a]Main (1979), [b]Vilarem et al. (1989), [c]Herzsprung et al. (1992), [d]Hetnarski and O'Brien (1975a), [e]Dawson (1994), [f]Yu et al. (1972b), [g]Rotenberg and Almog (1995), [h]Dawson (1995).

Table 13. Kinetic Constants for the Inhibition of BChE by Carbamate Pesticides.

Carbamate/metabolite	Dissociation constant K_d, (M)	Carbamylation constant k_2, (min^{-1})	Bimolecular rate constant k_i, ($M^{-1} min^{-1}$)
Aldicarb[a]	8.33×10^{-5}	2.0	2.4×10^4
Aldicarb sulfoxide[a]	3.33×10^{-6}	2.0	6.0×10^5
Aldicarb sulfone[a]	8.0×10^{-6}	2.0	2.5×10^5
Carbaryl[a]	1.05×10^{-5}	2.0	1.9×10^3
Carbofuran[a]	3.33×10^{-6}	2.0	6.0×10^5
3-OH Carbofuran[a]	1.82×10^{-4}	2.0	1.1×10^4
Formetanate[b]	—	—	—
Methiocarb[a]	1.54×10^{-4}	2.0	1.3×10^4
Methiocarb sulfoxide[a]	4.65×10^{-5}	2.0	4.3×10^4
Methiocarb sulfone[a]	6.25×10^{-4}	2.0	3.2×10^3
Methomyl[a]	5.13×10^{-4}	2.0	3.9×10^3
Oxamyl[a]	6.67×10^{-5}	2.0	3.0×10^4
Pirimicarb[b]	—	—	—
Propoxur[a]	2.5×10^{-3}	2.0	0.8×10^3
2-OH Propoxur[a]	0.022	2.0	0.9×10^2
Thiodicarb[b]	—	—	—

[a]Herzsprung et al. (1992).
[b]Not available.

Aldicarb-AChE and BChE Inhibition

Herzsprung et al. (1992) and Vilarem et al. (1989) determined k_i values for the inhibition of AChE by aldicarb, aldicarb sulfoxide, and aldicarb sulfone (see Table 12). Aldicarb sulfoxide is more inhibitory than aldicarb or its sulfone. Main (1979) reported a k_i value for aldicarb similar to the ones reported by Vilarem et al. (1989) and Herzsprung et al. (1992). According to Herzsprung et al. (1992), aldicarb sulfoxide is more inhibitory toward BChE than either aldicarb or its sulfone (see Table 13).

Carbaryl-AChE and BChE Inhibition

Hetnarski and O'Brien (1975a) determined K_d, k_2, and k_i for carbaryl using bovine erythrocyte AChE and the zero-time method (Hart and O'Brien 1973; O'Brien 1968). The values were 13×10^{-6}M, 1.83 min^{-1}, and $1.408 \times 10^5 M^{-1} min^{-1}$, respectively, as shown in Table 12. Herzsprung et al. (1992) determined the k_i for eel AChE to be $3.3 \times 10^4 M^{-1} min^{-1}$. Similar values were determined by Dawson (1994) for AChE. Herzsprung et al. (1992) determined k_i values for carbaryl of 1.9×10^3 and $0.7 \times 10^4 M^{-1} min^{-1}$ against human and horse BChE, respectively (see Table 13).

Carbofuran-AChE and BChE Inhibition

According to Herzsprung et al. (1992), the inhibition k_i values for carbofuran and 3-OH carbofuran in Table 12 were 1.7×10^6 and $3.5 \times 10^5 M^{-1} min^{-1}$, respectively, for eel AChE, with carbofuran being more inhibitory than 3-OH carbofuran. The K_d and k_2 values for the inhibition of bovine erythrocyte AChE by carbofuran at concentrations of 2.5×10^{-6}, 2.5×10^{-5}, and $2.5 \times 10^{-4} M$ were 3.5×10^{-6}, 3.9×10^{-6}, and $4.0 \times 10^{-6} M$, and 1.63, 1.55, and 1.32 min^{-1}, respectively. The corresponding k_i values were 4.66×10^5, 3.97×10^5, and $3.3 \times 10^5 M^{-1} min^{-1}$.

Herzsprung et al. (1992) determined the k_i values for carbofuran and 3-OH carbofuran using human (see Table 13) and horse BChE. The molar K_d values were calculated from the k_i values using a k_2 value of 2.0 min^{-1}. The K_d and k_2 values for the inhibition of equine plasma BChE by carbofuran at concentrations of 2.5×10^{-5} and $2.5 \times 10^{-4} M$ were 3.3×10^{-6} and $4.7 \times 10^{-6} M$, and 1.07 and 0.96 min^{-1}, respectively, with k_i values of 3.2×10^5 and $2.06 \times 10^5 M^{-1} min^{-1}$ (Barthova et al. 1989).

Formetanate-AChE and BChE Inhibition

No k_i values for AChE or BChE were found in the literature for formetanate.

Methiocarb-AChE and BChE Inhibition

Herzsprung et al. (1992) determined the k_i values for methiocarb, methiocarb sulfoxide, and methiocarb sulfone using eel and bovine AChE. The values shown in Table 12 are for eel AChE with methiocarb being more inhibitory than either the sulfoxide or sulfone. Main (1979) reported a similar value for methiocarb (see Table 12). Buronfosse et al. (1995) reported on the inhibitory properties of methiocarb and racemic methiocarb sulfoxide. The racemic sulfoxide was slightly less inhibitory than methiocarb ($k_i = 2.16 \times 10^5 M^{-1} min^{-1}$). A 10-fold difference was observed between the k_i for the (A) and (B) sulfoxide enantiomers ($5.4 \times 10^4 M^{-1} min^{-1}$ and $5.02 \times 10^5 M^{-1} min^{-1}$, respectively). Hetnarski and O'Brien (1975a) determined K_d, k_2, and k_i for methiocarb using bovine erythrocyte AChE and the zero-time method (Hart and O'Brien 1973; O'Brien 1968). The values were $1.5 \times 10^{-5} M$, 1.93 min^{-1}, and $1.3 \times 10^5 M^{-1} min^{-1}$, respectively.

Herzsprung et al. (1992) determined the bimolecular rate constants (k_i) for methiocarb, methiocarb sulfoxide, and methiocarb sulfone using human (see Table 13) and horse BChE. The K_d values were calculated from k_i using a value of 2.0 min^{-1} for k_2. The K_d and k_2 values for the inhibition of equine plasma BChE by methiocarb at concentrations of 2.5×10^{-4} and $2.5 \times 10^{-3} M$ were 1.0×10^{-3} and $8.2 \times 10^{-4} M$ and 1.21 and 1.13 min^{-1}, respectively (Barthova et al. 1989). The k_i values were 1.21×10^5 and $1.38 \times 10^5 M^{-1} min^{-1}$, respectively.

Methomyl-AChE and BChE Inhibition

Table 12 gives the k_i values for methomyl using AChE (Dawson 1995; Herzsprung et al. 1992; Vilarem et al. 1989). The Z isomer was more inhibitory than racemic methomyl (Vilarem et al. 1989). Methomyl was less inhibitory against human BChE than eel AChE as indicated in Table 13 (Herzsprung et al. 1992).

Oxamyl-AChE and BChE Inhibition

The k_i for oxamyl is similar to that of the Z isomer of methomyl (i.e., $1.31 \times 10^5 M^{-1} min^{-1}$ vs. $1.14 \times 10^5 M^{-1} min^{-1}$, respectively), as shown in Table 12 (Vilarem et al. 1989). Oxamyl is more inhibitory toward human BChE than methomyl, as shown in Table 13 (Herzsprung et al. 1992).

Pirimicarb-AChE and BChE Inhibition

No k_i values for AChE or BChE were found in the literature for pirimicarb.

Propoxur-AChE and BChE Inhibition

Hetnarski and O'Brien (1975a) determined K_d, k_2, and k_i for propoxur using bovine erythrocyte AChE and the zero-time method (Hart and O'Brien 1973; O'Brien 1968). The values were $2.0 \times 10^{-5} M$, $1.47 min^{-1}$, and $7.35 \times 10^4 M^{-1} min^{-1}$, respectively (see Table 12). According to Herzsprung et al. (1992), propoxur is considerably more inhibitory than 3-OH propoxur to eel AChE (Table 12). Propoxur and 2-OH propoxur are less inhibitory to human BChE as shown by their k_i values (see Table 13).

Thiodicarb-AChE and BChE Inhibition

No k_i values for AChE or BChE were found in the literature for thiodicarb. However, thiodicarb is metabolized to methomyl in the liver, and the k_i value for this carbamate may be used in PBPK/PD models.

VII. Discussion

Metcalf and March (1950), Gysin (1954), and Gubler et al. (1968) were largely responsible for starting the development of carbamates as insect control agents in the U.S. and Europe. Gysin (1954) concentrated on the development of dimethylcarbamates (Isolan, Dimetan, and Dimetilan) while Metcalf and March (1950) concentrated on the less stable but more active methylcarbamates. The research activities in the laboratories of Metcalf, March and Gysin, and others were ultimately responsible for the development, registration, and marketing of the 10 carbamate insecticides (i.e., aldicarb, carbaryl, carbofuran, formetanate, methiocarb, methomyl, oxamyl, pirimicarb, propoxur and thiodicarb) by chemical corporations that established agricultural chemical companies such as Union Carbide, Dow, Dupont, Bayer, and others. The WHO's alternative insecticide program supported the work of Metcalf at the University of Illinois and that of Metcalf and March at the University of California (UC) at Riverside along with a large number of cooperating laboratories in the U.S. and throughout the world (Wright 1971).

Oxime carbamates were developed by Union Carbide (aldicarb) and Dupont (methomyl and oxamyl). The structural rigidity of the C=N bond led to the formation of pairs of syn- and anti isomers in which the oximino oxygen is either cis or trans to the aldehyde H. The syn or Z isomer is more toxic to the housefly. The toxicity of these two isomers has not been considered in environmental studies. QSAR equations were developed by Metcalf and Fukuto at UC Riverside to assist them in understanding the relationship between the structure of the various carbamates and their insecticidal properties, ChE inhibition.

Goldblum et al. (1981) used the fly head I_{50} ChE data from UC Riverside to construct new models. The synthesis and testing of new carbamates encouraged others to examine the relationship between structure and ChE inhibition. Hastings et al. (1970), Main and Dauterman (1963), O'Brien (1960), and O'Brien (1976), and others showed that carbamates may be considered strong inhibitors of AChE, BChE, and CaE or substrates with very slow turnover rates. The dissociation rate of the inhibitor–enzyme complex is increased when it is diluted with saline, making it difficult to measure the initial amount of inhibited enzyme. Methods involving long incubation periods (i.e., Michel or Δ pH method; O'Brien 1960) also allow the inhibited enzyme to recover resulting in poor estimates of inhibition. On the basis of blood ChE measurements made according to Williams et al. (1957) during a 2-yr carbaryl feeding study, carbaryl was considered to be a poor ChE inhibitor (Carpenter et al. 1961). In another study in the same laboratory, aldicarb antiChE activity was analyzed using an automatic recording titrator (pH stat method; Knaak et al. 1966). ACh bromide was used as a substrate immediately after a pH adjustment was made in the reaction vessel by the titrator. Washing red cells with saline before analysis resulted in less inhibition and was avoided (Knaak et al. 1966).

A poor understanding of the chronic toxicity of aldicarb resulted in lowering the dose levels several times in a 2-yr feeding study, even with AChE inhibition data being available from 30- and 90-d studies. A NOAEL of 0.125 $mg\,kg^{-1}$ bw was reported in a 6-mon feeding study (USEPA 1988). A reference dose (RfD) of 0.0013 $mg\,kg^{-1}$ bw was established by EPA using an uncertainty factor of 100. A well-constructed PBPK/PD model would have been useful in setting up dose levels for a 2-yr study and predicting a NOAEL. A QSAR model, using TOPKAT, provides information on the oral, but not dermal, toxicity of the carbamate pesticides. A considerable amount of the *in vitro* work necessary for modeling carbamates [i.e., k_i, specific content of AChE, BChE, and CaE in tissues, skin permeation constants (K_p), $PC_{t:b}$, V_{max}, and K_m] including model development, e.g., with ACSL and ERDEM, was carried out after the carbamates were registered for use.

The pesticide registration requirements (Knaak et al. 1993) enforced by EPA via FIFRA require the development of considerable toxicological and metabolic data for assessing risks to the environment and to human and animal health. Human health risk assessments are the most difficult, involving the integration of toxicological, pathological, and metabolic data into a well-connected seamless story. Pathological data are easily integrated into toxicological reports involving the effects of chronic exposure to pesticides, but no one knows exactly what to do with ADME. Consequently, the development of metabolic data is still considered to be supplementary and not very useful until it is placed in the framework of a predictive PBPK/PD model. FIFRA does not require the development of the parameters (i.e., metabolic V_{max}, K_m values, tissue:blood partition coefficients, oral, skin, and inhalation parameters, plasma binding measurements, etc.) necessary for modeling insecticides. The reason for not requiring these values in the past was based on the belief that the science of PBPK modeling was immature, the work too costly, and not enough scientists were available to develop data or models or both. This is certainly not the case today, based on a monograph by Reddy et al. (2005) in which the authors presented the results of PBPK modeling in more than 1,000 publications.

Major problems in model development involve obtaining useful V_{max} and K_m values for metabolic transformations and tissue:blood partition coefficients for parent materials and their metabolites. Methods currently exist for obtaining V_{max} and K_m from studies involving human liver microsomal P450 and the individual CYPs (Foxenberg et al. 2006, 2007). QSAR models for predicting metabolic rates are currently being developed by Enslein et al. (2007). The Poulin and Theil (2002a,b) mechanistic model provides reasonable partition coefficients for nonvolatile metabolites when using log $D_{pH\ 7.4}$ values (ACD, Toronto, Ontario) in place of log P values.

In this review on insecticidal carbamates, a decision was made to develop preliminary PBPK/PD models for the 10 carbamates by using published metabolic pathway data and the multipathway PBPK/PD model of isofenphos (Knaak et al. 2002) as a partial template for the carbamate models. The aldicarb PBPK/PD model is given in Fig. 9 (see Appendix C). The conceptual models were translated into ACSL-based mathematical models incorporating the preliminary V_{max} and K_m values given in Tables 24–33 (see Appendix B), and the tissue:blood partition coefficients listed in Tables 14–23 (see Appendix A) in the model. The ACSL model was compiled using Compaq Visual Fortran, Version 6.6A (Compaq Computer Corporation, Houston, TX) before running the model.

The output (i.e., time tissue concentration curves, etc.) from each model was examined to make sure the model(s) performed as desired. The metabolic pathways in Tables 24–33 (Appendix B) were taken from the PBPK/PD models and are suitable for direct entry into the front end of ERDEM (exports command file to ERDEM202x) (Graphic Model). The molecular weights for parent carbamates and their metabolites are required by ERDEM along with $PC_{t:b}$ values, V_{max} and K_m values. The models used the k_i values for AChE and BChE in Tables 11 and 12. k_i values for CaEs were not found in the literature. ERDEM models were developed for carbofuran (Zhang et al. 2006) and carbaryl (Okino et al. 2005a,b; Power et al. 2005) using the metabolic pathways and preliminary V_{max} and K_m values presented in Tables 24–33 (Appendix B), along with the corresponding $PC_{t:b}$ values and molecular weights given in Tables 14–23 (see Appendix A).

As of 2006, under the Food Quality Protection Act (FQPA) of 1996, the EPA must consider available information concerning the cumulative effects on human health resulting from exposure to multiple chemicals that have a common mechanism of action, such as the *N*-methyl carbamates. A cumulative risk assessment also includes exposure data from multiple pathways. EPA used the relative potency factor (RPF) method to determine the joint risk associated with exposure to *N*-methylcarbamates. Oxamyl was selected as the index chemical for standardizing the toxic potencies and calculating the relative potency factor for each carbamate. Brain ChE inhibition from rats, measured at or near peak inhibition following a single oral exposure, was used as the endpoint for measuring RPF.

In this review, preliminary PBPK/PD models were developed for the 10 *N*-methylcarbamates of interest. The hydrolysis of aryl *N*-methylcarbamates (i.e., carbaryl, carbofuran, methiocarb, propoxur, etc.) by hydrolases was considered to be the primary detoxification pathway in humans as opposed to the P450-catalyzed oxidative pathway leading to hydroxylated derivatives in the rat. Hydrolytic enzymes catalyzing this pathway have not been well characterized despite the fact that it has been more than 40 yr since the carbamate insecticides were developed and registered. Rat and human microsomal P450 and individual CYP studies designed to obtain V_{max} and K_m data were carried out by Tang et al. (2002) and Usmani et al. (2004a) involving carbaryl and carbofuran, respectively. Metabolic rates were not determined for the hydrolysis of these two carbamates. The oxime carbamates are predominantly hydrolysed to their respective oximes by human or rat liver hydrolases. Metabolic rate constants for these enzymes (e.g., first-order or Michaelis–Menten) were not found in the literature. Less metabolic data were found on oxamyl than on the other carbamoyloximes (i.e., aldicarb, methomyl, and thiodicarb). Oxamyl is metabolized by hydrolysis to its oxime, des *N*-methyl oxamyl, or by enzymatic conversion via oxamyl nitrile to oxamyl acid. On the basis of the differences in structure between the oxime and aryl carbamates, the metabolic enzymes involved in their metabolism and their AChE k_i values, it would seem logical to use two index chemicals to represent the 10 carbamates. In terms of oral toxicity, the oxime carbamates vary as follows: aldicarb >

oxamyl > methomyl > thiodicarb, with aldicarb being the most toxic and thiodicarb the least toxic. Carbofuran is the most toxic of the aryl carbamates, followed by methiocarb > formetanate > propoxur > pirimicarb, with carbaryl being the least toxic.

VIII. Recommendations

A research program is needed to develop liver microsomal metabolic rate constants (V_{max}, K_m) for the hydroxylation of the aryl *N*-methylcarbamates and the hydrolysis of the hydroxylated products and the oxime carbamates. So far, data exist only on the formation of 3-hydroxycarbofuran from carbofuran and 4-hydroxycarbaryl from carbaryl. Studies involving the metabolism of the OP pesticides parathion and chlorpyrifos strongly suggest that information on the activity of the individual CYPs against these insecticides, and the content of these CYPs in HLMs by age, may be combined to give the metabolic rate constants for HLMs.

A significant amount of work has been carried out on the nature of the active sites of AChE and BChE as well as the CaEs. Information regarding the content of these enzymes in tissues needs to be improved. Current estimates involve one study on tissue enzyme activity carried out by Maxwell et al. (1987). A tabulation of ACh turnover numbers for AChE gives a coefficient of variation of 28.6%. The results [esterase binding site = (enzyme activity)/turnover rate] suggest that there may be a 28% variation in the inhibition estimates from PBPK/PD models.

Tissue:blood partition coefficients estimated by mechanistic models need to be validated based on experimental data using a high-throughput automated system based on the work of Jepson et al. (1994). To date, a system of this nature does not exist.

Summary

A review of the scientific literature was carried out for parameters [i.e., IC_{50}, LD_{50}, K_p (cm/hr) for percutaneous absorption, skin:water and tissue: blood partition coefficients, inhibition k_i values, and metabolic parameters such as V_{max} and K_m] suitable for building QSAR and PBPK/PD models [using e.g., ACSL or ERDEM] for assessing human risk to 10 *N*-methylcarbamate insecticides: aldicarb, carbaryl, carbofuran, formetanate, methiocarb, methomyl, oxamyl, pirimicarb, propoxur, and thiodicarb. While searching for the parameters, PBPK/PD models were developed for these carbamates based on published metabolic pathways. ACSL-based models were written for each carbamate and were run using command files containing the parameters or parameter estimates. During this process, tissue: blood partition coefficients were developed for parent carbamates and metabolites using published mechanistic models and $\log D_{pH\ 7.4}$ values. New estimates of tissue AChE, BChE, and CaE were made using substrate turnover numbers found in the literature. In the case of AChE, sufficient turnover data were available to calculate a coefficient of variation (28.6%). Bimolecular inhibition rate constants (k_i) involving AChE and BChE were found for all the carbamates except pirimicarb and formetanate. Inhibition rate constants were not found for CaEs. Several metabolic rate constants, V_{max} and K_m, determined using liver microsomes and individual CYPs (P450 isozymes), were found for one or more biotransformations (i.e., hydroxylation, etc) involving carbaryl and carbofuran. Limited information is available on the enzymes carrying out hydrolysis or their rate constants. A program is needed to develop enzyme (P450, carbamate hydrolases, etc.) activity and enzyme content to support risk assessment methodologies involving QSAR and PBPK/PD models.

Acknowledgments

Our thanks are extended to Prof. Herbert N. Nigg, University of Florida, and Prof. James R. Olson, The State University of New York at Buffalo, for reviewing the manuscript of this article and making helpful suggestions that improve its quality. We also thank Peter Mueller and Dr. Robert W. Gerlach, General Dynamics Information Technology, for their skilled work in preparing this manuscript for publication. The work on this review was funded by the U.S. EPA through General Services Administration contract GS-09K-99-BHD-0001, tasks 9T3Z061TMA-C and EP06HO00393 with General Dynamics Information Technology (GDIT), Henderson, NV. Although this work was reviewed by EPA and approved for publication, it may not necessarily reflect official Agency policy, nor does it represent the official views of GDIT. Mention of trade names or commercial products does not constitute endorsement or recommendation for use.

References

Abd-Elroaf TK, Fahmy MAH, Fukuto TR, Elsebae AH (1977) Anticholinesterase activity and toxicity of substituted phenyl methylcarbamates to honey bee. J Econ Entomol 70:78–82.

Aeschlimann JA, Reinert M (1931) The pharmacological action of some analogues of physostigmine. J Pharmacol Exp Ther 43:413.

Ahmad S, Knowles CO (1971) Formamidase involvement in the metabolism of chlorphenamidine and formetanate acaricides. J Econ Entomol 64:792–795.

Aldridge WN, Reiner E (1972) Kinetics of reaction of B-esterases with carbamates. In: Neuberger A, Tatum EL (eds) Frontiers in Biology. North Holland, Amsterdam.

ACGIH (1991) Documentation of the Threshold Limit Values and Biological Exposure Indices, 6th Ed, vols I, II, III. ACGIH, Cincinnati, OH.

Amidon GL, Lennernas H, Shah VP, Crison JR (1995) A theoretical basis for a biopharmaceutics drug classification: the correlation of *in vitro* drug product dissolution and *in vivo* bioavailability. Pharm Res 12:413–420.

Andrawes NR, Dorough HW, Lindquist DA (1967) Degradation and elimination of Temik in rats. J Econ Entomol 60:979–987.

Arpagaus M, Kott M, Vatsis KP, Bartels CF, La Du BN, Lockridge O (1990) Structure of the gene for human butyrylcholinesterase. Evidence for a single copy. Biochemistry 29:124–131.

Austin L, Berry WK (1953) Two selective inhibitors of cholinesterase. Biochem J 54:695–700.

Baron RL (1994) A carbamate insecticide: a case study of aldicarb. Environ Health Perspect 102 (suppl 11):23–28.

Baron RL Jr (1991) Carbamate insecticides. In: Hayes, WJ Jr, Laws ER Jr (eds) Handbook of Pesticide Toxicology, vol 3. Classes of Pesticides. Academic Press, New York, pp 1125–1189.

Barthova J, Mekonnen T, Woldemarian G, Leblova S (1989) The effect of carbamate insecticides on cholinesterases of nontarget organisms. Biologia 44: 353–358.

Benet LZ, Cummins CL, Wu CY (2003) Transporter–enzyme interactions: implications for predicting drug–drug interactions from *in vitro* data. Curr Drug Metab 4:393–398.

Beraud M, Forgue MF, Pipy B, Didier A, Carre P (1989) Lipid peroxidation in rat liver microsomes: influence of carcinogenic *N*-nitrosocarbaryl. Toxicology 58: 299–311.

Blancato JN, Knaak JB, Power F (2000) Use of PBPK models for assessing absorbed dose and ChE inhibition from aggregate exposure of infants and children to organophosphorus insecticides. Presented at 10th Annual Meeting of the International Society of Exposure Analysis, Monterey, CA (abstract 3F-09o).

Blancato JN, Power FW, Brown RN, Dary CC (2006) Exposure related dose estimating model (ERDEM): a physiologically-based pharmacokinetic and pharmacodynamic (PBPK/PD) model for assessing human exposure and risk. EPA/600/R-06/061 (NTIS PB2006-114712). U.S. Environmental Protection Agency, Washington, DC.

Boyd AE, Dunlop CS, Wong L, Radic Z, Taylor P, Johnson DA (2004) Nanosecond dynamics of acetylcholinesterase near the active center gorge. J Biol Chem 279:26612–26618.

Branch RA, Jacqz E (1986a) Is carbaryl as safe as its reputation? Does it have a potential for causing chronic neurotoxicity in humans? Am J Med 80:659–664.

Branch RA, Jacqz E (1986b) Subacute neurotoxicity following long-term exposure to carbaryl. Am J Med 80:741–745.

Bronaugh RL, Maibach HI (eds) (1985) Dermatology, vol. 6. Percutaneous Absorption: Mechanisms, Methodology and Drug Delivery. Dekker, New York.

Buronfosse T, Moroni P, Benoit E, Riviere JL (1995) Stereoselective sulfoxidation of the pesticide methiocarb by flavin-containing monooxygenase and CYP-dependent monooxygenases of rat liver microsomes. Anticholinesterase activity of the two sulfoxide enantiomers. J Biochem Toxicol 10:179–189.

Cambon C, Declume C, Derache R (1979) Effect of the insecticidal carbamate derivatives (Carbofuran, Pirimicarb, Aldicarb) on the activity of acetylcholinesterase in tissues from pregnant rats and fetuses. Toxicol Appl Pharmacol 40:203–208.

Cardona RA, Dorough HW (1973) Syntheses of the beta-D-glucosides of 4- and 5-hydroxy-1-naphthyl *N*-methylcarbamate. J Agric Food Chem 21:1065–1071.

Carpenter CP, Weil CS, Palm PE, Woodside MW, Nair JH III, Smyth HF Jr (1961) Mammalian toxicity of 1-naphthyl-*N*-methylcarbamate (Sevin insecticide). J Agric Food Chem 9:30–39.

Casale GP, Vennerstrom JL, Bavari S, Wang TL (1993) Inhibition of interleukin 2 driven proliferation of mouse CTLL2 cells, by selected carbamate and organophosphate insecticides and congeners of carbaryl. Immunopharmacol Immunotoxicol 15:199–215.

Chang JH, Benet LZ (2005) Glucuronidation and the transport of the glucuronide metabolites in LLC-PK1 cells. Mol Pharmacol 2:428–434.

Chang K-M, Knowles CO (1979) Metabolism of oxamyl in mice and twospotted spider mites. Arch Environ Contam Toxicol 8:499–508.

Chatonnet A, Lockridge O (1989) Comparison of butyrylcholinesterase and acetylcholinesterase. Biochem J 260:625–634.

Chen KC, Dorough HW (1979) Glutathione and mercapturic acid conjugations in the metabolism of naphthalene and 1-naphthyl *N*-methylcarbamate (carbaryl). Drug Chem Toxicol 2:331–354.

Cho TM, Rose RL, Hodgson E (2006) In vitro metabolism of naphthalene by human liver microsomal cytochrome P450 enzymes. Drug Metab Dispos 34: 176–183.

Christian TS, Beasley JG (1968) Michaelis constants for isolated cholinesterase systems. J Pharmacol Sci 6:1025–1027.

Clark DE, Pickett SD (2000) Computational methods for the prediction of "druglikeness." Drug Discov Today 5:49–58.

Cobb GP, Harper FD, Weisskopf CP (2001) Nonlethal method of forensic evaluation of aldicarb exposure in wildlife. Arch Environ Contam Toxicol 40: 77–88.

Cohen JA, Oosterbann RA, Warringa MGPJ (1955) Turnover number of ali-esterase, pseudo- and true cholinesterase and the combination of these enzymes with diisopropyl fluorophosphates (DFP). Biochim Biophys Acta 18:228–235.

Convard T, Dubost JP, Le Solleu H, Kummer E (1994) SmilogP: a program for a fast evaluation of theoretical log P from the SMILES code of a molecule. Quant Struct-Act Relat 13:34–37.

Cool M, Jankowski CK (1985) A survey of the metabolism of carbamate insecticides. In: Hutson DH, Roberts TR (eds) Progress in Pesticide Biochemistry and Toxicology, vol 5 Wiley. Chichester, pp 87–142.

Crop Protection Handbook (2007) MeisterPRO Information Resources, Willoughby, OH.

Cygler M, Schrag JD, Susman JL, Harel M, Silman I, Gentry MK, Doctor DP (1993) Relationship between sequence conservation and three-dimensional structure in a large family of esterases, lipases, and related proteins. Protein Sci 2:366–382.

Dawson RM (1994) Rate constants of carbamylation and decarbamylation of acetylcholinesterase for physostigmine and carbaryl in the presence of an oxime. Neurochem Int 24:173–182.

Dawson RM (1995) Oxime effects on the rate constants of carbamylation and decarbamylation of acetylcholinesterase for pyridostigmine, physostigmine and insecticidal carbamates. Neurochem Int 26:643–654.

Dawson RM, Poretski M (1985) Carbamylated acetylcholinesterase: acceleration of decarbamylation by bispyridinium oximes. Biochem Pharmacol 34:4337–4340.

de Coninck O (1886) On the nature and properties of alkaloids. J Am Chem Soc 8:23–24.
De La Hoz D, Doctor BP, Ralston JS, Rush RS, Wolfe AD (1986) A simplified procedure for the purification of large quantities of fetal bovine serum acetylcholinesterase. Life Sci 39:195–199.
Department of Health Education and Welfare, National Cancer Institute CD HEW/NCD (1979) Toxicology and Carcinogenesis Studies of Aldicarb, Report 136. Publication #79-1391. National Institutes of Health, Washington, DC.
Dorough HW (1968) Metabolism of furadan (NIA-10242) in rats and houseflies. J Agric Food Chem 16:319–325.
Dorough HW, Casida JE (1964) Nature of certain carbamate metabolites of the insecticide sevin. J Agric Food Chem 12:294–304.
Dorough HW, Leeling NC, Casida JE (1963) Nonhydrolytic pathway in metabolism of *N*-methylcarbamate insecticides. Science 140:170–171.
Durden JA, Bartley WJ, Stephan JF (1970) The preparation of standards for Aldicarb (Temik) metabolism study. J Agric Food Chem 18:454–458.
ECETOC (1993) Percutaneous absorption. In: Stringer DA (ed) Monograph No. 20. Brussels, Belgium.
Ecobichon DJ (1970) Characterization of the esterases of canine serum. Can J Biochem 48:1359–1367.
Ekins S, Rose J (2002) In silico ADME/TOX: the state of the art. J Mol Graph Model 20:305–309.
Ellenhorn MJ, Barceloux DG (1988) Medical Toxicology: Diagnosis and Treatment of Human Poisoning. Elsevier, New York.
Ellman GL, Courtney KD, Andres V, Featherstone RM (1961) A new and rapid colorimetric determination of acetylcholinesterase activity. Biochem Pharmacol 7:88–95.
Enslein K (1988) An overview of structure–activity relationships as an alternative to testing in animals for carcinogenicity, mutagenicity, dermal and eye irritation, and acute oral toxicity. Toxicol Ind Health 4:479.
Enslein K, Gombar VK, Shapiro D, Blake BW (1998) Prediction of rat oral LD_{50} values of organophosphates by QSAR equations. Topkat Health Designs, Rochester, NY.
Enslein K, Knaak JB, VanNostrand K, Susa R, Koestler D (2007) CYP3A4 hydroxylation kinetics: QSAR models with application to PBPK modeling. J Comp-Aided Mol Design (submitted).
Eriksson H, Augustinsson KB (1979) A mechanistic model for butyrylcholinesterase. Biochim Biophys Acta 567:161–173.
Fahmy MAH, Metcalf RL, Fukuto TR, Hennessy DJ (1966) Effects of deuteration, fluorination, and other structural modifications of the carbamyl moiety upon the anticholinesterase and insecticidal activities of phenyl *N*-methylcarbamates. J Agric Food Chem 14:79–83.
Fahmy MAH, Fukuto TR, Myers RO, March RB (1970) The selective toxicity of new N-phosphorothioylcarbamate esters. J Agric Food Chem 18:793–796.
Felton JC (1968) Insecticidal activity of some oxime carbamates. J Sci Food Agric (Suppl) 32–38.
Ferguson PW, Dey MS, Jewell SA, Krieger RI (1984) Carbofuran metabolism and toxicity in the rat. Fundam Appl Toxicol 4:14–21.

Ferro A, Masson P (1987) Kinetic evidence for thermally induced conformational change of butyrylcholinesterase. Biochim Biophys Acta 916:193–199.

Foss W, Krechniak J (1980) The fate of propoxur in rat. Arch Toxicol (Suppl) 4:346–349.

Fouchecourt M-O, Beliveau M, Krishnan K (2001) Quantitative structure-pharmacokinetic relationship modeling. Sci Total Environ 274:125–135.

Foxenberg RJ, Knaak JB, McGarrigle BP, Kostyniak PJ, Olson JR (2006) Utilizing human cytochrome P450 specific content and activity to better model parathion and chlorpyrifos metabolism in infants, children and adults. Presented at 45th Annual Meeting of the Society of Toxicology, San Diego, CA (abstract #1207).

Foxenberg RJ, Knaak JB, McGarrigle BP, Kostyniak PJ, Olson JR (2007) Human hepatic cytochrome P450-specific metabolism of parathion and chlorpyrifos. Drug Metab Dispos 35:189–193.

Fraser J, Harrison IR, Wakerly SB (1968) Synthesis and insecticidal activity of *N*-acyl-*N*-methylcarbamates. J Sci Food Agric (Suppl):8–12.

Fukuto TR (1972) Metabolism of carbamate insecticides. Drug Metab Rev 1:117–151.

Fukuto TR, Fahmy MAH, Metcalf RL (1967) Alkaline hydrolysis, anticholinesterase, and insecticidal properties of some nitro-substituted phenyl carbamates. J Agric Food Chem 15:273–281.

Fukuto TR, Metcalf RL, Jones RL, Myers RO (1969) Structure, reactivity, and biological activity of O-(diethyl phosphoryl) oximes and O-(methylcarbamoyl) oximes of substituted acetophenones and á-substituted benzaldehydes. J Agric Food Chem 17:923–930.

Gagua DI (1976) Acute toxicity of pirimor for laboratory animals. Sbornik Trudov-Nauchno-Issledovatel'skii Institut Gigieny Truda I Profzabolevanii im. N.I. Makhviladze.

Gearhart JM, Jepson GW, Clewell HJ 3rd, Andersen ME, Conolly RB (1990) Physiologically based pharmacokinetic and pharmacodynamic model for the inhibition of acetylcholinesterase by diisopropylfluorophosphate. Toxicol Appl Pharmacol 106:295–310.

Gearhart JM, Jepson GW, Clewell, HJ, Andersen ME, Connolly RB (1994) Physiologically based pharmacokinetic model for the inhibition of acetylcholinesterase by organophosphate esters. Environ Health Perspect 102:51–60.

Goldblum A, Yoshimoto M, Hansch C (1981) Quantitative structure–activity relationship of phenyl *N*-methylcarbamate inhibition of acetylcholinesterase. J Agric Food Chem 29:277–288.

Golicnik M, Sinko G, Simeon-Rudolf V, Grubic Z, Stojan J (2002) Kinetic model of ethopropazine interaction with horse serum butyrylcholinesterase and its docking into the active site. Arch Biochem Biophys 398:23–31.

Gordon MA, Carpenter DE, Wilson IB (1978) The turnover numbers of acetylcholinesterase forms. Mol Pharmacol 14:266–270.

Groff WA, Kaminskia A, Ellin RI (1976) Interconversion of cholinesterase enzyme activity units by the manual pH method and a recommended automated method. Clin Toxicol 9:353–358.

Grubic Z, Sketel J, Kinar B, Brzin M (1981) Recovery of acetylcholinesterase in the diaphragm, brain, and plasma of the rat after irreversible inhibition by

soman: a study of cytochemical localization and molecular forms of the enzyme in the motor end plate. J Neurochem 37:909–916.
Gubler K, Margot A, Gysin H (1968) Alicyclic and heterocyclic carbamates. J Sci Food Agric (Suppl):13–18.
Guengerich FP, Martin MV (1998) Purification of cytochromes P450. Rat and human hepatic forms. Methods Mol Biol 107:35–53.
Gupta RC, Dettbarn WD (1993) Role of carboxylesterases in the prevention and potentiation of *N*-methylcarbamate toxicity. Chem Biol Interact 87:295–303.
Gupta RC, Kadel WL (1990) Toxic interaction of tetraisopropylpyrophosphoramide and propoxur: some insights into the mechanisms. Arch Environ Contam Toxicol 19:917–920.
Gysin H (1954) Uber einige neue insektizide. Chimia 8:205–244.
Hansch C, Leo A (1979) Substituent Constants for Correlation Analysis in Chemistry and Biology. Wiley, New York.
Hansch C, Maloney RR, Fujita T (1962) Correlation of biological activity of phenoxyacetic acids with Hammett substituent constants and partition coefficients. Nature (Lond) 194:178–180.
Hansch C, Hoekman D, Leo A, Weininger D, Selassie CD (2002) Chem-bioinformatics: comparative QSAR at the interface between chemistry and biology. Chem Rev 102:783–812.
Hardt J, Angerer J (1999) Determination of pirimicarb in human urine by gas chromatography-mass spectrometry. J Chromatogr B Biomed Sci Appl 730:229–238.
Hardt J, Appl E, Angerer J (1999) Biological monitoring of exposure to pirimicarb: hydroxypyrimidines in human urine. Toxicol Lett 107:89–93.
Harel M, Sussman JL, Krejei E, Bon S, Chanal P, Massoulie J, Silman JL (1992) Conversion of acetylcholinesterase to butylcholinesterase: modeling and mutagenesis. Proc Natl Acad Sci U S A 89:10827–10831.
Hart GJ, O'Brien RD (1973) Recording spectrophotometric method for determination of dissociation and phosphorylation constants for the inhibition of acetylcholinesterase by organophosphates in the presence of substrate. Biochemistry 12:2940–2944.
Harvey J Jr, Han JC-Y (1978) Metabolism of oxamyl and selected metabolites in the rat. J Agric Food Chem 26:902–910.
Harvey J Jr, Jelinek AG, Sherman H (1973) Metabolism of methomyl in the rat. J Agric Food Chem 21:769–774.
Hastings FL, Main AR, Iverson F (1970) Carbamylation and affinity constants of some carbamate inhibitors of acetylcholinesterase and their relation to analogous substrate constants. J Agric Food Chem 18:497–502.
Hayes WJ Jr, Laws ER Jr (eds) (1991) Handbook of Pesticide Toxicology, vol 3. Classes of Pesticides. Academic Press, New York, pp 1137, 1162.
Hendley P, Lam J (1981) Stable isotopes as MS tracers in a study of the metabolites and residues from a pesticide. Method Surv 10:148–149.
Herzsprung P, Weil L, Niessner R (1992) Measurement of bimolecular rate constants k_i of the cholinesterase inactivation reaction by 55 insecticides and of the influence of various pyridimium oximes on k_i. Int J Environ Anal Chem 47:181–200.
Hetnarski B, O'Brien RD (1973) Charge transfer in cholinesterase inhibition. Role of the conjugation between carbamyl and aryl groups of aromatic carbamates. Biochemistry 12:3883–3887.

Hetnarski B, O'Brien RD (1975a) Electron-donor and affinity constants and their application to the inhibition of acetylcholinesterase by carbamates. J Agric Food Chem 23:709–713.

Hetnarski B, O'Brien RD (1975b) The charge-transfer constant. A new substituents constant for structure–activity relationships. J Med Chem 18:29–33.

Honeycutt RC, Day EW Jr (eds) (2001) Worker Exposure to Agrochemicals: Methods for Monitoring and Assessment. Lewis, Boca Raton, FL.

Honeycutt RC, Zweig G, Ragsdale NN (eds) (1985) Dermal Exposure Related to Pesticide Use: Discussion of Risk Assessment. ACS Symposium Series 273, American Chemical Society, Washington, DC.

Hood SR, Shah G, Jones P (1998) Expression of cytochromes P450 in a *Baculovirus* system. In: Phillips IR, Shephard EA (eds) Methods in Molecular Biology: CYP Protocols. Humana Press, Totowa, NJ, Ch 24, pp 203–218.

Hosokawa M, Maki T, Satoh T (1987) Multiplicity and regulation of hepatic microsomal carboxylesterases in rats. Mol Pharmacol 31:579–584.

Hosokawa M, Maki T, Satoh T (1990) Characterization of molecular species of liver microsomal carboxylesterases of several animal species and humans. Arch Biophys 277:219–227.

Huhtanen K, Dorough HW (1976) Isomerization and Beckmann rearrangement reactions in the metabolism of methomyl in rats. Pestic Biochem Physiol 6:571–583.

Humerickhouse R, Lohrbach K, Li L, Bosron WF, Dolan ME (2000) Characterization of CPT-11 hydrolysis by human liver carboxylesterase isoforms hCE-1 and hCE-2. Cancer Res 60:1189–1192.

Hunter DL, Marshall RS, Padilla S (1997) Automated instrument analysis of cholinesterase activity in tissues from carbamate-treated animals: a cautionary note. Toxicol Methods 7:43–53.

Ikeda Y, Okamura K, Fujjii S (1977) Purification and characterization of rat liver microsomal monoacylglycerol lipase in comparison to the other esterases. Biochim Biophys Acta 488:128–139.

Iverson F, Main AR (1969) Effect of charge on the carbamylation and binding constants of eel acetylcholinesterase in reaction with neostigmine and related carbamates. Biochemistry 8:1889–1895.

Ivie GW, Dorough HW (1968) Furadan-^{14}C metabolism in a lactating cow. J Agric Food Chem 16:849–855.

Jepson GW, Hoover DK, Black RK, McCafferty JD, Mahle DA, Gearhart JM (1992) Partition coefficient determination for nonvolatile and intermediate volatility chemicals in biological tissues. Toxicologist 12:262.

Jepson GW, Hoover DK, Black RK, McCafferty JD, Mahle DA, Gearhart JM (1994) A partition coefficient determination method for non-volatile chemicals in biological tissues. Fundam Appl Toxicol 22:519–524.

Kalow W (1964) The influence of pH on the hydrolysis of benzoylcholine by pseudocholinesterase of human plasma. Can J Physiol Pharmacol 42:161–168.

Kaplan D, Ordentlich A, Barak D, Ariel N, Kronman C, Velan B, Shafferman A (2001) Does "butyrylization" of acetylcholinesterase through substitution of the six divergent aromatic amino acids in the active center gorge generate an enzyme mimic of butyrylcholinesterase? Biochemistry 40:7433–7445.

Kennedy GL (1986) Acute toxicity studies with oxamyl. Fundam Appl Toxicol 6:423–429.

Khoury GF, Brill J, Walts L, Busuttil RW (1987) Atypical serum cholinesterase eliminated by orthotopic liver transplantation. Anesthesiology 67:273–274.

Klopman G, Li J, Wang S, Dimayuga MJ (1994) Computer automated log P calculations based on an extended group contribution approach. J Chem Inform Comput Sci 34:752–781.

Klopman G, Stefan L, Saiakhov RD (2000) New approach to the evaluation of bioavailability of drugs. Proceedings of the 220th American Chemical Society National Meeting, American Chemical Society, Washington, DC.

Knaak JB (1971) Biological and nonbiological modifications of carbamates. Bull WHO 44:121–131.

Knaak JB, Sullivan LJ (1967) Metabolism of carbaryl in the dog. J Agric Food Chem 15:1125–1126.

Knaak JB, Wilson BW (1985) Dermal dose-cholinesterase response and percutaneous absorption studies with several cholinesterase inhibitors. In: Honeycutt RC, Zweig G, Ragsdale NN (eds) Dermal Exposure Related to Pesticide Use: Discussion of Risk Assessment. ACS Symposium Series 273. American Chemical Society, Washington, DC, pp 63–79.

Knaak JB, Tallant MJ, Bartley MJ, Sullivan LJ (1965) The metabolism of carbaryl in the rat, guinea pig, and man. J Agric Food Chem 13:537–543.

Knaak JB, Tallant MJ, Sullivan LJ (1966) The metabolism of 2-methyl-2-(methylthio) propionaldehyde *O*-(methylcarbamoyl) oxime in the rat. J Agric Food Chem 14:573–578.

Knaak JB, Eldridge J, Sullivan LJ (1967) Systematic approach to preparation and identification of glucuronic acid conjugates. J Agric Food Chem 15: 605–609.

Knaak JB, Tallant MJ, Kozbelt SJ, Sullivan LJ (1968) The metabolism of carbaryl in man, monkey, pig and sheep. J Agric Food Chem 16:465–470.

Knaak JB, Munger DM, McCarthy JF, Satter LD (1970) Metabolism of carbofuran alfalfa residues in the dairy cow. J Agric Food Chem 18:832–837.

Knaak JB, Maddy KT, Jackson T, Fredrickson AS, Peoples SA, Love R (1978) Cholinesterase activity in blood samples collected from field workers and nonfield workers in California. Toxicol Appl Pharmacol 45:755–770.

Knaak JB, Schlocker P, Ackerman CR, Sieber JN (1980) Reentry research: establishment of safe pesticide levels on foliage. Bull Environ Contam Toxicol 24:796–804.

Knaak JB, Yee K, Ackerman CR, Zweig G, Fry DM, Wilson BW (1984) Percutaneous absorption and dermal dose-cholinesterase response studies with parathion and carbaryl in the rat. Toxicol Appl Pharmacol 76:252–263.

Knaak JB, Iwata Y, Maddy KT (1989) The worker hazard posed by reentry into pesticide-treated foliage: development of safe reentry times, with emphasis on chlorthiophos and carbosulfan. In: Paustenbach D (ed) The Risk Assessment of Environmental Hazards: A Textbook of Case Studies. Wiley, New York, Ch 24, pp 797–842.

Knaak JB, Wagner B, Boutchyard H Jr, Smith LW, Jones T, Wang RG (1993) Computerization of toxicological data by government agencies, chemical, and information industries. In: Jolley RL, Wang RGM (eds) Effective and Safe Waste Management: Interfacing Sciences and Engineering with Monitoring and Risk Analysis. CRC Press, Boca Raton, FL, Ch 19, pp 227–264.

Knaak JB, Al-Bayati MA, Raabe OG (1995) Development of partition coefficients, V_{max} and K_m values, and allometric relationships. Toxicol Lett 79:87–98.

Knaak JB, Dary C, Patterson GT, Blancato J (2002) The worker hazard posed by reentry into pesticide-treated foliage: reassessment of reentry levels/intervals using foliar residue transfer-percutaneous absorption PBPK/PD models, with emphasis on isofenphos and parathion. In: Paustenbach D (ed) Human and Ecological Risk Assessment: Theory and Practice. Wiley, New York, pp 673–731.

Knaak JB, Dary CC, Power F, Thompson CB, Blancato JN (2004) Physicochemical and biological data for the development of predictive organophosphorus pesticide QSARs and PBPK/PD models for human risk assessment. Crit Rev Toxicol 34:143–207.

Knaak JB, Chang D, Dary CC, Tornero-Velez R, Okino M, Power F (2005) Estimation of adipose tissue: blood partition coefficients of carbamates and OP pesticides using relative lipid content and log $P_{o:w}$ as the mechanistic determinants. Presented at 231st Annual Meeting of the American Chemical Society, Washington, DC, Aug 28–Sept 1.

Kolbezen MJ, Metcalf RL, Fukuto TR (1954) Insecticidal activity of carbamate cholinesterase inhibitors. J Agric Food Chem 2:864–870.

Kousba A, Sultatos LG (2002) Continuous system modeling of equilibrium dialysis for determinations of tissue partitioning of parathion and paraoxon. Toxicol Lett 133:153–159.

Krechniak J, Foss W (1982) Cholinesterase activity in rats treated with propoxur. Bull Environ Contam Toxicol 29:599–604.

Krechniak J, Foss W (1983) Fate of propoxur in the experimental animal after repeated administration. Bromat Chem Toksykol 16:205–208.

Krishna JG, Casida JE (1966) Fate of the radiocarbon from ten variously labeled methyl- and dimethylcarbamate-C14 insecticide chemicals and their hydrolysis products. J Agric Food Chem 14:98–105.

Kuhr RJ, Dorough HW (1976) Carbamate Insecticides: Chemistry, Biochemistry and Toxicology. CRC Press, Cleveland, OH.

La Motta RV, McComb RB, Wetstone HJ (1965) Isozymes of serum cholinesterase: a new polymerization sequence. Can J Physiol Pharmacol 43:313–318.

Lam JL, Benet LZ (2004) Hepatic microsome studies are insufficient to characterize *in vivo* hepatic metabolic clearance and metabolic drug-drug interactions: studies of digoxin metabolism in primary rat hepatocytes versus microsomes. Drug Metab Dispos 32:1311–1316.

Lau YY, Wu C-Y, Okochi H, Benet LZ (2003) Ex situ inhibition of hepatic uptake and efflux significantly changes metabolism: hepatic enzyme-transporter interplay. J Pharmacol Exp Ther 308:1040–1045.

Leeling NC, Casida JE (1966) Metabolites of carbaryl (1-naphthyl methylcarbamate) in mammals and enzymatic systems for their formation. J Agric Food Chem 14:281–290.

Leo AJ (1993) Calculating log P_{oct} from structures. Chem Rev 93:1281–1306.

Leo AJ, Hansch C, Elkins D (1971) Partition coefficients and their uses. Chem Rev 71:525–616.

Levy D, Ashani Y (1986) Synthesis and *in vitro* properties of a powerful quanternary methylphosphonate inhibitor of acetylcholinesterase. Biochem Pharmacol 35:1079–1085.

Lewis RJ (1996) Sax's Dangerous Properties of Industrial Materials, 9th Ed, vol 1–3. Van Nostrand Reinhold, New York, pp 653–654.
Lewis DFV (2000) Structural characteristics of human P450s involved in drug metabolism: QSARs and lipophilicity profiles. Toxicology 144:197–203.
Li Y, Barefoot AC, Reiser RW, Fogiel AJ, Sabourin PJ (1997) Identification of thiocyanate as the principal metabolite of oxamyl in lactating goat. J Agric Food Chem 45:962–966.
Lipinski CA (2000) Drug-like properties and the causes of poor solubility and poor permeability. J Pharmacol Toxicol Methods 44:235–249.
Lipinski CA, Lombardo F, Dominy BW, Feeney PJ (2001) Experimental and computational approaches to estimate solubility and permeability in drug discovery and development settings. Adv Drug Delivery Rev 46:3–26.
Liu HX, Yao XJ, Zhang RS, Liu MC, Hu ZD, Fan BT (2005) Prediction of the tissue/blood partition coefficients of organic compounds based on the molecular structure using least-squares support vector machines. J Comput-Aided Mol Design 19:499–508.
Lockridge O (1990) Genetic variants of human serum cholinesterase influence metabolism of the muscle relaxant succinylcholine. Pharmacol Ther 47: 35–60.
Lockridge O, La Du BN (1978) Comparison of atypical and usual human serum cholinesterase. J Biol Chem 253:361–366.
Lockridge O, La Du BN (1991) Structure of human butylcholinesterase gene and expression in mammalian cells. In: Massoulie J, Bacou F, Barnard E, Chatonnet A, Doctor BP (eds) Cholinesterases: Structure, Function, Mechanism, Genetics, and Cell Biology. Conference Proceedings Series, American Chemical Society, Washington, DC.
Lockridge O, Bartel CF, Vaughan TA, Wong CK, Norton SE, Johnson LL (1987) Complete amino acid sequence of human serum cholinesterase. J Biol Chem 262:549–557.
Lu AYH, West SB (1980) Multiplicity of mammalian microsomal cytochrome P450. Pharmacol Rev 31:277–291.
MacPherson SE, Scott RC, Williams FM (1991) Fate of carbaryl in rat skin. Arch Toxicol 65:594–598.
Main AR (1964) Affinity and phosphorylation constants for the inhibition of esterases by organophosphates. Science 144:992–993.
Main AR (1969a) Kinetics of cholinesterase inhibition by organophosphate and carbamate insecticides. Can Med Assoc J 100:161–167.
Main AR (1969b) Kinetic evidence of multiple reversible cholinesterases based on inhibition by organophosphates. J Biol Chem 244:829–840.
Main AR (1979) Mode of action of anticholinesterases. Pharmacol Ther 6:579–629.
Main AR, Dauterman WC (1963) Determination of the bimolecular rate constant for the reaction between organophosphorus inhibitors and esterases in the presence of substrate. Nature (Lond) 198:551–553.
Main AR, Tarkon E, Aull J, Soucie W (1972) Purification of horse serum cholinesterase by preparative polyacrylamide gel electrophoresis. J Biol Chem 247: 566–571.
Marshall TC, Dorough HW (1977) Bioavailability in rats of bound and conjugated plant carbamate insecticide residues. J Agric Food Chem 25:1003–1009.

Marshall TC, Dorough HW (1979) Biliary excretion of carbamate insecticides in the rat. Pestic Biochem Physiol 11:56–63.

Masson P (1979) Multiple molecular forms of human plasma butylcholinesterase. I. Apparent molecular parameters and broad pattern of the quaternary structure. Biochim Biophys Acta 578:493–504.

Masson P, Goasdoue JL (1986) Evidence that the conformational stability of 'aged' organophosphate-inhibited cholinesterase is altered. Biochim Biophys Acta 869:304–313.

Matsumura F (1985) Toxicology of Insecticides, 2nd Ed. Plenum Press, New York.

Mattie DR, Bates GD, Jepson GW, Fisher JW, McDougal JN (1994) Determination of skin: air partition coefficients for volatile chemicals: experimental method and applications. Toxicol Appl Pharmacol 22:51–57.

Maulet Y, Camp S, Gibney G, Rachinsky TL, Ekstrom TJ, Taylor P (1990) Single gene encodes glycophospholipid-anchored and asymmetric acetylcholinesterase forms: alternative coding exons contain inverted repeat sequences. Neuron 4:289–301.

Maxwell DM (1992) The specificity of carboxylesterase protection against the toxicity of organophosphorus compounds. Toxicol Appl Pharmacol 114:306–312.

Maxwell DM, Lenz DE, Groff WA, Kaminskis A, Froehlich HL (1987) The effects of blood flow and detoxification on *in vivo* cholinesterase inhibition by Soman in rats. Toxicol Appl Pharmacol 88:66–76.

McTiernan C, Adkins S, Chatonnet A, Vaughan TA, Bartels CF, Kott M, Rosenberry TL, La Du BN (1987) Brain cDNA clone for human cholinesterase. Proc Natl Acad Sci USA 86:953–957.

Menzie CM (1974) Metabolism of Pesticides, An Update. Special Scientific Report: Wildlife No. 184. U.S. Department of the Interior, Fish, Wildlife Service, Washington, DC.

Metcalf RL (1971) Structure–activity relationships for insecticidal carbamates. Bull WHO 44:43–78.

Metcalf RL, Fukuto TR (1962) *m*-Sulfurpentafluorophenyl diethyl phosphate and *m*-sulfurpentafluorophenyl *N*-methylcarbamate as insecticides and anticholinesterases. J Econ Entomol 55:340–341.

Metcalf RL, Fukuto TR (1965a) Effects of chemical structure on intoxication and detoxication of phenyl *N*-methylcarbamates. J Agric Food Chem 13:220–231.

Metcalf RL, Fukuto TR (1965b) Silicon-containing carbamate insecticides. J Econ Entomol 58:1151.

Metcalf RL, Fukuto TR (1967) Some effects of molecular structure upon Anticholinesterase and insecticidal activity of substituted phenyl *N*-methylcarbamates. J Agric Food Chem 15:1022–1029.

Metcalf RL, Fukuto TR, Frederickson M, Peak L (1965) Insecticidal activity of alkylthiophenyl *N*-methylcarbamates. J Agric Food Chem 13:473.

Metcalf RL, Fukuto TR, Collins C, Borck K, El-Aziz SA, Munoz R, Cassil CC (1968) Metabolism of 2,2-dimethyl-2,3-dihydrobenzofurnayl-7 *N*-methylcarbamate (Furadan) in plants, insects, and mammals. J Agric Food Chem 16: 300–311.

Metcalf RL, March RB (1950) Properties of acetylcholine esterases from the bee, the fly, and the mouse and their relation to insecticide action. J Econ Entomol 43:670–677.

Meylan WM, Howard PH (1994a) Upgrade of PCGEMS water solubility estimation method (draft). U.S. Environmental Protection Agency, Office of Pollution Prevention and Toxics, Washington, DC.
Meylan WM, Howard PH (1994b) Validation of water solubility estimation methods using log K_{ow} for application in PCGEMS & EPI. Final report. U.S. Environmental Protection Agency, Office of Pollution Prevention and Toxics, Washington, DC.
Meylan WM, Howard PH (1995) Atom/fragment contribution method for estimating octanol-water partition coefficients. J Pharm Sci 84:83–92.
Meylan WM, Howard PH (2000) Estimating log P with atom/fragments and water solubility with log P. Perspect Drug Discov Design 19:67–84.
Michalek H, Nemesio R, Menguz A, Bisso GM (1983) Interactions between anticholinesterase agents and neuroleptics in terms of cholinesterase inhibition in brain and other tissues of rats. Arch Toxicol (Suppl) 6:383–390.
Montesissa C, Huveneers MBM, Hoogenboom LAP, Amorena M, De Liguoro M, Lucisano A (1994) The oxidative metabolism of aldicarb in pigs: *in vivo–in vitro* comparison. Drug Metab Drug Interact 11:127–138.
Moser VC, MacPhail RC (1987) Cholinergic involvement in the action of formetanate on operant behavior in rats. Pharmacol Biochem Behav 26:119–121.
Muensch H, Goedde HW, Yoshida A (1976) Human-serum cholinesterase subunits and number of active sites of the major component. Eur J Biochem 70: 217–223.
Myers RC, DePass LR (1993) Acute toxicity testing by the dermal route. In: Wang RGM, Knaak JB, Maibach HI (eds) Health Risk Assessment: Dermal and Inhalation Exposure and Absorption of toxicants. CRC Press, Boca Raton, FL, pp 167–199.
Nigg HN, Knaak JB (2000) Blood cholinesterases as human biomarkers of organophosphorus pesticide exposure. Rev Environ Contam Toxicol 163:29–112.
Nishioka T, Kitamura K, Fujita T, Nakajima M (1976) Kinetic constants for the inhibition of acetylcholinesterase by phenyl carbamates. Pestic Biochem Physiol 6:320–327.
Nishioka T, Fujita T, Kamoshita K, Nakajima M (1977) Mechanism of inhibition reaction of acetylcholinesterase by phenyl *N*-methylcarbamates. Separation of hydrophobic, electronic, hydrogen bonding and proximity effects of aromatic substituents. Pestic Biochem Physiol 7:107–121.
Nostrandt AC, Duncan JA, Padilla S (1993) A modified spectrophotometric method appropraiate for measuring cholinesterase activity in tissue from carbaryl-treated animals. Fundam Appl Toxicol 21:196–203.
O'Brien RD (1960) The reaction with cholinesterase *in vitro*. In: Toxic Phosphorus Esters: Chemistry, Metabolism and Biological Effects. Academic Press, New York. Ch 3, pp 76–84.
O'Brien RD (1968) Kinetics of the carbamylation of cholinesterase. Mol Pharmacol 4:121–130.
O'Brien RD (1976) Acetylcholinesterase and its inhibition. In: Wilkinson CF (ed) Insecticide Biochemistry and Physiology. Plenum, New York, pp 271–296.
Okino MS, Power FW, Tornero-Velez R, Blancato JN, Dary CC (2005a) Assessment of carbaryl exposure following turf application using a physiologically

based pharmacokinetic model. EPA/HQ/OPP/2004/0405/0039. U.S. Environmental Protection Agency, Washington, DC.

Okino MS, Power FW, Knaak JB, Tornero-Velez R, Lunchick C, Lowit A, Blancato JN, Dary CC (2005b) Use of a physiologically-based pharmacokinetic model to estimate absorbed carbaryl dose in children after turf application. Presented at 44th Annual Meeting of the Society of Toxicology, New Orleans, LA (abstract 870). March 6–10.

Omura T, Sato R (1964) the carbon monoxide-binding pigment of liver microsomes. I. Evidence for its hemoprotein nature. J Biol Chem 239:2370–2378.

Oonnithan ES, Casida JE (1968) Oxidation of methyl- and dimethylcarbamate insecticide chemicals by microsomal enzymes and anticholinesterase activity of the metabolites. J Agric Food Chem 16:28–44.

Ordentlich A, Kronman C, Barak D, Stein D, Ariel N, Marcus D, Velan B, Shafferman A (1993) Engineering resistance to 'aging' of phosphorylated human acetylcholinesterase. FEBS J 334:215–220.

Parham FM, Kohn MC, Matthews HB, DeRosa C, Portier CJ (1997) Using structural information to create physiologically based pharmacokinetic models for all polychlorinated biphenyls. 1. Tissue: blood partition coefficients. Toxicol Appl Pharmacol 144:340–347.

Parke DV (1968) The Biochemistry of Foreign Compounds. Oxford Pergamon Press, Oxford, p 205.

Payne LK Jr, Stansbury JA Jr, Weiden MHJ (1966) The synthesis and insecticidal properties of some cholinergic trisubstituted acetaldehyde *O*-(methylcarbamoyl) oxime. J Agric Food Chem 14:356–365.

Pelekis M, Krishnan K (1997) Determination of the rate of aldicarb sulphoxidation in rat liver, kidney and lung microsomes. Xenobiotica 27:1112–11120.

Perkins EJ, El-Alfy A, Schlenk D (1999) *In vitro* sulfoxidation of aldicarb by hepatic microsomes of channel catfish, Ictalurus punctatus. Toxicol Sci 48:67–73.

Petrauskas AA, Kolovanov EA (2000) ACD/Log P method description. Perspect Drug Discov Design 19:99–116.

Petrovskii VV (1970) Level of Sevin content and duration of its excretion in milk. Veterinariia 8:39–40.

Pintore M, van de Waterbeemd H, Piclin N, Chretien JR (2003a) Prediction of oral bioavailability by adaptive fuzzy partitioning. Eur J Med Chem 38:427–431.

Pintore M, Piclin N, Benfenati E, Gini G, Chretien JR (2003b) Database mining with adaptive fuzzy partition: application to the prediction of pesticide toxicity on rats. Environ Toxicol Chem 22:983–991.

Poulin P, Krishnan K (1996) Molecular structure-based predictions of the partition coefficients of organic chemicals for physiological pharmacokinetic models. Toxicol Methods 6:117–137.

Poulin P, Theil F-P (2000) A priori prediction of tissue: plasma partition coefficients of drugs to facilitate the use of physiologically-based pharmacokinetic models in drug discovery. J Pharm Sci 89:16–35.

Poulin P, Theil F-P (2002a) Prediction of pharmacokinetics to *in vivo* studies. 1. Mechanism-based prediction of volume of distribution. J Pharm Sci 91:129–156.

Poulin P, Theil F-P (2002b) Prediction of pharmacokinetics prior to *in vivo* studies. II. Generic physiologically based pharmacokinetic models of drug disposition. J Pharm Sci 91:1358–1370.

Poulin P, Schoenlein K, Theil F-P (2001) Prediction of adipose tissue: plasma partition coefficients for structurally unrelated drugs. J Pharm Sci 90:436–447.

Power FW, Knaak JB, Okino MS, Tornero-Velez R, Lowit A, Blancato JN, Dary CC (2005) Use of exposure-related dose estimating model (ERDEM) for assessment of aggregate exposure of infants and children to N-methyl carbamate insecticides. Presented at 44th Annual Meeting of the Society of Toxicology, New Orleans, LA (abstract 1298). March 6–10.

Pryor AN, Selwood T, Lee-Shan L, Andracki MA, Lee Bong Ho, Rao M, Rosenberry T, Doctor BP, Silman I, Quinn DM (1992) Simple general acid-base catalysis of physiological acetylcholinesterase reactions. J Am Chem Soc 114:3896–3900.

Radic Z, Gibney G, Kawamoto S, MacPhee-Quigley K, Bongiorno C, Taylor P (1992) Expression of recombinant acetylcholinesterase in a *Baculovirus* system: kinetic properties of glutamate 199 mutants. Biochemistry 31:9760–9767.

Radic Z, Pickering NA, Vellom DC, Camp S, Taylor P (1993) Three distinct domains in the cholinesterase molecule confer selectivity for acetylcholinesterase and butylcholinesterase inhibitors. Biochemistry 32:12074–12084.

Raevsky OA (1997) Hydrogen bond strength estimation by means of HYBOT. In: Waterbeemd H, van de Testa B, Folkers G (eds) Computer-Assisted Lead Finding and Optimization. Helvetica Chimica Acta, Basel, pp 367–378.

Randall WR, Tsim KWK, Lai J, Barnard EA (1987) Monoclonal antibodies against chicken brain acetylcholinesterase. Their use in immunopurification and immunochemistry to demonstrate allelic variants of the enzyme. Eur J Biochem 164: 95–102.

Reddy MB, Yang RS, Clewell HJ, Andersen ME (eds) (2005) Physiologically Based Pharmacokinetic Modeling: Science and Applications. Wiley, New York.

Redinbo MR, Bencharit S, Potter PM (2003) Human carboxylesterase 1: from drug metabolism to drug discovery. Biochem Soc Trans 31:620–624.

Reiser RW, Dietrich RF, Djanegara TK, Fogiel AJ, Payne WG (1997) Identification of a novel animal metabolite of methomyl insecticides. J Agric Food Chem 45:2309–2313.

Riddles PW, Richards LJ, Bowles MR, Pond SM (1991) Cloning and analysis of a cDNA encoding a human liver carboxylesterase. Gene (Amst) 108:389–292.

Roberts TR, Hutson DH (1999) Carbamates. In: Jewess PJ, Lee PW, Nicholls PH, Plimmer JR (eds) Metabolic Pathways of Agrochemicals, Part 2. Insecticides and Fungicides. The Royal Society of Chemistry, Information Services, London.

Robertson A (1863) The calabar bean as a new agent in ophthalmic medicine. Edinb Med J 8:815.

Rosenberry TL, Scoggin DM (1984) Structure of human erythrocyte acetylcholinesterase. Characterization of intersubunit bonding and detergent interaction. J Biol Chem 259:5643–5652.

Ross MK, Borazjani A, Edwards CC, Potter PM (2006) Hydrolytic metabolism of pyrethroids by human and other mammalian carboxylesterasess. Biochem Pharmacol 71:657–669.

Rotenberg M, Almog S (1995) Evaluation of the decarbamylation process of cholinesterase during assay of enzyme activity. Clin Chim Acta 240:107–116.

Rotundo RL (1988) Biogenesis of acetylcholinesterase molecular forms in muscle. Evidence for a rapidly turning over, catalytically inactive precursor pool. J Biol Chem 263:19398–19406.

Rubinstein JS (1978) Abuse of anticholinergic drugs. N Engl J Med 299: 834.

Ruppert PH, Cook LL, Dean KF, Reiter LW (1983) Acute behavioral toxicity of carbaryl and propoxur in adult rats. Pharmacol Biochem Behav 18:579–584.

Sacher RM, Olin JF (1972) The insecticidal and anticholinesterase activity of meta-acylamidophenyl and meta-thioureidophenyl N-methylcarbamates. J Agric Food Chem 20:354–359.

Sanghani SP, Quinney SK, Fredenburg TB, Davis WI, Murry DJ, Bosron WF (2004) Hydrolysis of irinotecan and its oxidative metabolites, 7-ethyl-10-[4-*N*-(1-piperidino)-1-amino]-carbonyloxycamptothecin, by human carboxylesterases CES1A1, CES2, and a newly expressed carboxylesterase isoenzymes, CES3. Drug Metab Dispos 32:505–511.

Satoh T, Hosokawa M (1998) The mammalian carboxylesterases: from molecules to functions. Annu Rev Pharmacol Toxicol 38:257–288.

Saxena A, Maxwell DM, Quinn DM, Radic Z, Taylor P, Doctor BP (1997) Mutant acetylcholinesterases as potential detoxification agents for organophosphate poisoning. Biochem Pharmacol 54:269–274.

Schmitt H, Altenburger R, Jastroff B, Schuurmann G (2000) Quantitative structure–activity analysis of the algae toxicity of nitroaromatic compounds. Chem Res Toxicol 13:441–450.

Schwer H, Langmann T, Diag R, Becker A, Aslanidis C, Schmitz G (1997) Molecular cloning and characterization of a novel putative carboxylesterase, present in human intestine and liver. Biochem Biophys Res Commun 233: 117–120.

Scott RC, Guy RH, Hadgraft J (eds) (1990) Prediction of Percutaneous Penetration: Methods, Measurements, Modelling. IBC Technical Services, London, Great Britain.

Segal IH (1993) Enzyme Kinetics: Behavior and Analysis of Rapid Equilibrium and Steady-State Enzyme System. Wiley, New York.

Selwood T, Feaster SR, States MJ, Pryor AN, Quinn DM (1993) Parallel mechanisms in acetylcholinesterase-catalyzed hydrolysis of choline esters. J Am Chem Soc 115:10477–10482.

Sen-Gupta AK, Knowles CO (1970) Fate of formetanate-^{14}C acaricide in the rat. J Econ Entomol 63:10–14.

Shafferman A, Kronman C, Flashner Y, Leitner S, Grosfeld H, Ordentlich A, Gozes Y, Cohen S, Ariel N, Barak D, Harel M, Silman I, Sussman JL, Velan B (1992) Mutagenesis of human acetylcholinesterase. Identification of residues involved in catalytic activity and in polypeptide folding. J Biol Chem 267: 17640–17648.

Siewierski M (ed) (1984) Studies in Environmental Science, vol 24. Determination and Assessment of Pesticide Exposure. Elsevier, New York.

Sikorav JL, Krejei E, Massoulie J (1987) cDNA sequences of *Torpedo marmorata* acetylcholinesterase: primary structure of the precursor of a catalytic subunit; existence of multiple 5′-untranslated regions. EMBO J 6:1865–1873.

Sinha S, Bano S, Agrawal VK, Khadikar PV (1999) QSAR studies on some antimalarials. Oxid Commun 22:479–486.

Sogorb MA, Carrera V, Benabent M, Vilanova E (2002) Chem Res Toxicol 15: 520–526.

Solomonson L (1981) Cyanide as metabolite inhibitor. In: Vennesland B, Conn EE, Knowles CJ, Westley J, Wissing F (eds) Cyanide in Biology. Academic Press, New York, pp 11–28.
Somani SM, Khalique A (1986) Distribution and pharmacokinetics of physostigmine in rat after intramuscular administration. Fundam Appl Toxicol 6: 327–334.
Sparkes RS, Field LL, Sparkes MC, Crist M, Spence MA, James K, Garry PJ (1984) Genetic linkage studies of transferring, pseudocholinesterase, and chromosome 1 loci. Hum Hered 34:96–100.
Stedman E, Barger G (1925) Physostigmine (eserine). III. J Am Chem Soc 127: 247–258.
Stemple A, Aeschlimann JA (1956) Synthetic analogues of physostigmine. In: Blacks FF, Suter CM (eds) Medicinal Chemistry, vol II. Chap. 4. Wiley and Sons, New York.
Stok JE, Goloshchapov A, Song C, Wheelock CE, Derbel MB, Morisseau C, Hammock BD (2004) Investigation of the role of a second conserved serine in carboxylesterases via site-directed mutagenesis. Arch Biochem Biophys 430: 247–255.
Strother A (1972) *In vitro* metabolism of methylcarbamate insecticides by human and rat liver fraction. Toxicol Appl Pharmacol 21:112–129.
Struble CB, Feil VJ, Pekas JC, Gerst JW (1983) Biliary secretion of ^{14}C and identification of 5,6-dihydro-5,6-dihydroxycarbaryl glucuronide as a biliary metabolite of (^{14}C) carbaryl in the rat. Pestic Biochem Physiol 19:85–94.
Sullivan LJ, Eldridge JM, Knaak JB (1967) Determination of carbaryl and some other carbamates by gas chromatography. J Agric Food Chem 15:927–930.
Sullivan LJ, Eldridge JM, Knaak JB, Tallant MJ (1972) 5,6-Dihydro-5,6-dihydroxycarbaryl glucuronide as a significant metabolite of carbaryl in the rat. J Agric Food Chem 20:980–985.
Sussman JL, Harel M, Frolow F, Varon L, Toker L, Futerman AH, Silman I (1988) Purification and crystallization of a dimeric form of acetylcholinesterase from *Torpedo californica* subsequent to solubilization with phosphatidylinositol-specifc phospholipase. J Mol Biol 203:821–823.
Sussman JL, Harel M, Frolow F, Oefner C, Goldman A, Toker L, Silman L (1991) Atomic structure of acetylcholinesterase from *Torpedo californica*: a prototypic acetylcholine-binding protein. Science 253:872–879.
Sutton LD, Froelich S, Hendrickson HS, Quinn DM (1991) Cholinesterase catalyzed hydrolysis of mixed micellar thiophosphatidylcholines: a possible charge-relay mechanism. Biochemistry 30:5888–5893.
Suzuki T, Kudo Y (1990) Automatic log P estimation based on combined additive modeling methods. J Comput-Aided Mol Design 4:155–198.
Tang J, Cao Y, Rose RL, Hodgson E (2002) *In vitro* metabolism of carbaryl by human CYP and its inhibition by chlorpyrifos. Chem-Biol Interact 141: 229–241.
Timchalk C, Nolan RJ, Mendrala AL, Dittenber DA, Brzak KA, Mattsson JL (2002) A physiologically based pharmacokinetic and pharmacodynamic (PBPK/PD) model for the organophosphate insecticide chlorpyrifos in rats and humans. Toxicol Sci 66:34–53.
Tobia A, Ayers A, Blacker A, Hodges L, Carmichael N (2004) Aldicarb study misrepresented in human testing debate. Environ Health Perspect 112:A155–156.

Tomlin CDS (ed) (1994) The Pesticide Manual: World Compendium, 10th Ed. British Crop Protection Council, Surrey, UK, pp. 680, 820, 983.

Tynes RE, Hodgson E (1985) Magnitude of involvement of the mammalian flavin-containing monooxygenase in the microsomal oxidation of pesticides. J Agric Food Chem 33:471–479.

USEPA (1988) Acceptable Daily Intake (ADI) Tracking Report. U.S. Environmental Protection Agency, Office of Pesticide Programs, Washington, DC.

USEPA (1992) Dermal Exposure Assessment: Principles and Applications. Interim Report. EPA 600/8–91/011B. U.S. Environmental Protection Agency, Office of Research and Development, Washington, DC.

USEPA (1994) Reregistration Eligibility Decision (RED) Methiocarb. EPA 738-R94-002. U.S. Environmental Protection Agency, Office of Prevention, Pesticides and Toxic Substances, Washington, DC.

USEPA (2004) Exposure Related Dose Estimating Model (ERDEM) for Assessing Human Exposure and Dose. EPA 600/X-04/060. U.S. Environmental Protection Agency, Office of Research and Development, Washington, DC. ERDEM web site: http://www.epa.gov/heasdweb/erdem/erdem.htm (accessed March 8, 2007).

USEPA (2005) Estimation of cumulative risk from N-methyl carbamate pesticides: preliminary assessment. U.S. Environmental Protection Agency, Office of Pesticide Programs, Washington, DC. Available at: http://www.epa.gov/scipoly/sap/meetings/2005/august/preliminarynmc.pdf (accessed November 28, 2006).

Usmani KA, Hodgson E, Rose RL (2004a) *In vitro* metabolism of carbofuran by human, mouse, and rat cytochrome P450 and interactions with chlorpyrifos, testosterone, and estradiol. Chem-Biol Interact 150:221–232.

Usmani KA, Karoly ED, Hodgson E, Rose RL (2004b) *In vitro* sulfoxidation of thioether compounds by human cytochrome P450 and flavin-containing monooxygenase isoforms with particular reference to the CYP2C subfamily. Drug Metab Dispos 32:333–339.

Van de Sandt JJ, Rutten AA, Van Ommen B (1993) Species-specific cutaneous biotransformation of the pesticide propoxur during percutaneous absorption *in vitro*. Toxicol Appl Pharmacol 123:144–150.

Van de Sandt JJ, Meuling WJ, Elliott GR, Cnubben NH, Hakkert BC (2000) Comparative *in vitro-in vivo* percutaneous absorption of the pesticide propoxur. Toxicol Sci 58:15–22.

Van Hoof F, Heyndrickx A (1975) Excretion in urine of four insecticidal carbamates and their phenolic metabolites after oral administration to rats. Arch Toxikol 34:81–88.

Velan B, Grosfeld H, Kronman C, Leitner M, Gozes Y, Lazar A, Marcus D, Sery T, Papier Y, Grosfeld H, Cohen S, Shafferman A (1991) The effect of elimination of intersubunit disulfide bonds on the activity, assembly, and secretion of recombinant human acetylcholinesterase. Expression of acetylcholinesterase Cys-580–Ala mutant. J Biol Chem 266:23977–23984.

Venkatapathy R, Moudgal CJ, Bruce RM (2004) Assessment of the oral rat chronic lowest observed adverse effect level model in TOPKAT, a QSAR software package for toxicity prediction. J Chem Inform Comput Sci 44:1623–1629.

Verschueren K (1983) Handbook of Environmental Data of Organic Chemicals, 2nd Ed. Van Nostrand Reinhold, New York, p 230.

Vilarem G, Mrlina G, Calmon JP (1989) Synthesis and Anticholinesterase activity of some alpha-thio-substituted *O*-(methylcarbamoyl) oximes. J Agric Food Chem 37:173–177.
Viswanadhan VN, Ghose AK, Wendoloski JJ (2000) Estimating aqueous salvation and lipophilicity of small organic molecules: a comparative overview of atom/group contribution methods. Perspect Drug Discov Design 19:85–98.
Wang C, Murphy SD (1982) The role of noncritical binding proteins in the sensitivity of acetylcholinesterase from different species to diisopropyl fluorophosphates (DFP) *in vitro*. Life Sci 31:139–149.
Wang R, Gao Y, Lai L (2000) Calculating partition coefficient by atom-additive method. Perspect Drug Discov Design 19:47–66.
Wang RGM, Franklin CA, Honeycutt RC, Reinert JC (eds) (1989) Biological Monitoring for Pesticide Exposure: Measurement, Estimation and Risk Reduction. ACS Symposium Series 382. American Chemical Society, Washington, DC.
Webb GD, Johnson RL (1969) Apparent dissociation constants for several inhibitors of acetylcholinesterase in the intact electroplax of the electric eel. Biochem Pharmacol 18:2153–2161.
Weiden MHJ (1968) Insecticidal carbamoyloximes. J Sci Food Agric (suppl): 19–31.
WHO (1996) Methomyl. In: Litchfield ML (ed) Environ Health Criteria Series 178. World Health Organization, Geneva, Switzerland, p 59.
WHO/FAO (2000) Working Groups. Thiodicarb. TA: Pesticide Residues in Food. Toxicology Evaluations: Pesticide Residues. WHO, Geneva.
Wildman SA, Crippen GM (1999) Prediction of physicochemical parameters by atomic contributions. J Chem Inform Comput Sci 39:868–873.
Williams MW, Frawley JP, Fuyat HH, Blake JR (1957) Modification of the Michel electrometric technique for dog and rat blood cholinesterase. J Assoc Off Agric Chem 40:1118–1120.
Wilson IB (1951a) Mechanism of enzymic hydrolysis. I. Role of the acidic group in the esteratic site of acetylcholinesterase. Biochim Biophys Acta 7:466–470.
Wilson IB (1951b) Mechanism of hydrolysis. II. New evidence for an acylated enzyme as intermediate. Biochim Biophys Acta 7:520–525.
Wilson BW, Seiber JN, Stelljes ME, Henderson JD, Archer TE, Pollock GA, Knaak JB (1989) Bioassays for detection of aldicarb in watermelon. Bull Environ Contam Toxicol 42:159–166.
Wilson BW, Sanborn JR, O'Malley MA, Henderson JD, Billitti JR (1997) Monitoring the pesticide-exposure worker. Occup Med State Art Rev 12:347–363.
Wilson BW, Henderson JD, Ramirez A, O'Malley MA (2002) Standardization of clinical cholinesterase measurements. Int J Toxicol 21:385–388.
Worthing CR (ed) (1979) Pesticide Manual, 6th Ed, British Crop Protection Council, Worcestershire, England, p 284.
Wright JW (1971) The WHO Programme for the evaluation and testing of new insecticides. Bull WHO 44:11–22.
Wu C-Y, Benet LZ (2005) Predicting drug disposition via application of BCS: transport/absorption/elimination interplay and development of a biopharmaceutics drug disposition classification system. Pharm Res 22:11–23.
Yoshida F, Topliss JG (2000) QSAR model for drug human oral bioavailability. J Med Chem 43:2575–2585.

Yu C-C, Kearns CW, Metcalf RL (1972a) Acetylcholinesterase inhibition by substituted phenyl *N*-alkyl carbamates. J Agric Food Chem 20:537–540.
Yu C-C, Metcalf RL, Booth GM (1972b) Inhibition of acetylcholinesterase from mammals and insects by carbofuran and its related compounds and their toxicities toward these animals. J Agric Food Chem 20:923–926.
Zakut H, Zamir R, Sindel L, Soreq H (1989) Gene mapping on chorionic villi chromosomes by hybridization in situ: localization of cholinesterase cDNA binding sites to chromosomes 3q21, 3q26-ter and 16q21. Hum Reprod 4:941–946.
Zhang D, Suen J, Zhang Y, Song Y, Radic Z, Taylor P, Holst MJ, Bajaj C, Baker NA, McCannon J (2005) Tetrameric mouse acetylcholinesterase: continuum diffusion rate calculations by solving the steady-state Smoluchowski equation using finite element methods. Biophys J 88:1659–1665.
Zhang H (2005) A new approach for the tissue–blood partition coefficients of neutral and ionized compounds. J Chem Inform Model 45:121–127.
Zhang X, Tsang AM, Okino MS, Power FW, Knaak JB, Dary CC (2006) Construction of a physiologically based pharmacokinetic/pharmacodynamic (PBPK/PD) model for carbofuran using the Exposure Related Dose Estimating Model (ERDEM). EPA 600/R-06/135. U.S. Environmental Protection Agency, Office of Research and Development, Washington, DC.
Zhang X, Tsang AM, Okino MS, Power FW, Knaak JB, Dary CC (2007) A physiologically based pharmacokinetic/pharmacodynamic (PBPK/PD) model for carbofuran in Sprague-Dawley rats using the Exposure Related Dose Estimating Model (ERDEM). Toxicol Sci (accepted).

Manuscript received April 16; accepted April 21, 2007.

Appendix A: Chemical structures, physical parameters, and tissue: partition coefficients of parent carbamates and metabolites.

Table 14. Chemical structure, physical and chemical properties, and tissue partition coefficients for aldicarb and the resulting metabolites.

No.[a]	Pesticide, pesticide metabolite, and chemical structure[b]	Physical and chemical properties[c,d]		Partition coefficient[e]		
				Tissue	Rat	Human
1	Aldicarb					
	Propanal, 2-methyl-2-(methylthio)-, *O*-[(methylamino) carbonyl] oxime					
	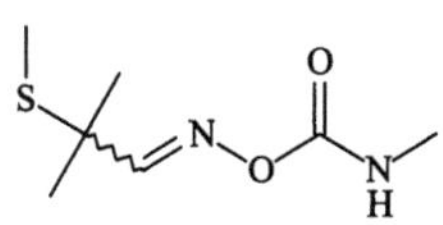	CAS No.	116-06-3	fat	0.97	1.00
		MW, $g\,mol^{-1}$	190.26	brain	1.81	1.98
		Exp K_{ow} @ pH 7.0	NA	rapid	1.18	1.14
		LogD @ pH 7.4	1.13	kidney	1.24	1.31
		LogP	1.13	liver	1.19	1.54
		p*K*a, MA	13.80	slow	1.08	1.28
		WS, $g\,L^{-1}$ (3.07)	5.31	skin	1.22	1.32
		Kp, $cm\,hr^{-1}$	NA			
		Log Kp	NA			
2	Aldicarb sulfoxide					
	Propanal, 2-methyl-2-(methylsulfinyl)-, *O*-[(methylamino) carbonyl] oxime					
	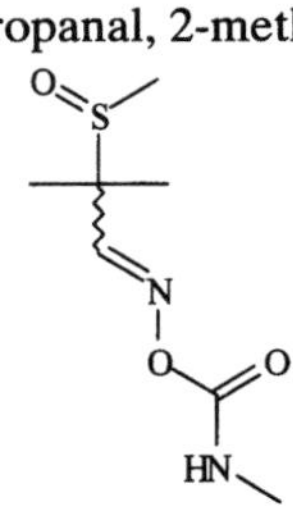	CAS No.	1646-87-3	fat	0.15	0.23
		MW, $g\,mol^{-1}$	206.26	brain	0.99	1.00
		Exp K_{ow} @ pH 7.0	NA	rapid	0.94	0.94
		LogD @ pH 7.4	−1.12	kidney	0.94	0.97
		LogP	−1.12	liver	0.87	0.94
		p*K*a, MA	13.57	slow	0.91	0.93
		WS, $g\,L^{-1}$ (77.5)	367.54	skin	0.79	0.89
3	Aldicarb sulfone					
	Propanal, 2-methyl-2-(methylsulfonyl)-, *O*-[(methylamino) carbonyl] oxime					
	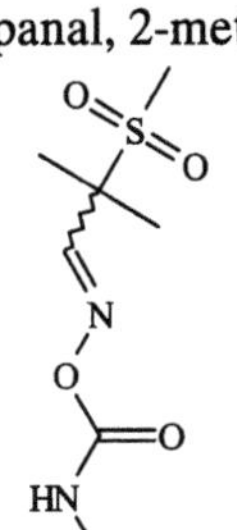	CAS No.	1646-88-4	fat	0.15	0.23
		MW, $g\,mol^{-1}$	222.26	brain	1.00	1.01
		Exp K_{ow} @ pH 7.0	NA	rapid	0.94	0.94
		LogD @ pH 7.4	−0.57	kidney	0.95	0.97
		LogP	−0.57	liver	0.87	0.95
		p*K*a, MA	13.44	slow	0.91	0.94
		WS, $g\,L^{-1}$ (7.09)	102.70	skin	0.80	0.89

Table 14. (*cont.*)

No.[a]	Pesticide, pesticide metabolite, and chemical structure[b]	Physical and chemical properties[c,d]		Partition coefficient[e]		
				Tissue	Rat	Human
4	Aldicarb oxime Propanal, 2-methyl-2-(methylthio)-, oxime					
	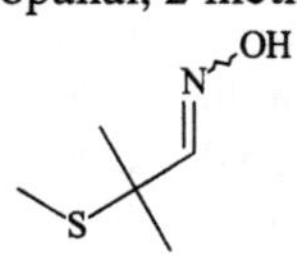	CAS No.	1646-75-9	fat	1.15	1.18
		MW, $g\,mol^{-1}$	133.21	brain	1.97	2.16
		Exp K_{ow} @ pH 7.0	NA	rapid	1.23	1.17
		LogD @ pH 7.4	1.21	kidney	1.30	1.37
		LogP	1.21	liver	1.26	1.65
		p*K*a, MA	11.52	slow	1.11	1.34
		WS, $g\,L^{-1}$ (12.6)	8.27	skin	1.31	1.40
5	Aldicarb oxime sulfoxide Propanal, 2-methyl-2-(methylsulfinyl)-, oxime					
	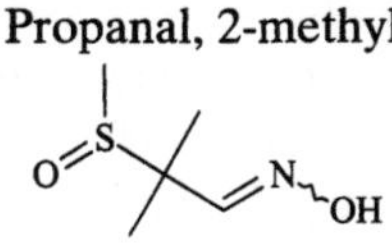	CAS No.	7635-32-7	fat	0.15	0.23
		MW, $g\,mol^{-1}$	149.21	brain	0.99	1.00
		Exp K_{ow} @ pH 7.0	NA	rapid	0.94	0.94
		LogD @ pH 7.4	–0.75	kidney	0.94	0.97
		LogP	–0.75	liver	0.86	0.94
		p*K*a, MA	10.38	slow	0.91	0.94
		WS, $g\,L^{-1}$ (58.29)	334.23	skin	0.80	0.89
6	Aldicarb oxime sulfone Propanal, 2-methyl-2-(methylsulfonyl)-, oxime					
	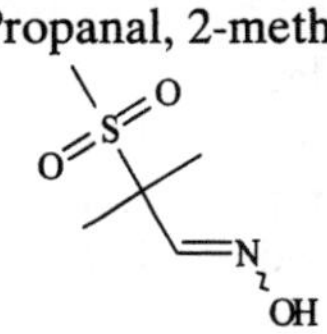	CAS No.	14357-44-9	fat	0.16	0.24
		MW, $g\,mol^{-1}$	165.21	brain	1.00	1.02
		Exp K_{ow} @ pH 7.0	NA	rapid	0.94	0.94
		LogD @ pH 7.4	–0.37	kidney	0.95	0.98
		LogP	–0.37	liver	0.87	0.96
		p*K*a, MA	9.93	slow	0.91	0.94
		WS, $g\,L^{-1}$ L (31.86)	134.06	skin	0.80	0.90
7	Aldicarb nitrile Propanenitrile, 2-methyl-2-(methylthio)-					
	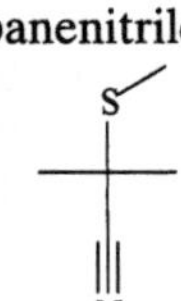	CAS No.	10074-86-9	fat	0.60	0.65
		MW, $g\,mol^{-1}$	115.19	brain	1.48	1.58
		Exp K_{ow} @ pH 7.0	NA	rapid	1.08	1.08
		LogD @ pH 7.4	0.90	kidney	1.12	1.17
		LogP	0.90	liver	1.06	1.30
		p*K*a	NA	slow	1.00	1.14
		WS, $g\,L^{-1}$ (11.78)	17.79	skin	1.05	1.15

Table 14. (*cont.*)

No.[a]	Pesticide, pesticide metabolite, and chemical structure[b]	Physical and chemical properties[c,d]		Partition coefficient[e]		
				Tissue	Rat	Human
8	Aldicarb nitrile sulfoxide Propanenitrile, 2-methyl-2-(methylsulfinyl)-					
	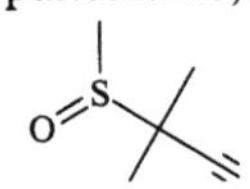	CAS No.	14668-28-1	fat	0.15	0.23
		MW, $g\,mol^{-1}$	131.20	brain	1.00	1.00
		Exp K_{ow} @ pH 7.0	NA	rapid	0.94	0.94
		LogD @ pH 7.4	−1.07	kidney	0.94	0.97
		LogP	−1.07	liver	0.87	0.94
		p*K*a	NA	slow	0.91	0.93
		WS, $g\,L^{-1}$ (248.99)	745.74	skin	0.80	0.89
9	Aldicarb nitrile sulfone Propanenitrile, 2-methyl-2-(methylsulfonyl)-					
	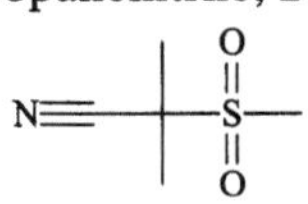	CAS No.	14668-29-2	fat	0.15	0.23
		MW, $g\,mol^{-1}$	147.20	brain	1.00	1.00
		Exp K_{ow} @ pH 7.0	NA	rapid	0.97	0.94
		LogD @ pH 7.4	−0.72	kidney	0.94	0.97
		LogP	−0.72	liver	0.87	0.94
		p*K*a	NA	slow	0.91	0.94
		WS, $g\,L^{-1}$ (27.52)	321.51	skin	0.80	0.89
10	Aldicarb aldehyde Propanal, 2-methyl-2-(methylthio)-					
	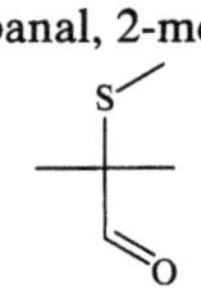	CAS No.	16042-21-0	fat	1.10	1.13
		MW, $g\,mol^{-1}$	118.20	brain	1.93	2.12
		Exp K_{ow} @ pH 7.0	NA	rapid	1.22	1.16
		LogD @ pH 7.4	1.19	kidney	1.28	1.35
		LogP	1.19	liver	1.24	1.62
		p*K*a	NA	slow	1.10	1.33
		WS, $g\,L^{-1}$ (12.88)	9.82	skin	1.28	1.38
11	Aldicarb aldehyde sulfoxide 2-methyl-2-(methylsulfinyl) propanal					
	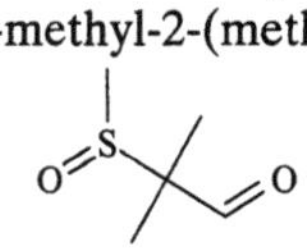	CAS No.	NA	fat	0.15	0.23
		MW, $g\,mol^{-1}$	134.20	brain	1.00	1.00
		Exp K_{ow} @ pH 7.0	NA	rapid	0.94	0.94
		LogD @ pH 7.4	−0.78	kidney	0.94	0.97
		LogP	−0.78	liver	0.87	0.94
		p*K*a	NA	slow	0.91	0.94
		WS, $g\,L^{-1}$ (271.30)	410.12	skin	0.80	0.89

Table 14. (*cont.*)

No.[a]	Pesticide, pesticide metabolite, and chemical structure[b]	Physical and chemical properties[c,d]		Partition coefficient[e]		
				Tissue	Rat	Human
12	Aldicarb aldehyde sulfone Propanal, 2-methyl-2-(methylsulfonyl)					
	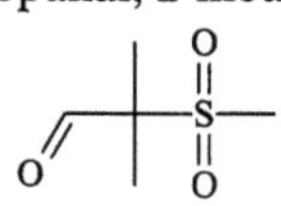	CAS No.	105531-86-0	fat	0.16	0.24
		MW, g mol^{-1}	150.20	brain	1.00	1.02
		Exp K_{ow} @ pH 7.0	NA	rapid	0.94	0.94
		LogD @ pH 7.4	−0.40	kidney	0.95	0.98
		LogP	−0.40	liver	0.87	0.95
		p*K*a	NA	slow	1.00	0.94
		WS, g L^{-1} (29.16)	166.28	skin	0.80	0.90
13	Aldicarb alcohol 1-Propanol, 2-methyl-2-(methylthio)-					
	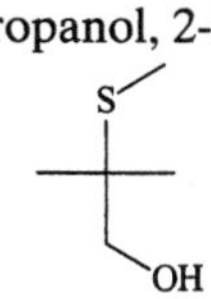	CAS No.	27874-69-7	fat	0.48	0.54
		MW, g mol^{-1}	120.21	brain	1.36	1.44
		Exp K_{ow} @ pH 7.0	NA	rapid	1.05	1.03
		LogD @ pH 7.4	0.78	kidney	1.07	1.12
		LogP	0.78	liver	1.01	1.21
		p*K*a MA	14.92	slow	0.98	1.10
		WS, g L^{-1} (61.52)	21.61	skin	0.99	1.10
14	Aldicarb alcohol sulfoxide 1-Propanol, 2-methyl-2-(methylsulfinyl)-					
	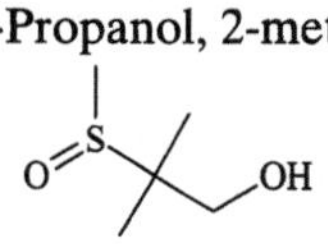	CAS No.	25841-37-6	fat	0.15	0.22
		MW, g mol^{-1}	136.21	brain	0.98	0.99
		Exp K_{ow} @ pH 7.0	NA	rapid	0.94	0.94
		LogD @ pH 7.4	−1.20	kidney	0.94	0.97
		LogP	−1.20	liver	0.87	0.94
		p*K*a, MA	14.02	slow	0.91	0.93
		WS, g L^{-1} (1000.0)	919.15	skin	0.79	0.89
15	Aldicarb alcohol sulfone 1-Propanol, 2-methyl-2-(methylsulfonyl)-					
	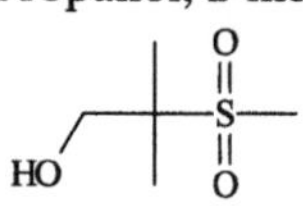	CAS No.	25841-38-7	fat	0.15	0.23
		MW, g mol^{-1}	152.21	brain	0.99	0.99
		Exp K_{ow} @ pH 7.0	NA	rapid	0.94	0.94
		LogD @ pH 7.4	−0.89	kidney	0.94	0.97
		LogP	−0.89	liver	0.87	0.94
		p*K*a, MA	13.67	slow	0.91	0.93
		WS, g L^{-1} (149.79)	426.98	skin	0.79	0.89

Table 14. (*cont.*)

No.[a]	Pesticide, pesticide metabolite, and chemical structure[b]	Physical and chemical properties[c,d]		Partition coefficient[e]		
				Tissue	Rat	Human
16	Aldicarb acid					
	Propanoic acid, 2-methyl-2-(methylthio)-					
	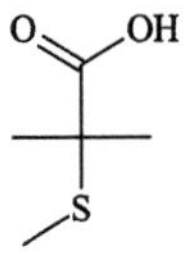	CAS No.	27868-56-0	fat	0.14	0.22
		MW, $g\,mol^{-1}$	134.20	brain	0.98	0.99
		Exp K_{ow} @ pH 7.0	NA	rapid	0.94	0.94
		LogD @ pH 7.4	−2.56	kidney	0.94	0.97
		LogP	0.87	liver	0.86	0.97
		p*Ka*, MA	3.68	slow	0.91	0.93
		WS, $g\,L^{-1}$ (1000.0)	13584.03	skin	0.79	0.88
17	Aldicarb acid sulfoxide					
	Propanoic acid, 2-methyl-2-(methylsulfinyl)-					
	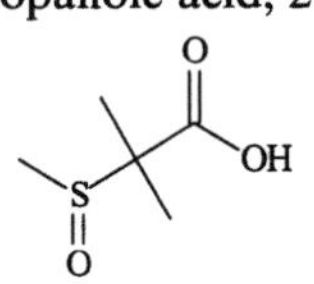	CAS No.	3680-05-5	fat	0.14	0.22
		MW, $g\,mol^{-1}$	150.20	brain	0.98	0.99
		Exp K_{ow} @ pH 7.0	NA	rapid	0.94	0.94
		LogD @ pH 7.4	−4.60	kidney	0.94	0.97
		LogP	−0.91	liver	0.86	0.94
		p*Ka*, MA	2.76	slow	0.91	0.93
		WS, $g\,L^{-1}$ (1000.0)	642142.60	skin	0.79	0.88
18	Aldicarb acid sulfone					
	Propanoic acid, 2-methyl-2-(methylsulfonyl)-					
	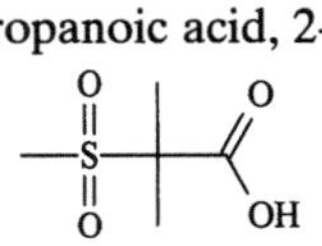	CAS No.	25841-43-4	fat	0.14	0.22
		MW, $g\,mol^{-1}$	166.20	brain	0.98	0.99
		Exp K_{ow} @ pH 7.0	NA	rapid	0.94	0.941
		LogD @ pH 7.4	−4.26	kidney	0.94	0.97
		LogP	−0.53	liver	0.86	0.94
		p*Ka*, MA	2.28	slow	0.91	0.93
		WS, $g\,L^{-1}$ (1000.0)	278457.10	skin	0.79	0.88

[a]Parent chemical or metabolite number for cross-reference to Table 24 (Appendix B).

[b]NA = Not Available. When CAS no. is available, CAS names are used. When CAS no. is not available, IUPAC names from ACD/ChemSketch, version 8.0 (Advanced Chemistry Development, Toronto, Ontario) are used.

[c]WS (g/L) was obtained using the Syracuse Equation in Accord 6 for Excel (http://www.accelrys.com/products/accord/accordproducts/acc_4excel.html) and Log $D_{pH7.4}$ above. Values in (...) after WS were calculated using ACD's LogD Solubility Suite, version 9.0 (http://www.acdlabs.com/products/phys_chem_lab/logd/suite.html).

[d]For p*Ka*, A = acidic, MA = more acidic, MB = more basic.

[e]Tissue: blood partition coefficients.

Table 15. Chemical Structure, Physical and Chemical Properties, and Tissue Partition Coefficients for Carbaryl and the Resulting Metabolites.

No.[a]	Pesticide, pesticide metabolite, and chemical structure[b]	Physical and chemical properties[c,d]		Partition coefficient[e]		
				Tissue	Rat	Human
1	Carbaryl 1-Napthalenyl methylcarbamate					
	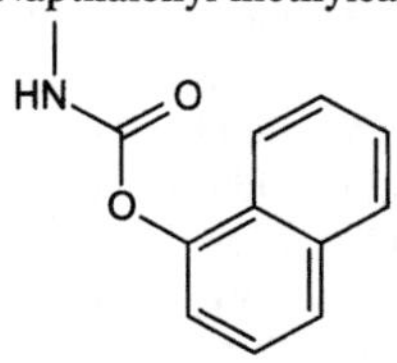	CAS No.	63-25-2	fat	20.65	18.77
		MW, $g\,mol^{-1}$	201.20	brain	11.14	10.07
		Exp Kow @ pH 7.0	2.36	rapid	3.94	2.76
		LogD @ pH 7.4	2.40	kidney	4.56	4.06
		LogP	2.40	liver	4.91	6.42
		p*K*a, MA	12.02	slow	2.99	4.10
		WS, $g\,L^{-1}$ (0.10)	0.38	skin	6.09	4.90
		Kp, $cm\,hr^{-1}$	0.04			
		Log Kp	−1.41			
2	4-OH carbaryl 1, 4-Napthalenediol, mono (methylcarbamate)					
	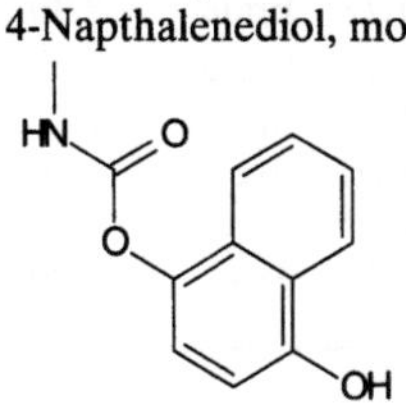	CAS No.	5266-97-7	fat	2.62	2.55
		MW, $g\,mol^{-1}$	217.22	brain	3.01	3.42
		Exp Kow @ pH 7.0	1.90	rapid	1.56	1.43
		LogD @ pH 7.4	1.56	kidney	1.70	1.80
		LogP	1.57	liver	1.71	2.40
		p*K*a, MA	9.20	slow	1.24	1.78
		WS, $g\,L^{-1}$ (0.32)	1.73	skin	1.90	1.96
3	5-OH carbaryl 1,5-Naphthalenediol, mono (methylcarbamate)					
	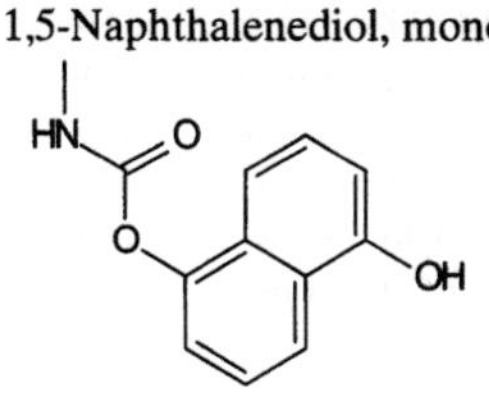	CAS No.	5721-72-2	fat	3.25	3.15
		MW, $g\,mol^{-1}$	217.22	brain	3.55	3.88
		Exp Kow @ pH 7.0	1.94	rapid	1.70	1.52
		LogD @ pH 7.4	1.66	kidney	1.86	1.95
		LogP	1.65	liver	1.89	2.68
		p*K*a, MA	9.02	slow	1.44	1.94
		WS, $g\,L^{-1}L$ (0.30)	1.48	skin	2.13	2.16
4	3,4-Dihydrodihydroxy carbaryl 1,2,4-Naphthalenetriol, 1,2-dihydro-, 4-(methylcarbamate)					
	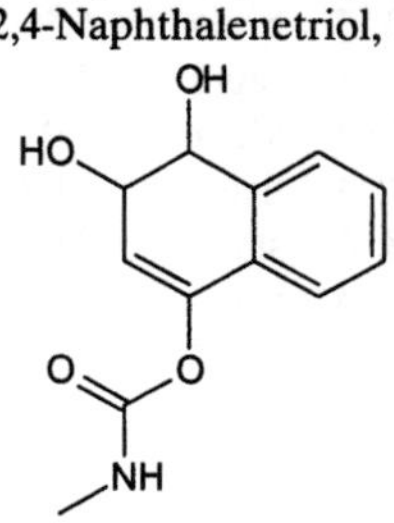	CAS No.	39647-39-7	fat	1.70	1.70
		MW, $g\,mol^{-1}$	235.24	brain	2.42	2.67
		Exp Kow @ pH 7.0	NA	rapid	1.36	1.28
		LogD @ pH 7.4	1.38	kidney	1.46	1.54
		LogP	1.38	liver	1.44	1.96
		p*K*a, MA	12.62	slow	1.20	1.52
		WS, $g\,L^{-1}$ (0.66)	1.89	skin	1.54	1.63

Table 15. Chemical Structure, Physical and Chemical Properties, and Tissue Partition Coefficients for Carbaryl and the Resulting Metabolites.

No.[a]	Pesticide, pesticide metabolite, and chemical structure[b]	Physical and chemical properties[c,d]		Partition coefficient[e] Tissue	Rat	Human
5	5,6-Dihydrodihydroxy carbaryl 1,2,5-Naphthalenetriol, 1,2-dihydro-, 5-(methylcarbamate)					
	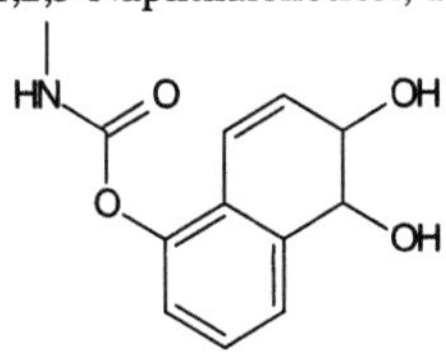	CAS No.	5375-49-5	fat	0.18	0.25
		MW, $g mol^{-1}$	235.24	brain	1.03	1.04
		Exp Kow @ pH 7.0	NA	rapid	0.95	0.95
		LogD @ pH 7.4	–0.14	kidney	0.96	0.99
		LogP	–0.14	liver	0.89	0.97
		p*K*a, MA	12.15	slow	0.92	0.95
		WS, $g L^{-1}$ (7.29)	37.54	skin	0.82	0.91
6	Hydroxymethyl carbaryl Carbamic acid, (hydroxymethyl)-, 1-naphthalenyl ester					
	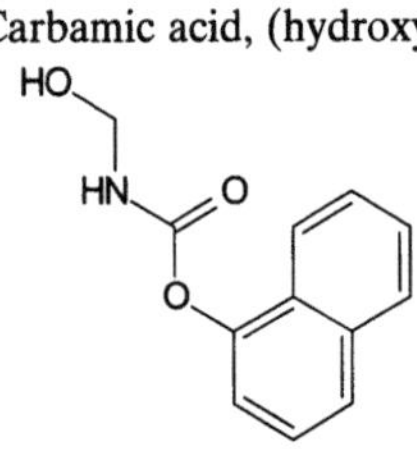	CAS No.	5266-96-6	fat	2.88	2.80
		MW, $g mol^{-1}$	217.22	brain	3.30	3.61
		Exp Kow @ pH 7.0	NA	rapid	1.62	1.47
		LogD @ pH 7.4	1.60	kidney	1.77	1.86
		LogP	1.60	liver	1.79	2.52
		p*K*a, MA	10.27	slow	1.38	1.85
		WS, $g L^{-1}$ (0.57)	1.53	skin	2.00	2.04
7	Naphthol 1-Naphthalenol					
	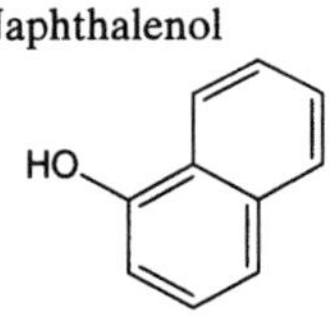	CAS No.	90-15-3	fat	43.20	37.31
		MW, $g mol^{-1}$	144.17	brain	16.04	12.87
		Exp Kow @ pH 7.0	NA	rapid	5.39	3.33
		LogD @ pH 7.4	2.71	kidney	6.32	5.02
		LogP	2.71	liver	6.87	8.12
		p*K*a, MA	9.40	slow	4.01	5.08
		WS, $g L^{-1}$ (0.51)	0.39	skin	8.64	6.14
8	1-Hydroxy-2-naphthol 1,2-Naphthalenediol					
	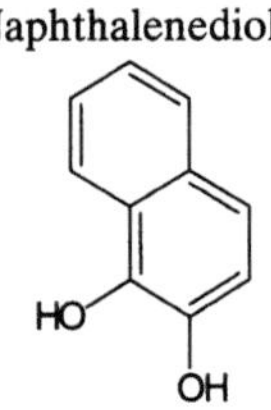	CAS No.	547-00-5	fat	8.51	8.00
		MW, $g mol^{-1}$	160.17	brain	6.55	6.67
		Exp Kow @ pH 7.0	NA	rapid	2.58	2.08
		LogD @ pH 7.4	2.04	kidney	2.93	2.91
		LogP	2.11	liver	3.08	4.37
		p*K*a, MA	8.42	slow	2.05	2.92
		WS, $g L^{-1} L$ (0.95)	1.24	skin	3.69	3.40

Table 15. (*cont.*)

No.[a]	Pesticide, pesticide metabolite, and chemical structure[b]	Physical and chemical properties[c,d]		Partition coefficient[e] Tissue	Rat	Human
9	4-OH Naphthol 1,4-Naphthalenediol					
	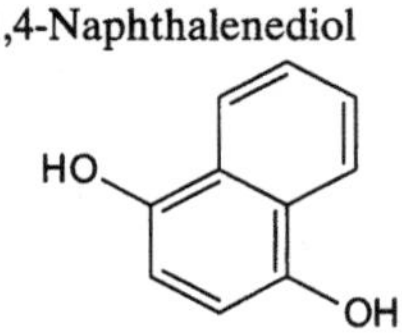	CAS No.	571-60-8	fat	5.59	5.32
		MW, $g\,mol^{-1}$	160.17	brain	5.01	5.30
		Exp Kow @ pH 7.0	NA	rapid	2.13	1.81
		LogD @ pH 7.4	1.87	kidney	2.38	2.44
		LogP	1.88	liver	2.47	3.54
		p*K*a, MA	10.56	slow	1.74	2.44
		WS, $g\,L^{-1}$ L (1.05)	1.73	skin	2.89	2.79
10	5-OH naphthol 1,5-Naphthalenediol					
	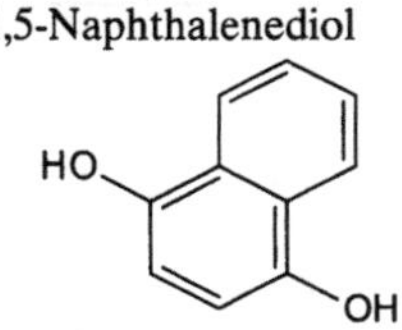	CAS No.	83-56-7	fat	6.81	6.44
		MW, $g\,mol^{-1}$	160.17	brain	5.68	5.92
		Exp Kow @ pH 7.0	NA	rapid	2.33	1.93
		LogD @ pH 7.4	1.95	kidney	2.62	2.65
		LogP	1.96	liver	2.74	3.92
		p*K*a, MA	9.58	slow	1.87	2.65
		WS, $g\,L^{-1}$ (0.97)	1.48	skin	3.24	3.06
11	3, 4-Dihydrodihydroxy naphthol 1, 2, 4-Naphthalenetriol, 1, 2-dihydro-					
	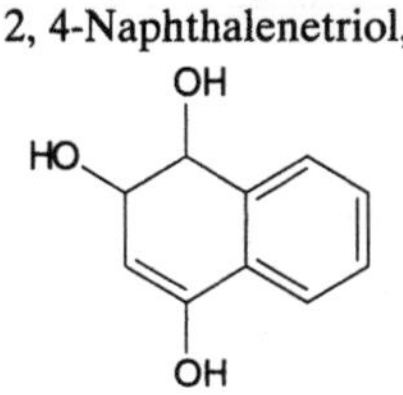	CAS No.	73092-92-9	fat	1.32	1.33
		MW, $g\,mol^{-1}$	178.18	brain	2.11	2.33
		Exp Kow @ pH 7.0	NA	rapid	1.27	1.21
		LogD @ pH 7.4	1.27	kidney	1.34	1.42
		LogP	1.27	liver	1.31	1.75
		p*K*a, MA	9.92	slow	1.14	1.40
		WS, $g\,L^{-1}$ (22.41)	4.63	skin	1.38	1.48
12	5, 6-Dihydrodihydroxy naphthol 1, 2, 5-Naphthalenetriol, 1, 2-dihydro-					
	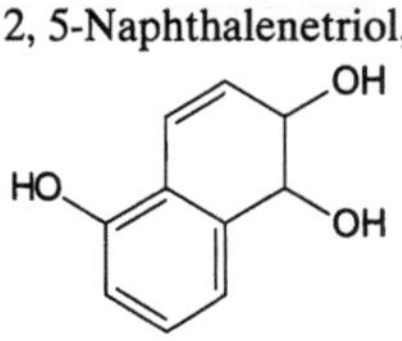	CAS No.	5536-39-0	fat	0.64	0.69
		MW, $g\,mol^{-1}$	178.18	brain	1.51	1.63
		Exp Kow @ pH 7.0	NA	rapid	1.09	1.07
		LogD @ pH 7.4	0.93	kidney	1.13	1.19
		LogP	0.93	liver	1.07	1.32
		p*K*a, MA	9.72	slow	1.02	1.15
		WS, $g\,L^{-1}$ (18.72)	9.03	skin	1.07	1.17
13	Naphthyl sulfuric acid 1-Naphthalenol, hydrogen sulfate					
	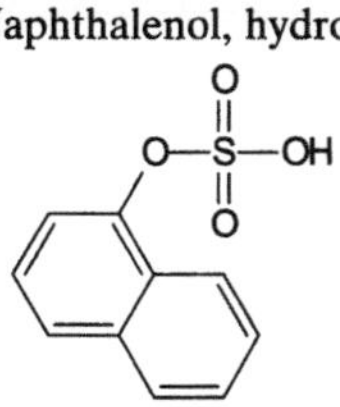	CAS No.	3197-94-2	fat	0.15	0.22
		MW, $g\,mol^{-1}$	224.23	brain	0.99	0.99
		Exp Kow @ pH 7.0	NA	rapid	0.94	0.94
		LogD @ pH 7.4	−0.97	kidney	0.94	0.97
		LogP	2.53	liver	0.87	0.94
		p*K*a, MA	−4.02	slow	0.91	0.93
		WS, $g\,L^{-1}$ (331.48)	220.13	skin	0.79	0.89

Table 15. (*cont.*)

No.[a]	Pesticide, pesticide metabolite, and chemical structure[b]	Physical and chemical properties[c,d]		Partition coefficient[e] Tissue	Rat	Human
14	Naphthyl glucuronic acid β-D-Glucopyranosiduronic acid, 1-naphthalenyl					
	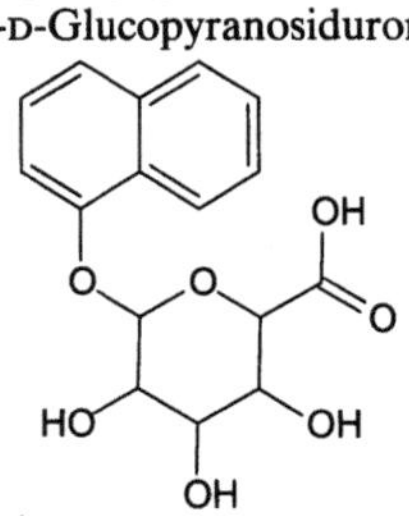	CAS No.	17238-47-0	fat	0.14	0.22
		MW, g mol^{-1}	320.29	brain	0.98	0.99
		Exp Kow @ pH 7.0	NA	rapid	0.94	0.94
		LogD @ pH 7.4	−3.31	kidney	0.94	0.97
		LogP	0.40	liver	0.86	0.94
		p*K*a, MA	2.78	slow	0.91	0.93
		WS, g L^{-1} (1000.0)	6259.11	skin	0.79	0.88
15	2-OH Naphthyl sulfuric acid 1,2-Naphthalenediol, 1-(hydrogen sulfate)					
	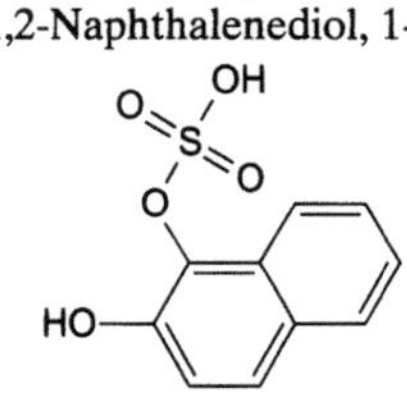	CAS No.	7234-05-1	fat	0.14	0.22
		MW, g mol^{-1}	240.23	brain	0.98	0.99
		Exp Kow @ pH 7.0	NA	rapid	0.94	0.94
		LogD @ pH 7.4	−2.23	kidney	0.94	0.97
		LogP	1.59	liver	0.86	0.94
		p*K*a, MA	−4.82	slow	0.91	0.93
		WS, g L^{-1} (1000.0)	2148.68	skin	0.79	0.88
16	2-OH Naphthyl glucuronic acid β-D-Glucopyranosiduronic acid, 2-hydroxy-1-naphthyl					
	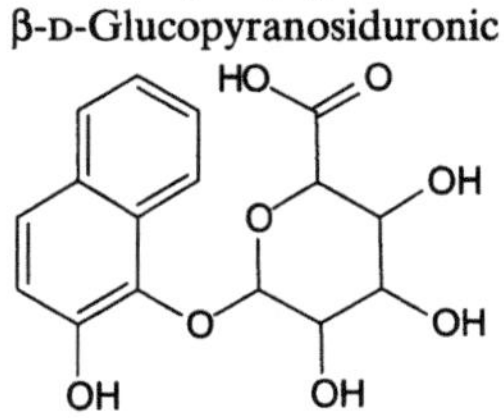	CAS No.	133102-07-5	fat	0.14	0.22
		MW, g mol^{-1}	336.29	brain	0.98	0.99
		Exp Kow @ pH 7.0	NA	rapid	0.94	0.94
		LogD @ pH 7.4	−4.39	kidney	0.94	0.97
		LogP	−0.55	liver	0.86	0.94
		p*K*a, MA	2.74	slow	0.91	0.93
		WS, g L^{-1} (1000.0)	42026.47	skin	0.79	0.88
17	4-OH Naphthyl sulfuric acid 1,4-Naphthalenediol, mono (hydrogen sulfate)					
	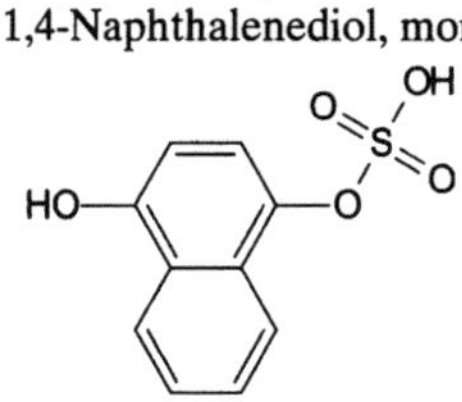	CAS No.	109841-40-9	fat	0.14	0.22
		MW, g mol^{-1}	240.23	brain	0.98	0.99
		Exp Kow @ pH 7.0	NA	rapid	0.94	0.94
		LogD @ pH 7.4	−1.80	kidney	0.94	0.97
		LogP	1.71	liver	0.86	0.94
		p*K*a, MA	−4.46	slow	0.91	0.93
		WS, g L^{-1} (1000.0)	922.47	skin	0.79	0.88

Table 15. (*cont.*)

No.[a]	Pesticide, pesticide metabolite, and chemical structure[b]	Physical and chemical properties[c,d]		Partition coefficient[e]		
				Tissue	Rat	Human
18	4-OH Naphthyl glucuronic acid 3,4,5-Trihydroxy-6-[(4-hydroxy-1-naphthyl)oxy]tetrahydro-2*H*-pyran-2-carboxylic acid					
	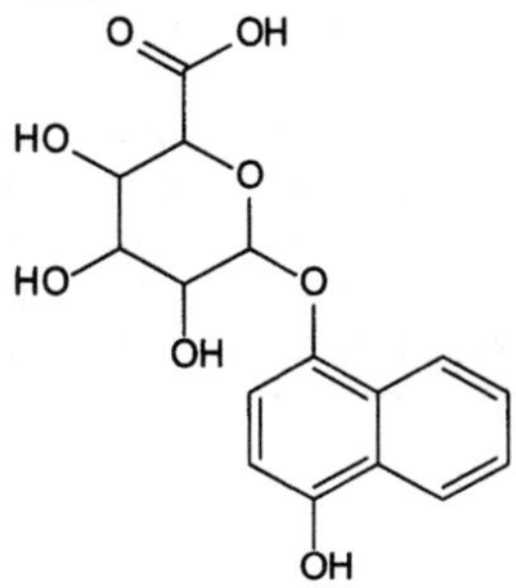	CAS No.	NA	fat	0.14	0.22
		MW, $g mol^{-1}$	336.30	brain	0.98	0.99
		Exp Kow @ pH 7.0	NA	rapid	0.94	0.94
		LogD @ pH 7.4	–4.14	kidney	0.94	0.97
		LogP	–0.42	liver	0.86	0.94
		p*K*a, MA	2.76	slow	0.91	0.93
		WS, $g L^{-1}$ (1000.0)	25705.31	skin	0.79	0.88
19	5-OH Naphthyl sulfuric acid 1,5-Naphthalenediol, mono (hydrogen sulfate)					
	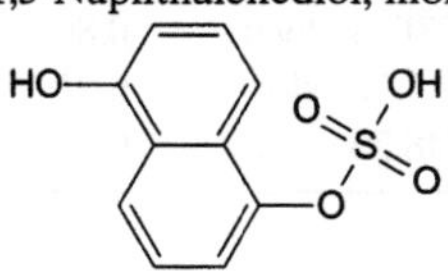	CAS No.	NA	fat	0.14	0.22
		MW, $g mol^{-1}$	240.23	brain	0.98	0.99
		Exp Kow @ pH 7.0	Na	rapid	0.94	0.94
		LogD @ pH 7.4	–1.71	kidney	0.94	0.97
		LogP	1.80	liver	0.86	0.94
		p*K*a, MA	–4.31	slow	0.91	0.93
		WS, $g L^{-1}$ (936.49)	772.84	skin	0.79	0.88
20	5-OH Naphthyl glucuronic acid β-D-Glucopyranosiduronic acid, 5-hydroxy-1-naphthalenyl					
	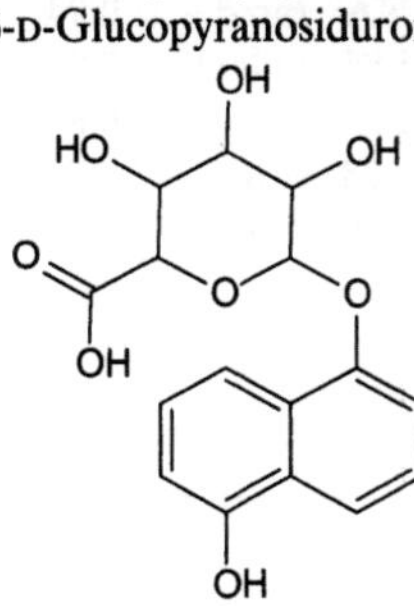	CAS No.	215609-66-8	fat	0.14	0.22
		MW, $g mol^{-1}$	336.29	brain	0.98	0.99
		Exp Kow @ pH 7.0	NA	rapid	0.94	0.94
		LogD @ pH 7.4	–4.05	kidney	0.94	0.97
		LogP	–0.34	liver	0.86	0.94
		p*K*a, A	2.77	slow	0.91	0.93
		WS, $g L^{-1}$ (1000.0)	21535.89	skin	0.79	0.88
21	3,4-Dihydrodihydroxy naphthyl sulfuric acid 3,4-Dihydroxy-3,4-dihydronaphthalen-1-yl hydrogen sulfate					
	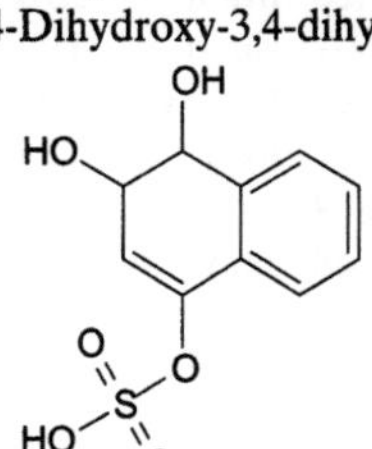	CAS No.	NA	fat	0.14	0.22
		MW, $g mol^{-1}$	258.25	brain	0.98	0.99
		Exp Kow @ pH 7.0	NA	rapid	0.94	0.94
		LogD @ pH 7.4	–2.30	kidney	0.94	0.97
		LogP	1.20	liver	0.86	0.94
		p*K*a, MA	–3.61	slow	0.91	0.93
		WS, $g L^{-1}$ (1000.0)	1959.66	skin	0.79	0.88

Table 15. (*cont.*)

No.[a]	Pesticide, pesticide metabolite, and chemical structure[b]	Physical and chemical properties[c,d]		Partition coefficient[e] Tissue	Rat	Human
22	3, 4-Dihydrodihydroxy naphthyl glucuronic acid 6-[(3,4-Dihydroxy-3,4-dihydronaphthalen-1-yl)oxy]-3,4,5-trihydroxytetrahydro-2*H*-pyran-2-carboxyl acid					
	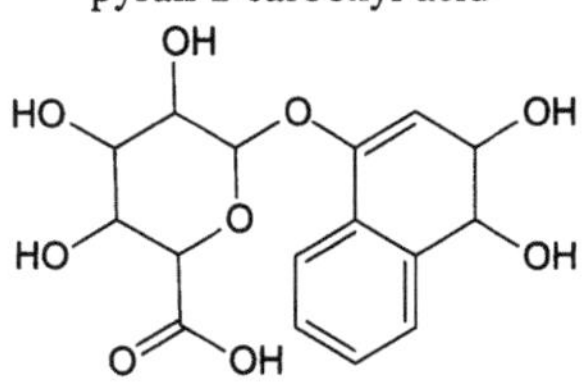	CAS No.	NA	fat	0.14	0.22
		MW, $g\,mol^{-1}$	354.30	brain	0.98	0.99
		Exp Kow @ pH 7.0	NA	rapid	0.94	0.94
		LogD @ pH 7.4	−2.94	kidney	0.94	0.97
		LogP	0.78	liver	0.86	0.94
		p*K*a, MA	2.81	slow	0.91	0.93
		WS, $g\,L^{-1}$ (1000.0)	1891.33	skin	0.79	0.88
23	5, 6-Dihydrodihydroxy naphthyl sulfuric acid 5,6-Dihydroxy-5,6-dihydronaphthalen-1-yl hydrogen sulfate					
	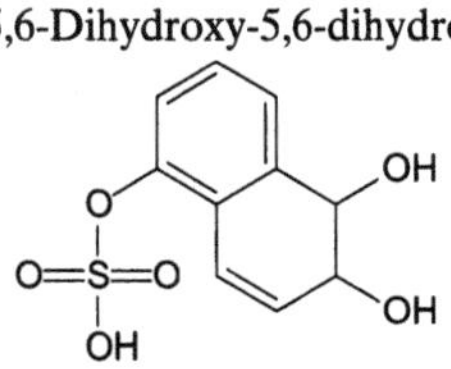	CAS No.	NA	fat	0.14	0.22
		MW, $g\,mol^{-1}$	258.25	brain	0.98	0.99
		Exp Kow @ pH 7.0	NA	rapid	0.94	0.94
		LogD @ pH 7.4	−3.50	kidney	0.94	0.97
		LogP	2.83E-4	liver	0.86	0.94
		p*K*a, MA	−4.10	slow	0.91	0.93
		WS, $g\,L^{-1}$ (1000.0)	20748.20	skin	0.79	0.88
24	5,6-Dihydro, 5,6-dihydroxy naphthyl glucuronic acid 6-[(5,6-Dihydroxy-5,6-dihydronaphthalen-1-yl)oxy]-3,4,5-trihydroxytetrahydro-2*H*-pyran-2-carboxylic acid					
	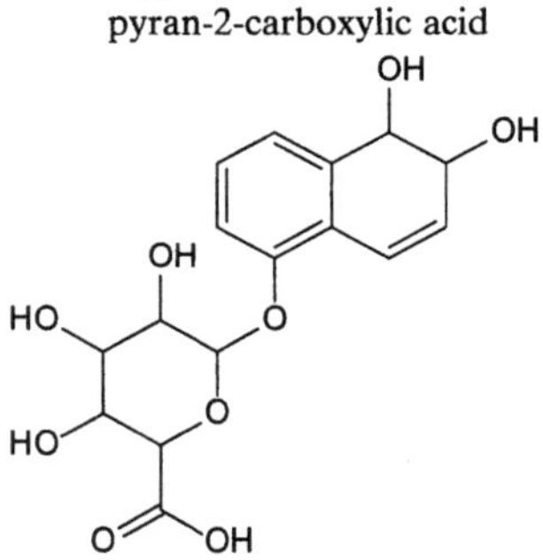	CAS No.	NA	fat	0.14	0.22
		MW, $g\,mol^{-1}$	354.31	brain	0.98	0.99
		Exp Kow @ pH 7.0	NA	rapid	0.94	0.94
		LogD @ pH 7.4	−5.85	kidney	0.94	0.97
		LogP	−2.13	liver	0.86	0.94
		p*K*a, MA	2.78	slow	0.91	0.93
		WS, $g\,L^{-1}$ (1000.0)	577971.70	skin	0.79	0.88
25	4-OH Carbaryl sulfuric acid 1,4-Naphthalenediol, hydrogen sulfate methylcarbamate					
	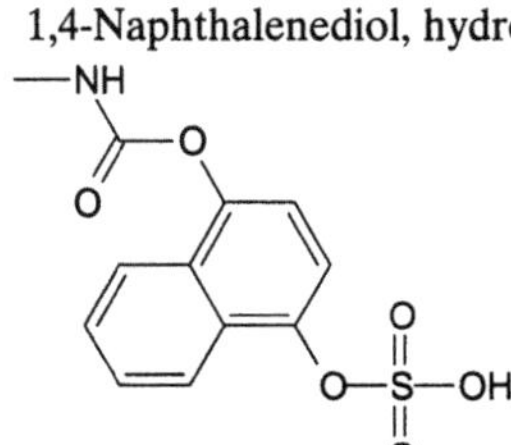	CAS No.	3197-93-1	fat	0.14	0.22
		MW, $g\,mol^{-1}$	297.28	brain	0.98	0.99
		Exp Kow @ pH 7.0	NA	rapid	0.94	0.94
		LogD @ pH 7.4	−1.93	kidney	0.94	0.97
		LogP	1.57	liver	0.86	0.94
		p*K*a, MA	−4.58	slow	0.91	0.93
		WS, $g\,L^{-1}$ (465.72)	566.48	skin	0.79	0.88

Table 15. (*cont.*)

No.[a]	Pesticide, pesticide metabolite, and chemical structure[b]	Physical and chemical properties[c,d]		Partition coefficient[e] Tissue	Rat	Human
26	4-OH Carbaryl glucuronic acid β-D-Glucopyranosiduronic acid, 4-[[(methylamino) carbonyl] oxy]-1-naphthalenyl					
	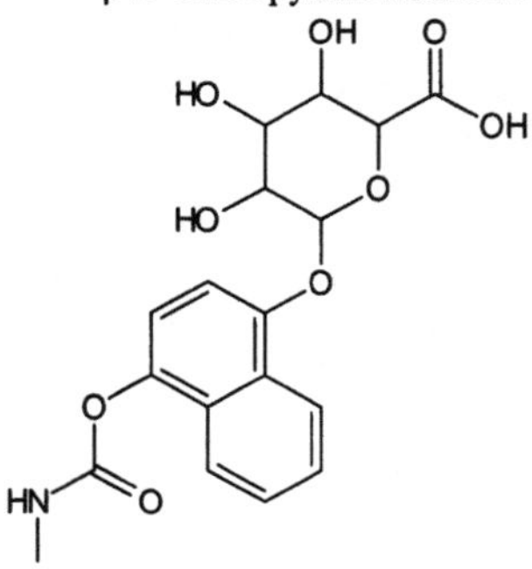	CAS No.	17236-46-9	fat	0.14	0.22
		MW, $g\,mol^{-1}$	393.34	brain	0.98	0.99
		Exp Kow @ pH 7.0	NA	rapid	0.94	0.94
		LogD @ pH 7.4	−4.28	kidney	0.94	0.97
		LogP	−0.57	liver	0.86	0.94
		p*Ka*, A	2.75	slow	0.91	0.93
		WS, $g\,L^{-1}$ (1000.0)	15214.21	skin	0.79	0.88
27	5-OH Carbaryl sulfuric acid 1, 5-Naphthalenediol, hydrogen sulfate methylcarbamate					
	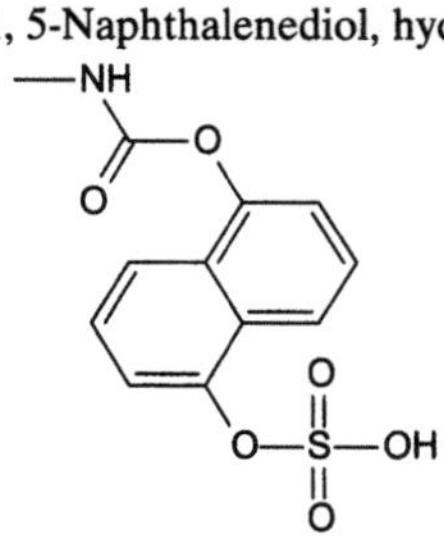	CAS No.	24281-28-5	fat	0.14	0.22
		MW, $g\,mol^{-1}$	297.28	brain	0.98	0.99
		Exp Kow @ pH 7.0	NA	rapid	0.94	0.94
		LogD @ pH 7.4	−2.02	kidney	0.94	0.97
		LogP	1.48	liver	0.86	0.94
		p*Ka*, MA	−4.40	slow	0.91	0.93
		WS, $g\,L^{-1}$ (518.41)	676.16	skin	0.79	0.88
28	5-OH Carbaryl glucuronic acid 3,4,5-Trihydroxy-6-[(5-{[(methylamino)carbonyl]oxy}-1-naphthyl)oxy]tetrahydro-2*H*-pyran-2-carboxylic acid					
	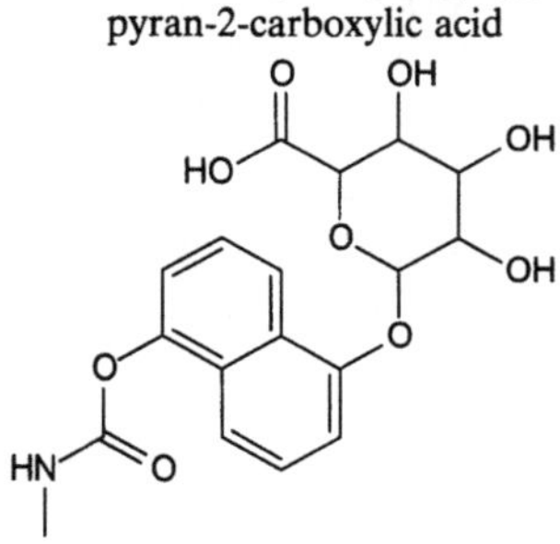	CAS No.	NA	fat	0.14	0.22
		MW, $g\,mol^{-1}$	393.34	brain	0.98	0.99
		Exp Kow @ pH 7.0	NA	rapid	0.94	0.94
		LogD @ pH 7.4	−4.37	kidney	0.94	0.97
		LogP	−0.65	liver	0.86	0.94
		p*Ka*, A	2.77	slow	0.91	0.93
		WS, $g\,L^{-1}$ (1000.0)	18159.73	skin	0.79	0.88

Table 15. (*cont.*)

No.[a]	Pesticide, pesticide metabolite, and chemical structure[b]	Physical and chemical properties[c,d]		Partition coefficient[e]		
				Tissue	Rat	Human
29	3, 4-Dihydrodihydroxy carbaryl sulfuric acid 3-Hydroxy-4-(sulfooxy)-3,4-dihyronaphthalen-1-yl methylcarbamate					
	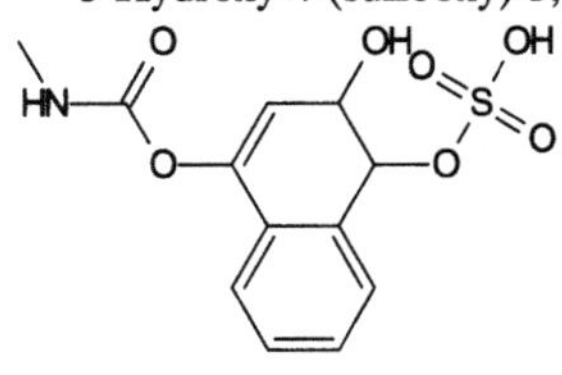	CAS No.	NA	fat	0.14	0.22
		MW, $g\,mol^{-1}$	315.30	brain	0.98	0.99
		Exp Kow @ pH 7.0	NA	rapid	0.94	0.94
		LogD @ pH 7.4	−2.20	kidney	0.94	0.97
		LogP	1.30	liver	0.86	0.94
		p*K*a, MA	−3.75	slow	0.91	0.93
		WS, $g\,L^{-1}$ (570.37)	755.34	skin	0.79	0.88
30	3, 4-Dihydrodihydroxy carbaryl glucuronic acid 3,4,5-Trihydroxy-6-[(2-hydroxy-4-{[(methylamino)carbonyl]oxy}-1,2-dihydronaphthalen-1-yl)oxy]tetrahydro-2*H*-pyran-2-carboxylic acid					
	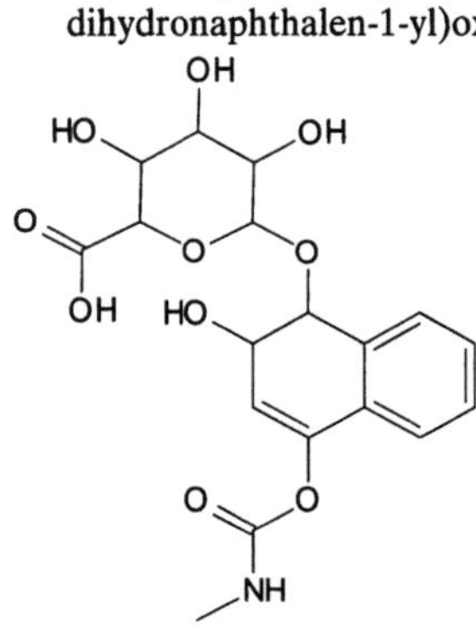	CAS No.	NA	fat	0.14	0.22
		MW, $g\,mol^{-1}$	411.36	brain	0.98	0.99
		Exp Kow @ pH 7.0	NA	rapid	0.94	0.94
		LogD @ pH 7.4	−4.70	kidney	0.94	0.97
		LogP	−0.99	liver	0.86	0.94
		p*K*a, MA	2.81	slow	0.91	0.93
		WS, $g\,L^{-1}$ (1000.0)	26873.32	skin	0.79	0.88
31	5,6-Dihydrodihydroxy carbaryl sulfuric acid 6-Hydroxy-5-(sulfooxy)-5,6-dihydronaphthalen-1-yl methyl carbamate					
	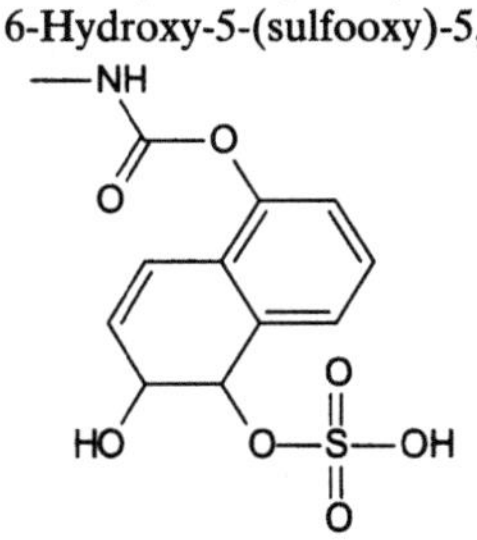	CAS No.	NA	fat	0.14	0.22
		MW, $g\,mol^{-1}$	315.30	brain	0.98	0.99
		Exp Kow @ pH 7.0	NA	rapid	0.94	0.94
		LogD @ pH 7.4	−3.71	kidney	0.94	0.97
		LogP	−0.21	liver	0.86	0.94
		p*K*a, MA	−3.82	slow	0.91	0.93
		WS, $g\,L^{-1}$ (1000.0)	14712.28	skin	0.79	0.88

Table 15. (*cont.*)

No.[a]	Pesticide, pesticide metabolite, and chemical structure[b]	Physical and chemical properties[c,d]		Partition coefficient[e]		
				Tissue	Rat	Human
32	5,6-Dihydrodihydroxy carbaryl glucuronic acid 3,4,5-Trihydroxy-6-[(2-hydroxy-5-{[(methylamino)carbonyl]oxy]oxy}-1,2-dihydronaphthalen-1-yl)oxy]tetrahydro-2*H*-pyran-2-carboxylic acid					
	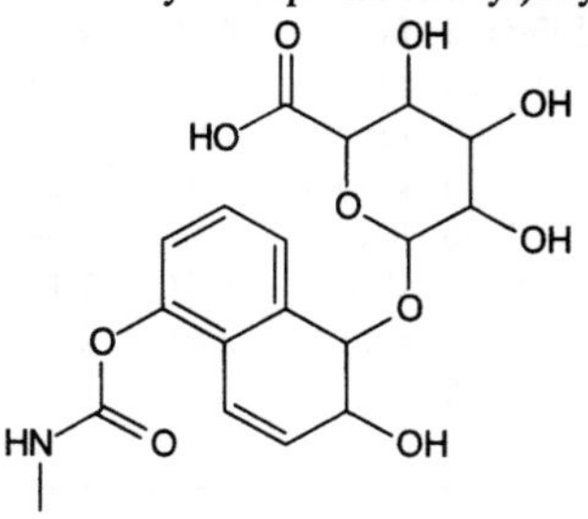	CAS No.	NA	fat	0.14	0.22
		MW, g mol^{-1}	411.36	brain	0.98	0.99
		Exp Kow @ pH 7.0	NA	rapid	0.94	0.94
		LogD @ pH 7.4	−6.21	kidney	0.94	0.97
		LogP	−2.50	liver	0.86	0.94
		p*K*a, MA	2.80	slow	0.91	0.93
		WS, g L^{-1} (1000.0)	523433.20	skin	0.79	0.88
33	Hydroxymethyl carbaryl sulfuric acid 1-Naphthyl[(sulfooxy)methyl]carbamate					
	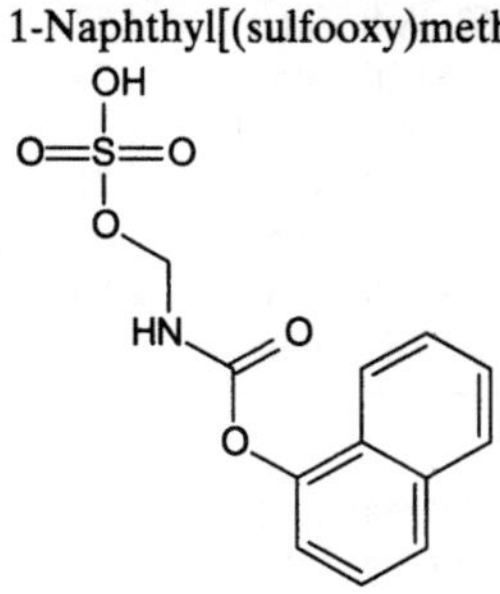	CAS No.	NA	fat	0.14	0.22
		MW, g mol^{-1}	297.28	brain	0.98	0.99
		Exp Kow @ pH 7.0	NA	rapid	0.94	0.94
		LogD @ pH 7.4	−1.64	kidney	0.94	0.97
		LogP	1.87	liver	0.86	0.94
		p*K*a, MA	−3.90	slow	0.91	0.93
		WS, g L^{-1} (312.32)	320.28	skin	0.79	0.88
34	Hydroxymethyl carbaryl glucuronic acid 3,4,5-Trihydroxy-6-({[(1-naphthyloxy)carbonyl]amino}methoxy)tetrahydro-2*H*-pyran-2-carboxylic acid					
	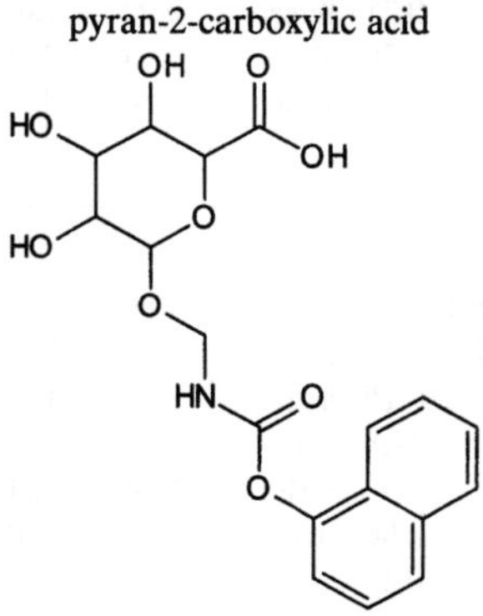	CAS No.	NA	fat	0.14	0.22
		MW, g mol^{-1}	393.34	brain	0.98	0.99
		Exp Kow @ pH 7.0	NA	rapid	0.94	0.94
		LogD @ pH 7.4	−4.14	kidney	0.94	0.97
		LogP	−0.42	liver	0.86	0.94
		p*K*a, MA	2.80	slow	0.91	0.93
		WS, g L^{-1} (1000.0)	11554.50	skin	0.79	0.88

[a]Parent chemical or metabolite number for cross-reference to Table 25 (Appendix B).
[b–e]See footnotes in Table 14 (Appendix A).

Table 16. Chemical Structure, Physical and Chemical Properties, and Tissue Partition Coefficients for Carbofuran and the Resulting Metabolites.

No.[a]	Pesticide, pesticide metabolite, and chemical structure[b]	Physical and chemical properties[c,d]		Partition coefficient[e] Tissue	Rat	Human
1	Carbofuran 7-Benzofuranol, 2,3-dihydro-2,2-dimethyl-, methylcarbamate					
	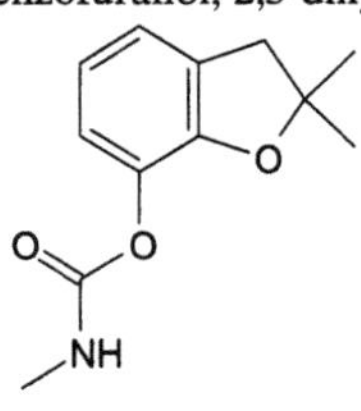	CAS No.	1563-66-2	fat	4.25	4.08
		MW, $g\,mol^{-1}$	221.25	brain	4.20	4.53
		Exp Kow @ pH 7.0	NA	rapid	1.89	1.65
		LogD @ pH 7.4	1.76	kidney	2.09	2.18
		LogP	1.76	liver	2.15	3.08
		p*K*a, MA	12.28	slow	1.57	2.17
		WS, $g\,L^{-1}$ (0.31)	1.06	skin	2.47	2.45
		Kp, $cm\,hr^{-1}$	NA			
		Log Kp	NA			
2	Carbofuran-7-phenol 7-Benzofuranol, 2,3-dihydro-2,2-dimethyl-					
	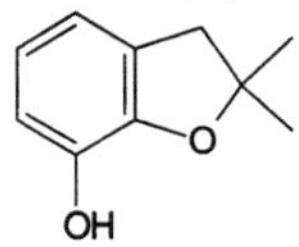	CAS No.	1563-38-8	fat	13.57	12.56
		MW, $g\,mol^{-1}$	164.20	brain	8.74	8.41
		Exp Kow @ pH 7.0	NA	rapid	3.23	2.43
		LogD @ pH 7.4	2.23	kidney	3.71	3.50
		LogP	2.23	liver	3.96	5.42
		p*K*a, MA	10.09	slow	2.50	3.52
		WS, $g\,L^{-1}L$ (1.23)	0.82	skin	4.84	4.16
3	3-OH carbofuran 3,7-Benzofurandiol, 2,3-dihydro-2,2-dimethyl-, 7-(methylcarbamate)					
	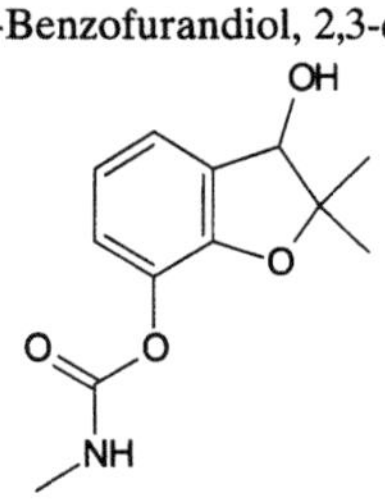	CAS No.	16655-82-6	fat	0.22	0.29
		MW, $g\,mol^{-1}$	237.25	brain	1.08	1.11
		Exp Kow @ pH 7.0	NA	rapid	0.97	0.96
		LogD @ pH 7.4	0.21	kidney	0.98	1.01
		LogP	0.21	liver	0.90	1.01
		p*K*a, MA	12.28	slow	0.93	0.98
		WS, $g\,L^{-1}$ (3.87)	18.38	skin	0.84	0.94
4	Hydroxymethyl carbofuran Carbamic acid, (hydroxymethyl)-, 2,3-dihydro-2,2-dimethyl-7-benzofuranyl ester					
	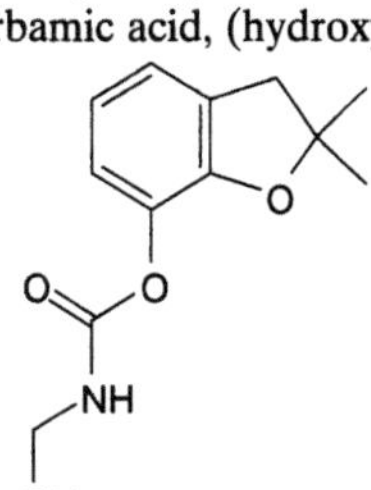	CAS No.	18999-70-7	fat	0.68	0.73
		MW, $g\,mol^{-1}$	237.25	brain	1.55	1.68
		Exp Kow @ pH 7.0	NA	rapid	1.11	1.08
		LogD @ pH 7.4	0.96	kidney	1.14	1.20
		LogP	0.96	liver	1.09	1.35
		p*K*a, MA	10.53	slow	1.02	1.17
		WS, $g\,L^{-1}$ (1.83)	4.21	skin	1.09	1.19

Table 16. (*cont.*)

No.[a]	Pesticide, pesticide metabolite, and chemical structure[b]	Physical and chemical properties[c,d]		Partition coefficient[e] Tissue	Rat	Human
5	Carbofuran-3,7-diol 3,7-Dihydroxy-2,2-dimethyl-2,3-dihydrobenzofuran					
	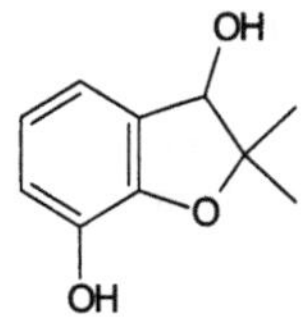	CAS No.	17781-15-6	fat	0.40	0.47
		MW, $g\,mol^{-1}$	180.20	brain	1.28	1.35
		Exp Kow @ pH 7.0	NA	rapid	1.02	1.01
		LogD @ pH 7.4	0.68	kidney	1.05	1.09
		LogP	0.68	liver	0.98	1.16
		p*K*a, MA	9.90	slow	0.97	1.06
		WS, $g\,L^{-1}$ (15.74)	14.43	skin	0.94	1.04
6	3-Keto carbofuran 3(2*H*)-Benzofuranone, 2,2-dimethyl-7-[[(methylamino) carbonyl] oxy]-					
	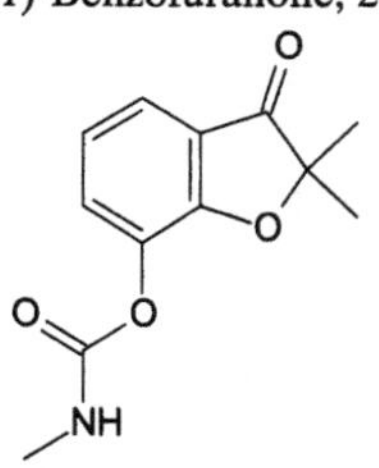	CAS No.	16709-30-1	fat	1.96	1.94
		MW, $g\,mol^{-1}$	235.24	brain	2.62	2.90
		Exp Kow @ pH 7.0	NA	rapid	1.42	1.32
		LogD @ pH 7.4	1.44	kidney	1.53	1.62
		LogP	1.44	liver	1.52	2.09
		p*K*a, MA	11.92	slow	1.24	1.60
		WS, $g\,L^{-1}$ (0.19)	1.68	skin	1.65	1.73
7	3-Keto-7-phenol 7-Hydroxy-2,2-dimethylbenzofuran-3-one					
	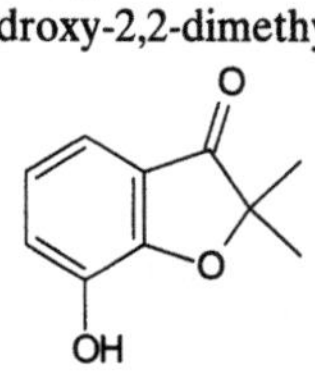	CAS No.	17781-16-7	fat	7.90	7.44
		MW, $g\,mol^{-1}$	178.18	brain	6.25	6.41
		Exp Kow @ pH 7.0	NA	rapid	2.49	2.03
		LogD @ pH 7.4	2.01	kidney	2.82	2.82
		LogP	2.05	liver	2.96	4.22
		p*K*a, MA	8.44	slow	1.99	2.83
		WS, $g\,L^{-1}$ (0.73)	1.08	skin	3.54	3.28
8	7-Phenyl sulfuric acid 7-Benzofuranol, 2,3-dihydro-2,2-dimethyl-, hydrogen sulfate					
	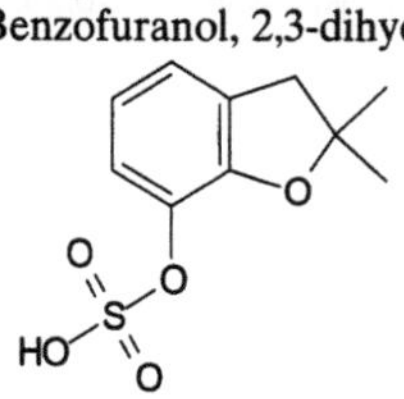	CAS No.	70988-92-0	fat	0.14	0.22
		MW, $g\,mol^{-1}$	244.27	brain	0.98	0.99
		Exp Kow @ pH 7.0	NA	rapid	0.94	0.94
		LogD @ pH 7.4	−1.59	kidney	0.94	0.97
		LogP	1.91	liver	0.86	0.94
		p*K*a, MA	−3.96	slow	0.91	0.93
		WS, $g\,L^{-1}$ (1000.0)	579.80	skin	0.79	0.88

Table 16. (*cont.*)

No.[a]	Pesticide, pesticide metabolite, and chemical structure[b]	Physical and chemical properties[c,d]		Partition coefficient[e] Tissue	Rat	Human
9	7-Phenyl glucuronic acid β-D-Glucopyranosiduronic acid, 2,3-dihydro-2,2-dimethyl -7-benzofuranyl					
	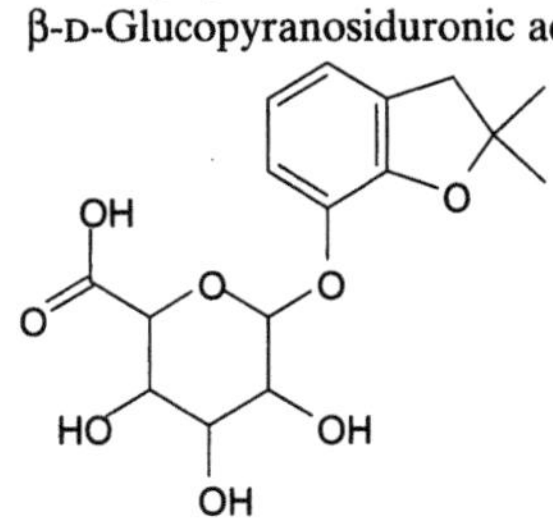	CAS No.	70988-91-9	fat	0.14	0.22
		MW, $g\,mol^{-1}$	340.33	brain	0.98	0.99
		Exp Kow @ pH 7.0	NA	rapid	0.94	0.94
		LogD @ pH 7.4	−3.94	kidney	0.94	0.97
		LogP	−0.23	liver	0.86	0.94
		p*Ka*, MA	2.79	slow	0.91	0.93
		WS, $g\,L^{-1}$ (1000.0)	16414.89	skin	0.79	0.88
10	3-OH carbofuran sulfuric acid 3,7-Benzofurandiol, 2,3-dihydro-2,2-dimethyl-, 3-(hydrogen sulfate) 7-(methylcarbamate)					
	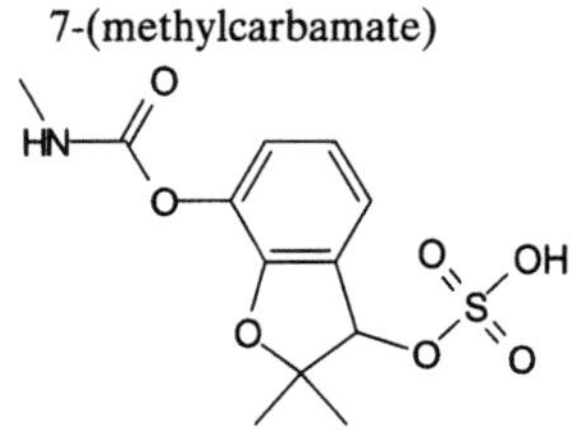	CAS No.	70988-90-8	fat	0.14	0.22
		MW, $g\,mol^{-1}$	317.32	brain	0.98	0.99
		Exp Kow @ pH 7.0	NA	rapid	0.94	0.94
		LogD @ pH 7.4	−3.00	kidney	0.94	0.97
		LogP	0.50	liver	0.86	0.94
		p*Ka*, MA	−4.03	slow	0.91	0.93
		WS, $g\,L^{-1}$ (1000.0)	3544.26	skin	0.80	0.88
11	3-OH carbofuran glucuronic acid β-D-Glucopyranosiduronic acid, 2,3-dihydro-2,2-dimethyl-7-[[(methylamino) carbonyl] oxy]-3-benzofuranyl					
	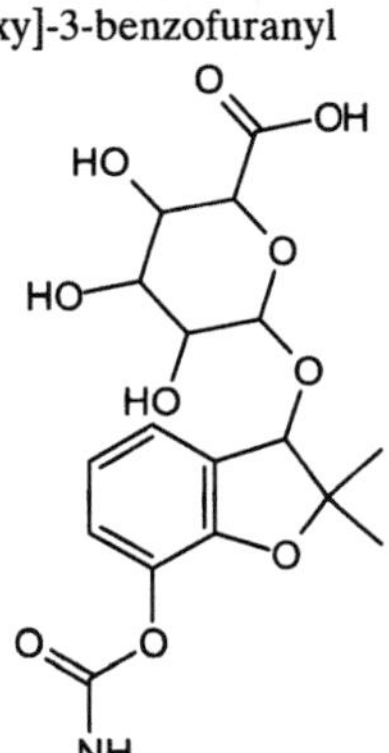	CAS No.	53305-32-1	fat	0.15	0.22
		MW, $g\,mol^{-1}$	413.38	brain	0.98	0.99
		Exp Kow @ pH 7.0	NA	rapid	0.94	0.94
		LogD @ pH 7.4	−5.50	kidney	0.94	0.97
		LogP	−1.79	liver	0.86	0.94
		p*Ka*, A	2.79	slow	0.91	0.93
		WS, $g\,L^{-1}$ (1000.0)	125839.90	skin	0.79	0.88

Table 16. (*cont.*)

No.[a]	Pesticide, pesticide metabolite, and chemical structure[b]	Physical and chemical properties[c,d]		Partition coefficient[e] Tissue	Rat	Human
12	Hydroxymethyl carbofuran sulfate 2,2-Dimethyl-2,3-dihydro-1-benzofuran-7-yl [(sulfooxy) methyl] carbamate 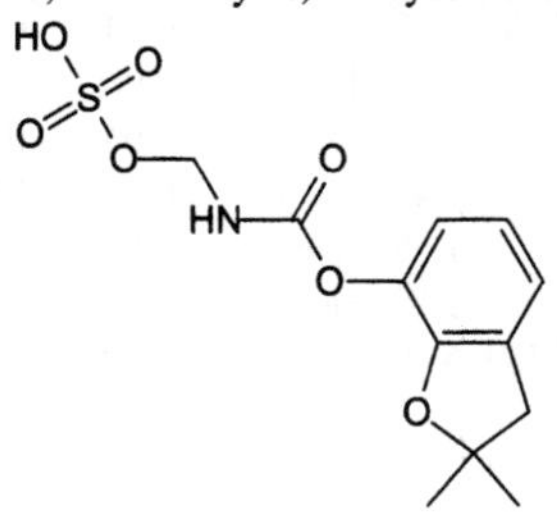	CAS No.	NA	fat	0.14	0.22
		MW, $g\,mol^{-1}$	317.32	brain	0.98	0.99
		Exp Kow @ pH 7.0	NA	rapid	0.94	0.94
		LogD @ pH 7.4	–2.27	kidney	0.94	0.97
		LogP	1.23	liver	0.86	0.94
		p*K*a, A	–3.87	slow	0.91	0.93
		WS, $g\,L^{-1}$ (1000.0)	843.54	skin	0.79	0.88
13	Hydroxymethyl carbofuran glucuronic acid ({[(2,2-dimethylhexahydro-2*H*-cyclopenta[b]furan-6-yl)oxy]carbonyl}amino) methyl hexopyranosiduronic acid 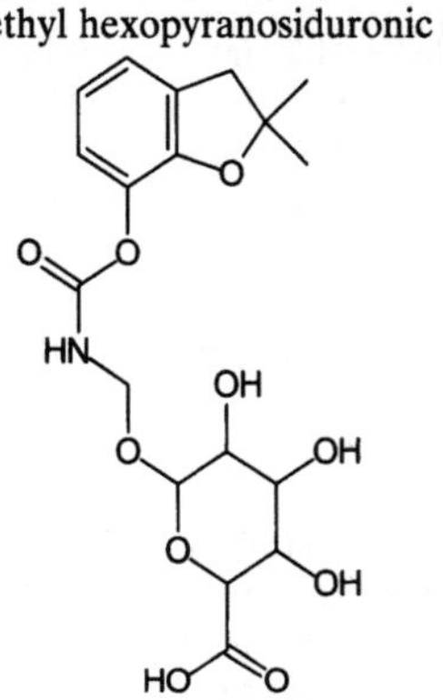	CAS No.	NA	fat	0.14	0.22
		MW, $g\,mol^{-1}$	413.38	brain	0.98	0.99
		Exp Kow @ pH 7.0	NA	rapid	0.94	0.94
		LogD @ pH 7.4	–4.77	kidney	0.94	0.97
		LogP	–1.06	liver	0.86	0.94
		p*K*a, MA	2.80	slow	0.91	0.93
		WS, $g\,L^{-1}$ (1000.0)	29959.0	skin	0.79	0.88
14	3,7-Diol sulfuric acid 3,7-Benzofurandiol, 2,3-dihydro-2,2-dimethyl-, 7-(hydrogen sulfate) 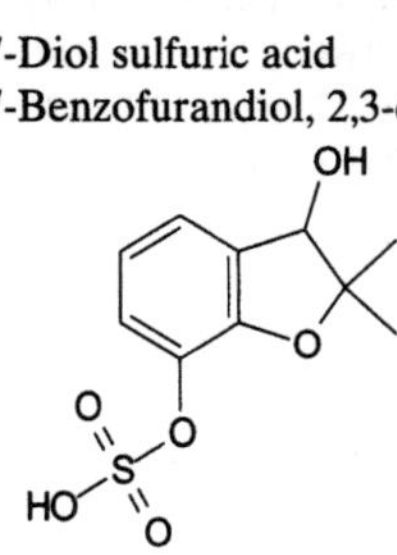	CAS No.	90433-38-8	fat	0.14	0.22
		MW, $g\,mol^{-1}$	260.26	brain	0.98	0.99
		Exp Kow @ pH 7.0	NA	rapid	0.94	0.94
		LogD @ pH 7.4	–3.14	kidney	0.94	0.97
		LogP	0.36	liver	0.86	0.94
		p*K*a, MA	–3.96	slow	0.91	0.93
		WS, $g\,L^{-1}$ (1000.0)	9955.20	skin	0.80	0.88

Table 16. (*cont.*)

No.[a]	Pesticide, pesticide metabolite, and chemical structure[b]	Physical and chemical properties[c,d]		Partition coefficient[e] Tissue	Rat	Human
15	3,7-Diol glucuronic acid β-D-Glucopyranosiduronic acid, 2,3-dihydro-3-hydroxy-2,2-dimethyl-7-benzofuranyl 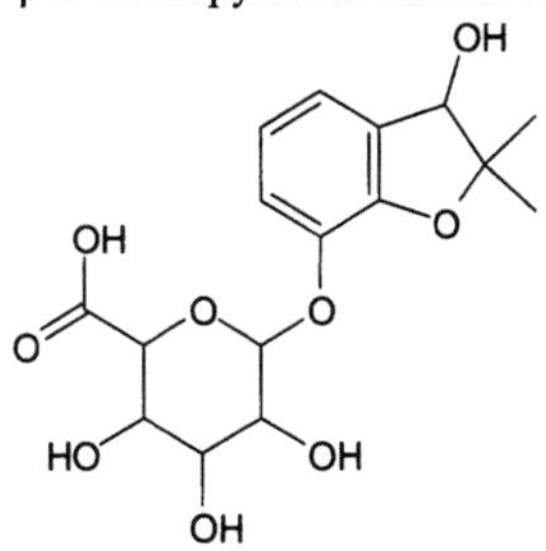	CAS No.	90433-37-7	fat	0.14	0.22
		MW, $g\,mol^{-1}$	356.32	brain	0.98	0.99
		Exp Kow @ pH 7.0	NA	rapid	0.94	0.94
		LogD @ pH 7.4	−5.49	kidney	0.94	0.97
		LogP	−1.78	liver	0.86	0.94
		p*K*a, MA	2.79	slow	0.91	0.93
		WS, $g\,L^{-1}$ (1000.0)	276956.70	skin	0.79	0.88
16	3-Keto-7-phenyl sulfuric acid 3(2*H*)-Benzofuranone, 2,2-dimethyl-7-(sulfooxy)- 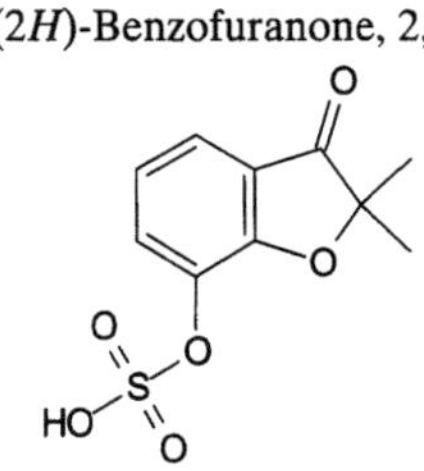	CAS No.	70988-94-2	fat	0.14	0.22
		MW, $g\,mol^{-1}$	258.25	brain	0.98	0.99
		Exp Kow @ pH 7.0	NA	rapid	0.94	0.94
		LogD @ pH 7.4	−1.98	kidney	0.94	0.97
		LogP	1.52	liver	0.86	0.94
		p*K*a, MA	−4.34	slow	0.91	0.93
		WS, $g\,L^{-1}$ (1000.0)	1045.15	skin	0.80	0.88
17	3-Keto-7-phenol glucuronic acid β-D-Glucopyranosiduronic acid, 2,3-dihydro-2,2-dimethyl-3-oxo-7-benzofuranyl 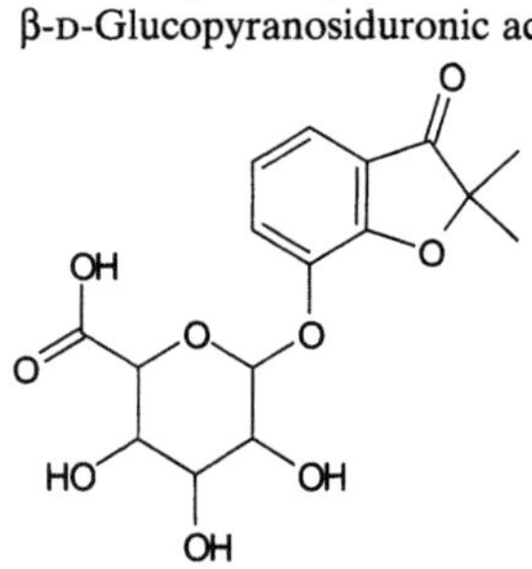	CAS No.	70988-93-1	fat	0.14	0.22
		MW, $g\,mol^{-1}$	354.31	brain	0.98	0.99
		Exp Kow @ pH 7.0	NA	rapid	0.94	0.94
		LogD @ pH 7.4	−4.33	kidney	0.94	0.97
		LogP	−0.61	liver	0.86	0.94
		p*K*a, MA	2.77	slow	0.91	0.93
		WS, $g\,L^{-1}$ (1000.0)	29099.61	skin	0.79	0.88

[a]Parent chemical or metabolite number for cross-reference to Table 26 (Appendix B).
[b–e]See footnotes in Table 14 (Appendix A).

Table 17. Chemical Structure, Physical and Chemical Properties, and Tissue Partition Coefficients for Formetanate and the Resulting Metabolites.

No.[a]	Pesticide, pesticide metabolite, and chemical structure[b]	Physical and chemical properties[c,d]		Partition coefficient[e]		
				Tissue	Rat	Human
1	Formetanate Methanimidamide, *N,N*-dimethyl-*N′*-[3-[[(methylamino) carbonyl] oxy]phenyl]-					
	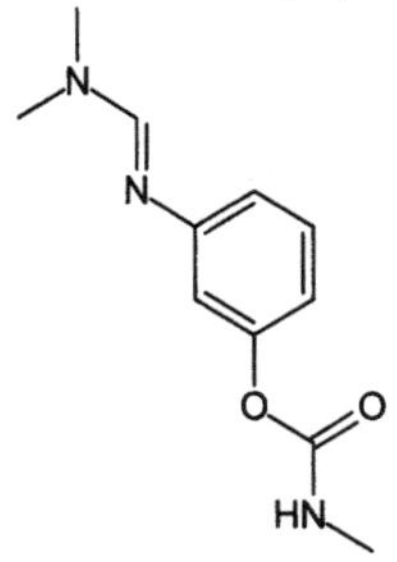	CAS No.	22259-30-9	fat	0.48	0.54
		MW, $g\,mol^{-1}$	221.26	brain	1.36	1.44
		Exp Kow @ pH 7.0	NA	rapid	1.05	1.03
		LogD @ pH 7.4	0.78	kidney	1.07	1.12
		LogP	0.86	liver	1.01	1.21
		p*K*a, MA	12.36	slow	0.98	1.09
		WS, $g\,L^{-1}$ (1.55)	7.31	skin	0.99	1.09
		Kp, $cm\,hr^{-1}$	NA			
		Log Kp	NA			
2	Demethylformetanate Methanimidamide, *N*-methyl-*N′*-[3-[[(methylamino) carbonyl] oxy] phenyl]-					
	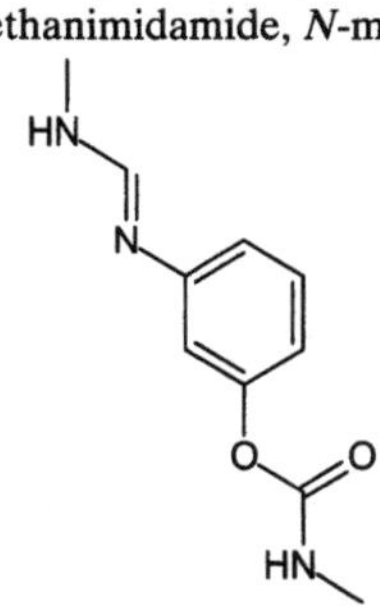	CAS No.	25636-15-1	fat	0.17	0.24
		MW, $g\,mol^{-1}$	207.23	brain	1.01	1.03
		Exp Kow @ pH 7.0	NA	rapid	0.95	0.95
		LogD @ pH 7.4	−0.29	kidney	0.95	0.98
		LogP	0.58	liver	0.88	0.96
		p*K*a, MA	12.38	slow	0.91	0.95
		WS, $g\,L^{-1}$ (35.16)	71.03	skin	0.81	0.90
3	3-Formaminophenyl N-methylcarbamate Formamide, *N*-[3-[[(methylamino) carbonyl] oxy] phenyl]-					
	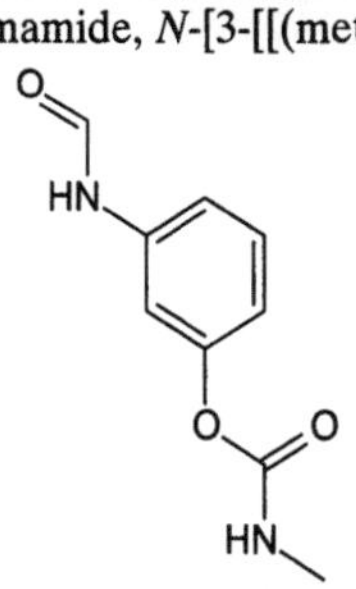	CAS No.	24891-34-7	fat	0.28	0.35
		MW, $g\,mol^{-1}$	194.19	brain	1.15	1.20
		Exp Kow @ pH 7.0	NA	rapid	0.99	0.98
		LogD @ pH 7.4	0.44	kidney	1.00	1.04
		LogP	0.44	liver	0.93	1.06
		p*K*a, MA	12.11	slow	0.94	1.01
		WS, $g\,L^{-1}$ (3.65)	19.71	skin	0.88	0.98
4	3-Formaminophenol Formamide, *N*-(3-hydroxyphenyl)-					
	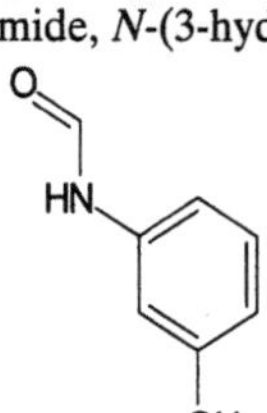	CAS No.	24891-35-8	fat	0.49	0.55
		MW, $g\,mol^{-1}$	137.14	brain	1.37	1.45
		Exp Kow @ pH 7.0	NA	rapid	1.05	1.03
		LogD @ pH 7.4	0.79	kidney	1.08	1.13
		LogP	0.80	liver	1.02	1.22
		p*K*a, MA	9.41	slow	0.99	1.10
		WS, $g\,L^{-1}$ (5.68)	18.20	skin	0.99	1.09

Table 17. (*cont.*)

No.[a]	Pesticide, pesticide metabolite, and chemical structure[b]	Physical and chemical properties[c,d]		Partition coefficient[e]		
				Tissue	Rat	Human
5	3-Aminophenol Phenol, 3-amino-					
	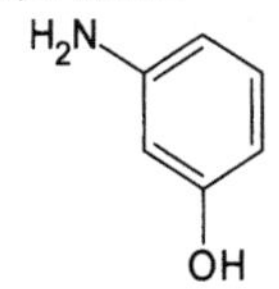	CAS No.	591-27-5	fat	0.25	0.32
		MW, $g\,mol^{-1}$	109.13	brain	1.12	1.15
		Exp Kow @ pH 7.0	NA	rapid	0.98	0.97
		LogD @ pH 7.4	0.34	kidney	0.99	1.02
		LogP	0.34	liver	0.92	1.04
		p*K*a, MA	10.01	slow	0.93	0.99
		WS, $g\,L^{-1}$ (21.55)	56.12	skin	0.86	0.96
6	3-Acetamidophenol Acetamide, *N*-(3-hydroxyphenyl)-					
	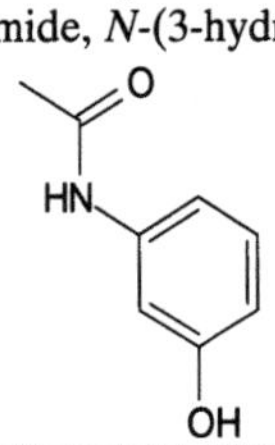	CAS No.	621-42-1	fat	0.44	0.50
		MW, $g\,mol^{-1}$	151.16	brain	1.32	1.40
		Exp Kow @ pH 7.0	NA	rapid	1.04	1.02
		LogD @ pH 7.4	0.73	kidney	1.06	1.11
		LogP	0.73	liver	1.00	1.18
		p*K*a, MA	9.50	slow	0.98	1.07
		WS, $g\,L^{-1}$ (12.92)	17.84	skin	0.97	1.06
7	3-Formaminophenyl glucuronic acid 6-[3-(Formylamino)phenoxy]-3,4,5-trihydroxytetrahydro-2*H*-pyran-2-carboxylic acid					
	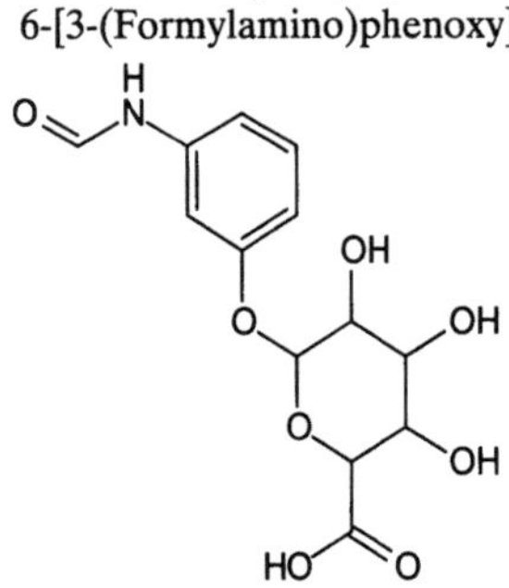	CAS No.	NA	fat	0.14	0.22
		MW, $g\,mol^{-1}$	313.26	brain	0.98	0.99
		Exp Kow @ pH 7.0	NA	rapid	0.94	0.94
		LogD @ pH 7.4	−5.27	kidney	0.94	0.97
		LogP	−1.56	liver	0.86	0.94
		p*K*a, MA	2.78	slow	0.91	0.93
		WS, $g\,L^{-1}$ (1000.0)	325042.00	skin	0.79	0.88
8	3-Formaminophenyl sulfuric acid 3-(Formylamino)phenyl hydrogen sulfate					
	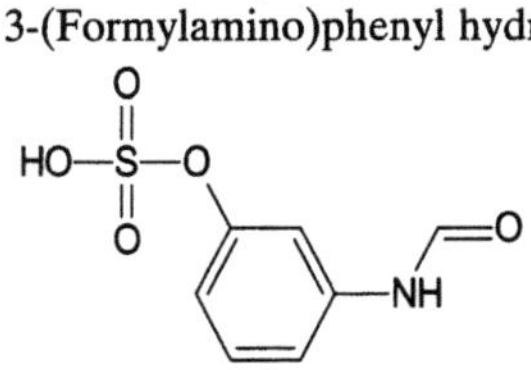	CAS No.	NA	fat	0.14	0.22
		MW, $g\,mol^{-1}$	217.20	brain	0.98	0.99
		Exp Kow @ pH 7.0	NA	rapid	0.94	0.94
		LogD @ pH 7.4	−2.92	kidney	0.94	0.97
		LogP	0.58	liver	0.86	0.94
		p*K*a, MA	−4.14	slow	0.91	0.93
		WS, $g\,L^{-1}$ (1000.0)	11100.29	skin	0.79	0.88

Table 17. (*cont.*)

No.[a]	Pesticide, pesticide metabolite, and chemical structure[b]	Physical and chemical properties[c,d]		Partition coefficient[e] Tissue	Rat	Human
9	3-Aminophenyl glucuronic acid β-D-Glucopyranosiduronic acid, *m*-aminophenyl 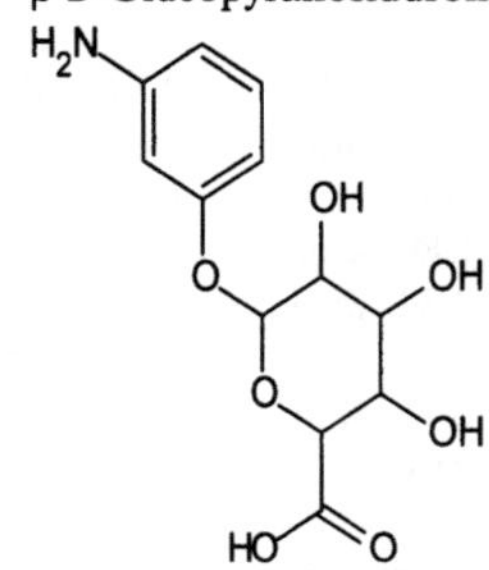	CAS No.	108595-12-6	fat	0.14	0.22
		MW, $g\,mol^{-1}$	285.25	brain	0.98	0.99
		Exp Kow @ pH 7.0	NA	rapid	0.94	0.94
		LogD @ pH 7.4	−5.84	kidney	0.94	0.97
		LogP	−2.13	liver	0.86	0.94
		p*K*a, MA	2.80	slow	0.91	0.93
		WS, $g\,L^{-1}$ (1000.0)	1451978.00	skin	0.79	0.88
10	3-Aminophenyl sulfuric aicd Phenol, 3-amino-, hydrogen sulfate 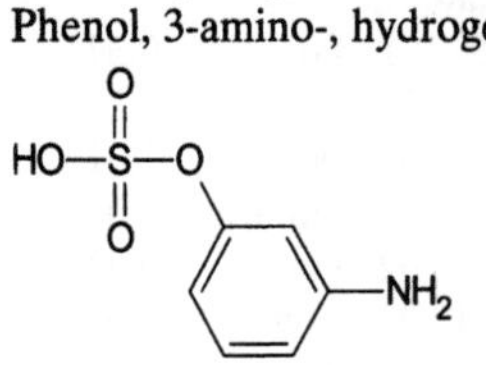	CAS No.	27991-69-1	fat	0.14	0.22
		MW, $g\,mol^{-1}$	189.19	brain	0.98	0.99
		Exp Kow @ pH 7.0	NA	rapid	0.94	0.94
		LogD @ pH 7.4	−3.5	kidney	0.94	0.97
		LogP	−0.003	liver	0.86	0.94
		p*K*a, MA	−3.78	slow	0.91	0.93
		WS, $g\,L^{-1}$ (1000.0)	48373.21	skin	0.79	0.88
11	3-Acetamidophenyl glucuronic acid β-D-Glucopyranosiduronic acid, 3-(acetylamino) phenyl 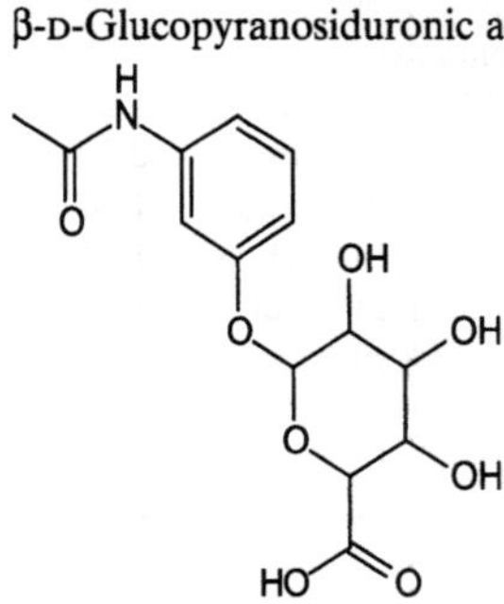	CAS No.	102989-23-1	fat	0.14	0.22
		MW, $g\,mol^{-1}$	327.79	brain	0.98	0.99
		Exp Kow @ pH 7.0	NA	rapid	0.94	0.94
		LogD @ pH 7.4	−5.48	kidney	0.94	0.97
		LogP	−1.62	liver	0.86	0.94
		p*K*a, MA	2.78	slow	0.91	0.93
		WS, $g\,L^{-1}$ (1000.0)	308058.70	skin	0.79	0.88
12	3-Acetamidophenyl sulfuric acid Acetamide, *N*-[3-(sulfooxy) phenyl]- 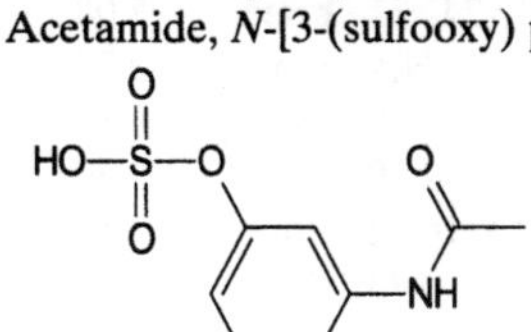	CAS No.	52175-03-8	fat	0.14	0.22
		MW, $g\,mol^{-1}$	231.23	brain	0.98	0.99
		Exp Kow @ pH 7.0	NA	rapid	0.94	0.94
		LogD @ pH 7.4	−2.99	kidney	0.94	0.97
		LogP	0.51	liver	0.86	0.94
		p*K*a, MA	−4.11	slow	0.91	0.93
		WS, $g\,L^{-1}$ (1000.0)	10719.42	skin	0.79	0.88

[a]Parent chemical or metabolite number for cross-reference to Table 27 (Appendix B).
[b–e]See footnotes in Table 14 (Appendix A).

Table 18. Chemical Structure, Physical and Chemical Properties, and Tissue Partition Coefficients for Methiocarb and the Resulting Metabolites.

No.[a]	Pesticide, pesticide metabolite, and chemical structure[b]	Physical and chemical properties[c,d]		Partition coefficient[e]		
				Tissue	Rat	Human
1	Methiocarb					
	Phenol, 3,5-dimethyl-4-(methylthio)-, methylcarbamate					
	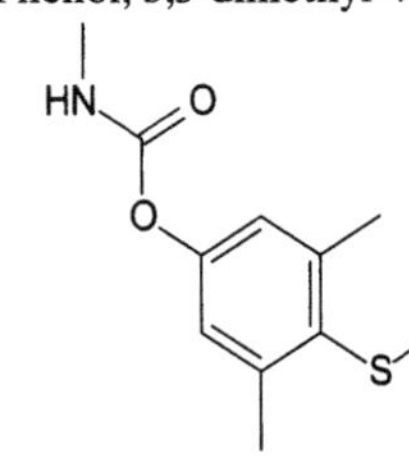	CAS No.	2032-65-7	fat	64.84	53.51
		MW, $g\,mol^{-1}$	225.31	brain	18.85	14.20
		Exp Kow @ pH 7.0	NA	rapid	6.22	3.60
		LogD @ pH 7.4	2.89	kidney	7.32	5.47
		LogP	2.89	liver	7.99	8.92
		p*K*a, MA	12.16	slow	4.58	5.55
		WS, $g\,L^{-1}$ (0.12)	0.11	skin	10.11	6.72
		Kp, $cm\,hr^{-1}$	NA			
		Log Kp	NA			
2	Methiocarb sulfoxide					
	Phenol, 3,5-dimethyl-4-(methylsulfinyl)-, methylcarbamate					
	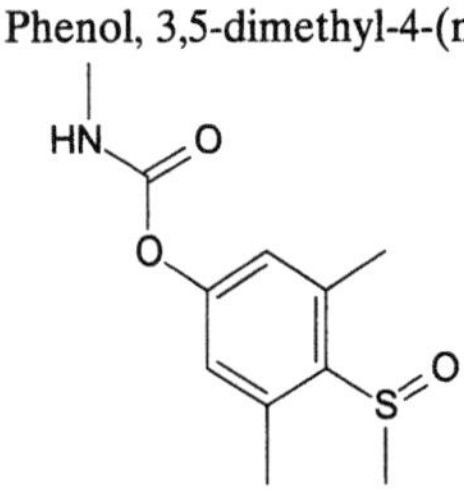	CAS No.	2635-10-1	fat	0.37	0.43
		MW, $g\,mol^{-1}$	241.31	brain	1.24	1.31
		Exp Kow @ pH 7.0	NA	rapid	1.01	1.00
		LogD @ pH 7.4	0.62	kidney	1.03	1.08
		LogP	0.62	liver	0.97	1.13
		p*K*a, MA	12.03	slow	0.96	1.04
		WS, $g\,L^{-1}$ (1.79)	7.81	skin	0.93	1.02
3	Methiocarb sulfone					
	Phenol, 3,5-dimethyl-4-(methylsulfonyl)-, methylcarbamate					
	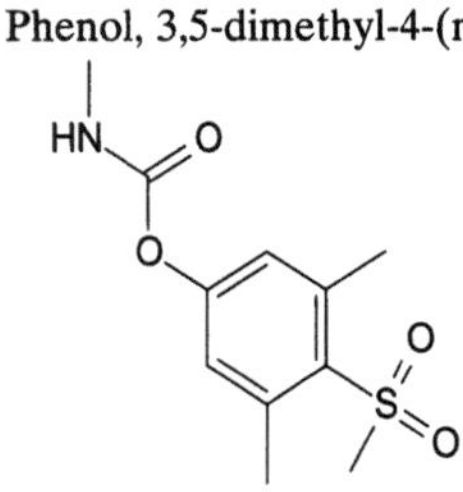	CAS No.	2179-25-1	fat	1.29	1.31
		MW, $g\,mol^{-1}$	257.31	brain	2.09	2.30
		Exp Kow @ pH 7.0	NA	rapid	1.26	1.20
		LogD @ pH 7.4	1.26	kidney	1.34	1.41
		LogP	1.26	liver	1.31	1.73
		p*K*a, MA	11.96	slow	1.13	1.39
		WS, $g\,L^{-1}$ (0.46)	1.81	skin	1.37	1.46
4	Hydroxymethyl methiocarb					
	3,5-Dimethyl-4-(methylthio) phenyl hydroxymethyl) carbamate					
	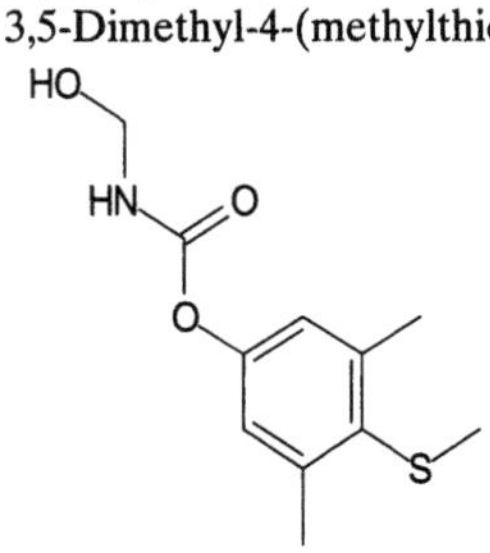	CAS No.	NA	fat	9.63	9.02
		MW, $g\,mol^{-1}$	241.31	brain	7.08	7.11
		Exp Kow @ pH 7.0	NA	rapid	2.74	2.17
		LogD @ pH 7.4	2.09	kidney	3.12	3.06
		LogP	2.09	liver	3.30	4.64
		p*K*a, MA	10.42	slow	2.16	3.07
		WS, $g\,L^{-1}$ (0.74)	0.43	skin	3.97	3.59

Table 18. (*cont.*)

No.[a]	Pesticide, pesticide metabolite, and chemical structure[b]	Physical and chemical properties[c,d]		Partition coefficient[e] Tissue	Rat	Human
5	Methiocarb phenol Phenol, 3,5-dimethyl-4-(methylthio)-					
	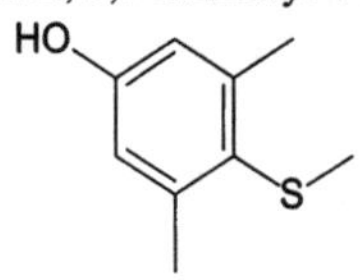	CAS No.	7379-51-3	fat	45.24	38.89
		MW, $g\,mol^{-1}$	168.26	brain	16.36	13.03
		Exp Kow @ pH 7.0	NA	rapid	5.48	3.36
		LogD @ pH 7.4	2.73	kidney	6.43	5.07
		LogP	2.73	liver	6.99	8.22
		p*K*a, MA	10.03	slow	4.07	5.14
		WS, $g\,L^{-1}$ (0.84)	0.29	skin	8.81	6.21
6	Methiocarb phenol sulfoxide Phenol, 3,5-dimethyl-4-(methylsulfinyl)-					
	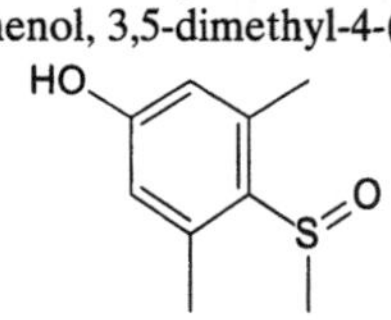	CAS No.	22454-92-8	fat	1.15	1.18
		MW, $g\,mol^{-1}$	184.26	brain	1.97	2.16
		Exp Kow @ pH 7.0	NA	rapid	1.23	1.17
		LogD @ pH 7.4	1.21	kidney	1.29	1.37
		LogP	1.23	liver	1.26	1.65
		p*K*a, MA	8.64	slow	1.11	1.34
		WS, $g\,L^{-1}$ (6.87)	4.86	skin	1.31	1.40
7	Methiocarb phenol sulfone Phenol, 3,5-dimethyl-4-(methylsulfonyl)-; 3,5-Xylenol, 4-(methylsulfonyl)-;					
	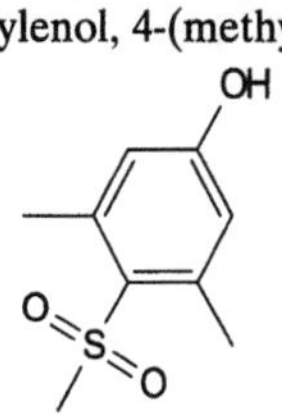	CAS No.	14763-62-3	fat	0.75	0.79
		MW, $g\,mol^{-1}$	200.26	brain	1.62	1.75
		Exp Kow @ pH 7.0	NA	rapid	1.12	1.09
		LogD @ pH 7.4	1.01	kidney	1.17	1.23
		LogP	1.15	liver	1.12	1.40
		p*K*a, MA	7.81	slow	1.04	1.20
		WS, $g\,L^{-1}$ (4.08)	5.98	skin	1.12	1.22
8	Hydroxymethyl methiocarb glucuronic acid ({[3,5-Dimethyl-4-(methylthio)phenoxy]carbonyl}amino)methyl hexopyranosiduronic acid					
	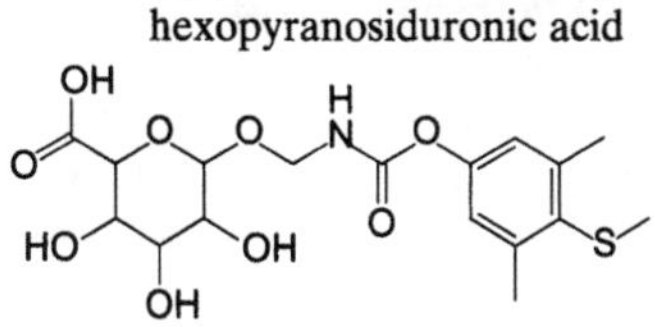	CAS No.	NA	fat	0.14	0.22
		MW, $g\,mol^{-1}$	417.43	brain	0.98	0.99
		Exp Kow @ pH 7.0	NA	rapid	0.94	0.94
		LogD @ pH 7.4	−3.65	kidney	0.94	0.97
		LogP	0.07	liver	0.86	0.94
		p*K*a, MA	2.8	slow	0.91	0.93
		WS, $g\,L^{-1}$ (1000.0)	3124.6	skin	0.79	0.85

Table 18. (*cont.*)

No.[a]	Pesticide, pesticide metabolite, and chemical structure[b]	Physical and chemical properties[c,d]		Partition coefficient[e] Tissue	Rat	Human
9	Hydroxymethyl methiocarb sulfuric acid 3,5-Dimethyl-4-(methylthio) phenyl [(sulfooxy) methyl] carbamate					
	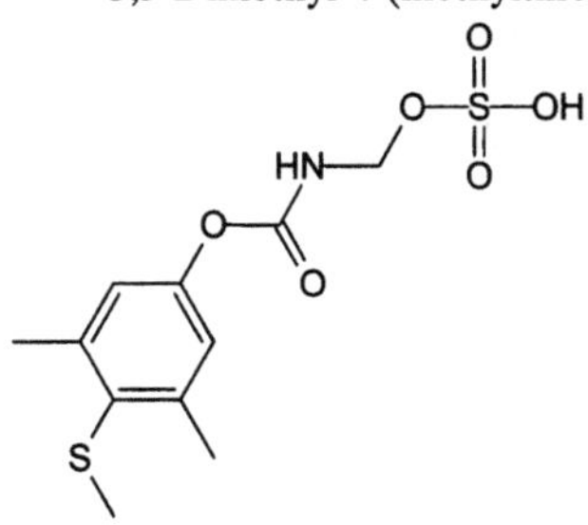	CAS No.	NA	fat	0.14	0.22
		MW, $g\,mol^{-1}$	321.37	brain	0.99	0.99
		Exp Kow @ pH 7.0	NA	rapid	0.94	0.94
		LogD @ pH 7.4	−1.15	kidney	0.94	0.97
		LogP	2.36	liver	0.86	0.94
		p*K*a, MA	−3.88	slow	0.91	0.93
		WS, $g\,L^{-1}$ (177.18)	88.21	skin	0.79	0.89
10	Methiocarb phenyl glucuronic acid 6-[3,5-Dimethyl-4-(methylthio) phenoxy]-3,4,5-trihydroxytetrahydro-2*H*-pyran-2-carboxylic acid					
	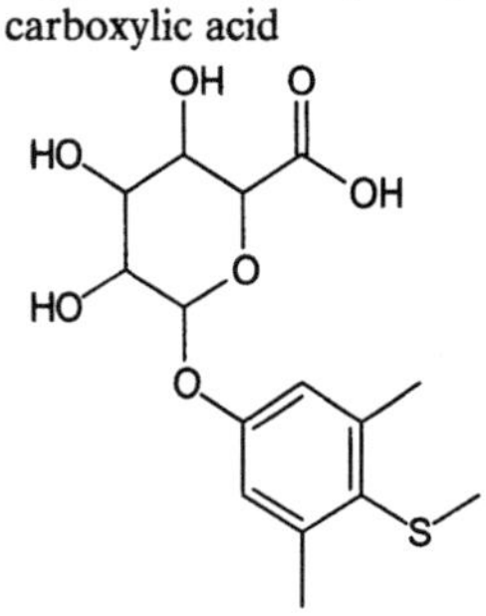	CAS No.	NA	fat	0.14	0.22
		MW, $g\,mol^{-1}$	344.39	brain	0.98	0.99
		Exp Kow @ pH 7.0	NA	rapid	0.94	0.94
		LogD @ pH 7.4	−3.11	kidney	0.94	0.97
		LogP	0.60	liver	0.86	0.94
		p*K*a, MA	2.78	slow	0.91	0.93
		WS, $g\,L^{-1}$ (1000.0)	3033.05	skin	0.79	0.88
11	Methiocarb phenyl sulfuric acid 3,5-Dimethyl-4-(methylthio) phenyl hydrogen sulfate					
	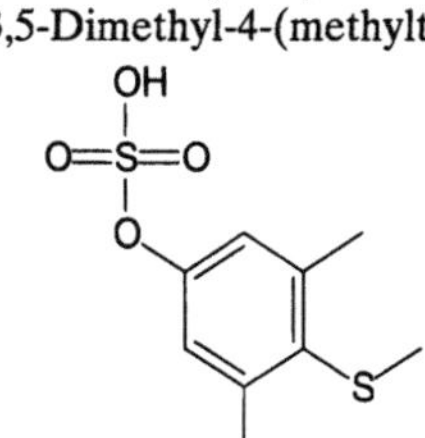	CAS No.	NA	fat	0.14	0.22
		MW, $g\,mol^{-1}$	248.32	brain	0.99	1.00
		Exp Kow @ pH 7.0	NA	rapid	0.94	0.94
		LogD @ pH 7.4	−0.76	kidney	0.94	0.97
		LogP	2.74	liver	0.87	0.94
		p*K*a, MA	−4.18	slow	0.91	0.94
		WS, $g\,L^{-1}$ (309.93)	107.72	skin	0.79	0.88

Table 18. (*cont.*)

No.[a]	Pesticide, pesticide metabolite, and chemical structure[b]	Physical and chemical properties[c,d]		Partition coefficient[e]		
				Tissue	Rat	Human
12	Methiocarb sulfoxide phenyl glucuronic acid 6-[3,5-Dimethyl-4-(methylsulfinyl)phenoxy]-3,4,5-trihyroxytetrahydro-2*H*-pyran-2-carboxylic acid 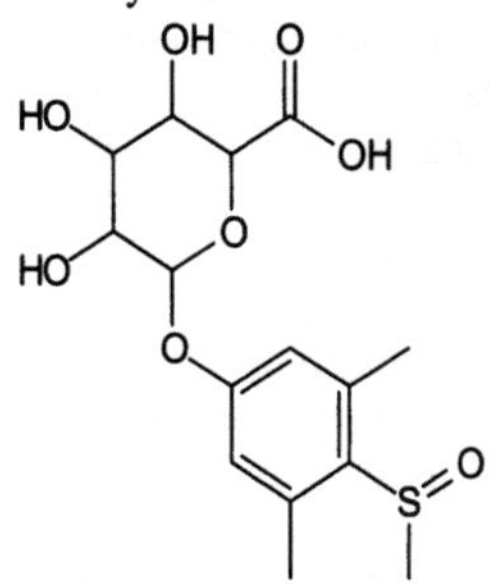	CAS No.	NA	fat	0.14	0.22
		MW, $g\,mol^{-1}$	360.38	brain	0.98	1.00
		Exp Kow @ pH 7.0	NA	rapid	0.94	0.94
		LogD @ pH 7.4	−5.09	kidney	0.94	0.97
		LogP	−1.38	liver	0.86	0.94
		p*K*a, MA	2.78	slow	0.91	0.93
		WS, $g\,L^{-1}$ (1000.0)	119139.10	skin	0.79	0.88
13	Methiocarb sulfoxide phenyl sulfuric acid 3,5-Dimethyl-4-(methylsulfinyl)phenyl hydrogen sulfate 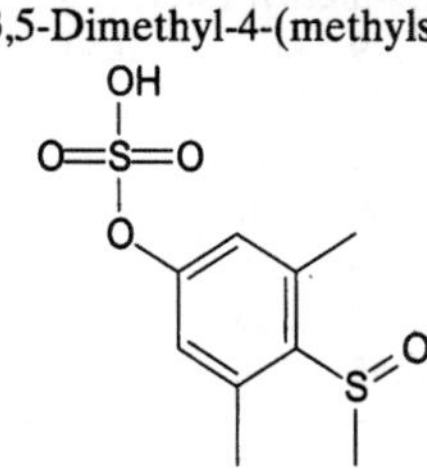	CAS No.	NA	fat	0.14	0.22
		MW, $g\,mol^{-1}$	264.32	brain	0.98	1.00
		Exp Kow @ pH 7.0	NA	rapid	0.94	0.94
		LogD @ pH 7.4	−2.75	kidney	0.94	0.97
		LogP	0.75	liver	0.86	0.94
		p*K*a, MA	−4.39	slow	0.91	0.93
		WS, $g\,L^{-1}$ (1000.0)	4389.26	skin	0.79	0.88
14	Methiocarb sulfone phenyl glucuronic acid 6-[3,5-Dimethyl-4-(methylsulfonyl) phenoxy]-3,4,5-trihydroxytetrahydro-2*H*-pyran-2-carboxylic acid 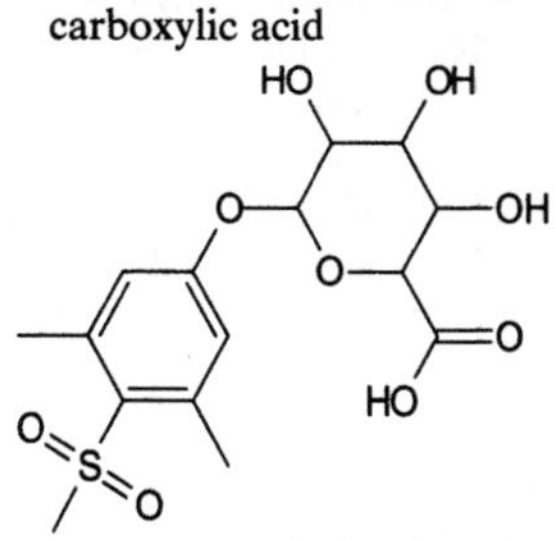	CAS No.	NA	fat	0.14	0.22
		MW, $g\,mol^{-1}$	376.38	brain	0.98	1.00
		Exp Kow @ pH 7.0	NA	rapid	0.94	0.94
		LogD @ pH 7.4	−4.74	kidney	0.94	0.97
		LogP	−1.03	liver	0.86	0.94
		p*K*a, MA	2.77	slow	0.91	0.93
		WS, $g\,L^{-1}$ (1000.0)	47813.03	skin	0.79	0.88
15	Methiocarb sulfone phenyl sulfuric acid 3,5-Dimethyl-4-(methylsulfonyl)phenyl hydrogen sulfate 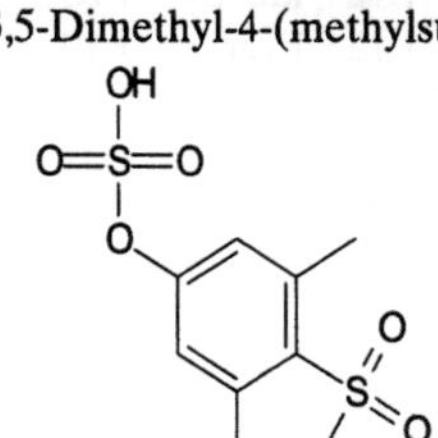	CAS No.	NA	fat	0.14	0.22
		MW, $g\,mol^{-1}$	280.32	brain	0.98	0.99
		Exp Kow @ pH 7.0	NA	rapid	0.94	0.94
		LogD @ pH 7.4	−2.39	kidney	0.94	0.97
		LogP	1.11	liver	0.86	0.94
		p*K*a, MA	−4.51	slow	0.91	0.93
		WS, $g\,L^{-1}$ (1000.0)	1753.91	skin	0.79	0.88

[a]Parent chemical or metabolite number for cross-reference to Table 28 (Appendix B).
[b–e]See footnotes in Table 14 (Appendix A).

Table 19. Chemical Structure, Physical and Chemical Properties, and Tissue Partition Coefficients for Methomyl and the Resulting Metabolites.

No.[a]	Pesticide, pesticide metabolite, and chemical structure[b]	Physical and chemical properties[c,d]		Partition coefficient[e] Tissue	Rat	Human
1	*syn*-(Z)-methomyl Ethanimidothioic acid, *N*-[[(methylamino)carbonyl]oxy]-methyl ester, (*Z*)-					
	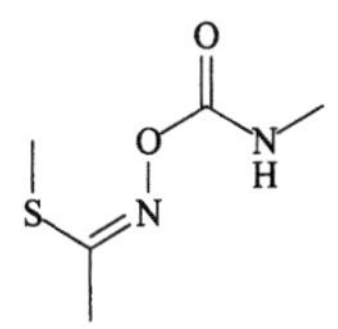	CAS No.	19928-37-1	fat	0.36	0.42
		MW, g mol^{-1}	162.21	brain	1.23	1.29
		Exp Kow @ pH 7.0	NA	rapid	1.01	1.00
		LogD @ pH 7.4	0.60	kidney	1.03	1.07
		LogP	0.60	liver	0.96	1.12
		p*K*a, MA	13.27	slow	0.96	1.04
		WS, g L^{-1} (11.38)	20.55	skin	0.92	1.02
		Kp, cm hr^{-1}	NA			
		Log Kp	NA			
2	*anti*-(*E*)-methomyl Ethanimidothioic acid, *N*-[[(methylamino) carbonyl] oxy]-methyl ester, (*E*)-					
	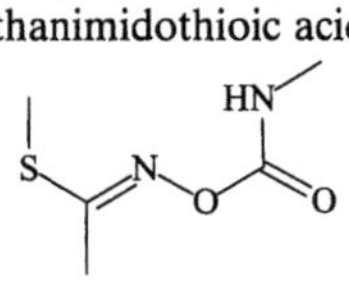	CAS No.	19928-35-9	fat	0.36	0.42
		MW, g mol^{-1}	162.21	brain	1.23	1.29
		Exp Kow @ pH 7.0	NA	rapid	1.01	1.00
		LogD @ pH 7.4	0.60	kidney	1.03	1.07
		LogP	0.60	liver	0.96	1.12
		p*K*a, MA	13.27	slow	0.96	1.04
		WS, g L^{-1} (11.38)	20.55	skin	0.92	1.02
3	*syn*-(*Z*)-methomyl oxime Ethanimidothioic acid, *N*-hydroxy-, methyl ester, (*Z*)-					
	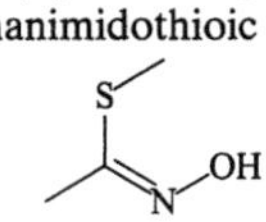	CAS No.	19125-12-3	fat	0.95	0.98
		MW, g mol^{-1}	105.16	brain	1.79	1.96
		Exp Kow @ pH 7.0	NA	rapid	1.18	1.13
		LogD @ pH 7.4	1.12	kidney	1.23	1.30
		LogP	1.12	liver	1.19	1.52
		p*K*a, MA	11.26	slow	1.07	1.27
		WS, g L^{-1} (16.69)	12.47	skin	1.21	1.31
4	*anti*-(*E*)-methomyl oxime Ethanimidothioic acid, *N*-hydroxy-, methyl ester, (*E*)-					
	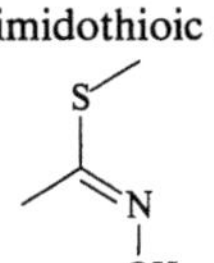	CAS No.	19145-16-5	fat	0.95	0.98
		MW, g mol^{-1}	105.16	brain	1.79	1.96
		Exp Kow @ pH 7.0	NA	rapid	1.18	1.13
		LogD @ pH 7.4	1.12	kidney	1.23	1.30
		LogP	1.12	liver	1.19	1.52
		p*K*a, MA	11.26	slow	1.07	1.27
		WS, g L^{-1} (16.69)	12.47	skin	1.21	1.31

Table 19. (*cont.*)

No.[a]	Pesticide, pesticide metabolite, and chemical structure[b]	Physical and chemical properties[c,d]		Partition coefficient[e]		
				Tissue	Rat	Human
5	S-Methyl, *N*-methyl carbamic acid					
	Carbamothioic acid, methyl-, *S*-methyl ester					
	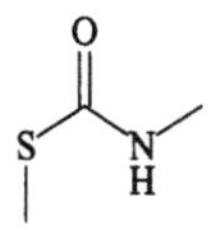	CAS No.	22013-97-4	fat	0.24	0.31
		MW, $g\,mol^{-1}$	105.16	brain	1.11	1.14
		Exp Kow @ pH 7.0	NA	rapid	0.97	0.97
		LogD @ pH 7.4	0.30	kidney	0.99	1.02
		LogP	0.30	liver	0.91	1.03
		p*K*a, MA	12.77	slow	0.93	0.99
		WS, $g\,L^{-1}$ (41.17)	62.53	skin	0.85	0.95
6	Ethylium					
	Ethylium, 1-[(methylthio) imino]-					
	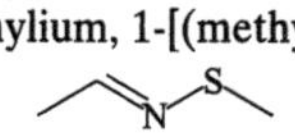	CAS No.	61599-22-2	fat	1.13	1.15
		MW, $g\,mol^{-1}$	89.16	brain	1.95	2.14
		Exp Kow @ pH 7.0	NA	rapid	1.22	1.17
		LogD @ pH 7.4	1.20	kidney	1.29	1.36
		LogP	1.20	liver	1.25	1.63
		p*K*a, MB	4.01	slow	1.11	1.33
		WS, $g\,L^{-1}$ (9.58)	11.81	skin	1.29	1.39
7	Carbon dioxide					
	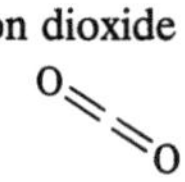	CAS No.	124-38-9	fat	0.53	0.58
		MW, $g\,mol^{-1}$	44.01	brain	1.40	1.50
		Exp Kow @ pH 7.0	NA	rapid	1.06	1.04
		LogD @ pH 7.4	0.83	kidney	1.09	1.14
		LogP	0.83	liver	1.03	1.25
		p*K*a	NA	slow	0.99	1.11
		WS, $g\,L^{-1}$ (358.82)	25.72	skin	1.01	1.11
8	Acetonitrile					
	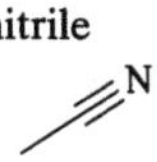	CAS No.	75-05-8	fat	0.16	0.23
		MW, $g\,mol^{-1}$	41.05	brain	1.00	1.01
		Exp Kow @ pH 7.0	NA	rapid	0.94	0.94
		LogD @ pH 7.4	−0.45	kidney	0.95	0.98
		LogP	−0.45	liver	0.87	0.95
		p*K*a	NA	slow	0.91	0.94
		WS, $g\,L^{-1}$ (122.06)	312.45	skin	0.80	0.90

Table 19. (*cont.*)

No.[a]	Pesticide, pesticide metabolite, and chemical structure[b]	Physical and chemical properties[c,d]		Partition coefficient[e]		
				Tissue	Rat	Human
9	Hydrogen cyanide					
	≡N	CAS No.	74-90-8	fat	0.15	0.23
		MW, g mol^{-1}	27.03	brain	1.02	1.03
		Exp Kow @ pH 7.0	NA	rapid	0.95	0.95
		LogD @ pH 7.4	−0.25	kidney	0.95	0.98
		LogP	−0.25	liver	0.88	0.96
		p*K*a	NA	slow	0.91	0.95
		WS, g L^{-1} (89.68)	175.60	skin	0.81	0.90
10	Glutathione conjugate of acetonitrile 2-Amino-5-({2-[(carboxylmethyl)amino]-1-[(ethanimidoylthio)methyl]-2-oxoethyl}amino)-5-oxopentanoic acid					
	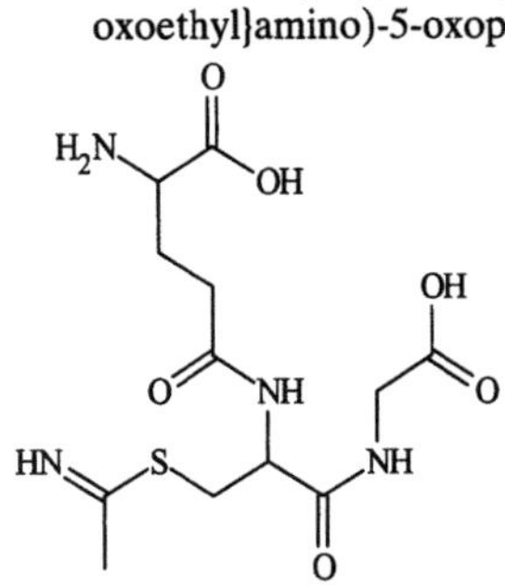	CAS No.	NA	fat	0.14	0.22
		MW, g mol^{-1}	348.38	brain	0.98	0.99
		Exp Kow @ pH 7.0	NA	rapid	0.94	0.94
		LogD @ pH 7.0	−3.98	kidney	0.94	0.97
		LogP	−1.11	liver	0.86	0.94
		p*K*a, MA, A	3.57	slow	0.91	0.93
			3.94			
		WS, g L^{-1} (3.56)	15877.38	skin	0.80	0.88
11	Glutamylcysteine conjugate of acetonitrile 2-Amino-5-{[1-carboxy-2-(ethanimidoylthio)ethyl]-5-oxopentanoic acid					
	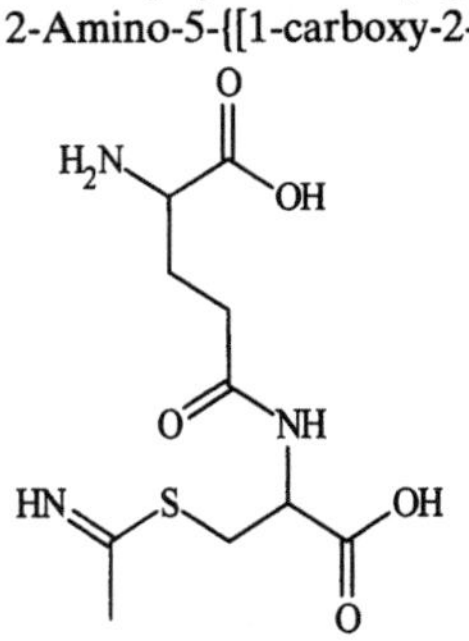	CAS No.	NA	fat	0.14	0.22
		MW, g mol^{-1}	291.33	brain	0.98	0.99
		Exp Kow @ pH 7.0	NA	rapid	0.94	0.94
		LogD @ pH 7.4	−3.67	kidney	0.94	0.97
		LogP	−0.92	liver	0.86	0.94
		p*K*a, MA, A	3.15	slow	0.91	0.93
			3.94			
		WS, g L^{-1} (4.3)	18780.45	skin	0.80	0.88

Table 19. (*cont.*)

No.[a]	Pesticide, pesticide metabolite, and chemical structure[b]	Physical and chemical properties[c,d]		Partition coefficient[e]		
				Tissue	Rat	Human
12	Cysteinylglycine conjugate of acetonitrile {[2-Amino-3-(ethanimidoylthio)propanoyl]amino}acetic acid					
	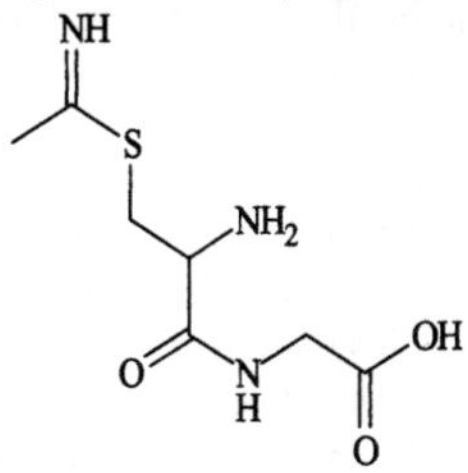	CAS No.	NA	fat	0.14	0.22
		MW, $g\,mol^{-1}$	219.26	brain	0.98	0.99
		Exp Kow @ pH 7.0	NA	rapid	0.94	0.94
		LogD @ pH 7.4	−3.62	kidney	0.94	0.97
		LogP	−0.85	liver	0.86	0.94
		pKa, MA	3.61	slow	0.91	0.93
		WS, $g\,L^{-1}$ (22.28)	42875.65	skin	0.80	0.88
13	Cysteine conjugate of acetonitrile 2-Amino-3-(ethanimidoylthio) propanoic acid					
	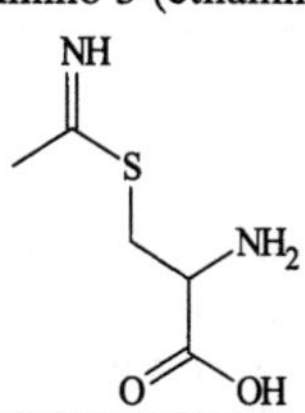	CAS No.	NA	fat	0.14	0.22
		MW, $g\,mol^{-1}$	162.21	brain	0.98	0.99
		Exp Kow @ pH 7.0	NA	rapid	0.94	0.94
		LogD @ pH 7.4	−2.67	kidney	0.94	0.97
		LogP	−0.025	liver	0.86	0.94
		pKa, MA	3.54	slow	0.91	0.93
		WS, $g\,L^{-1}$ (25.95)	12745.72	skin	0.80	0.88
14	*N*-Sulfate-cysteine conjugate of acetonitrile 3-Mercapto-2-{[(1*Z*)-*N*-(sulfinooxy)ethanimidoyl]amino}propanoic acid					
	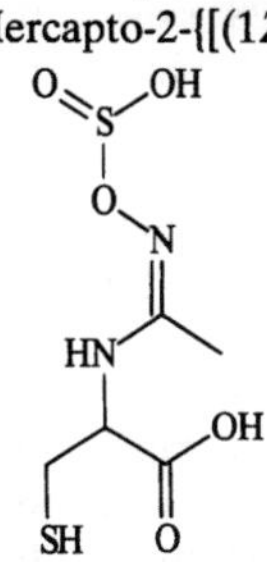	CAS No.	NA	fat	0.14	0.22
		MW, $g\,mol^{-1}$	242.27	brain	0.98	0.99
		Exp Kow @ pH 7.0	NA	rapid	0.94	0.94
		LogD @ pH 7.4	−4.73	kidney	0.94	0.97
		LogP	−0.083	liver	0.86	0.94
		PKa, MA, A	2.15 3.24	slow	0.91	0.93
		WS, $g\,L^{-1}$ (1000.0)	285646.80	skin	0.80	0.88

[a]Parent chemical or metabolite number for cross-reference to Table 29 (Appendix B).
[b–e]See footnotes in Table 14 (Appendix A).

Table 20. Chemical Structure, Physical and Chemical Properties, and Tissue Partition Coefficients for Oxamyl and the Resulting Metabolites.

No.[a]	Pesticide, pesticide metabolite, and chemical structure[b]	Physical and chemical properties[c,d]		Partition coefficient[e] Tissue	Rat	Human
1	Oxamyl Ethanimidothioic acid, 2-(dimethylamino)-*N*-[[(methylamino)] oxy]-oxo-, methyl ester; CAS No. for *Z* isomer, 32817-80-4, is shown; CAS No. for *E* isomer, 32817-79-1.					
	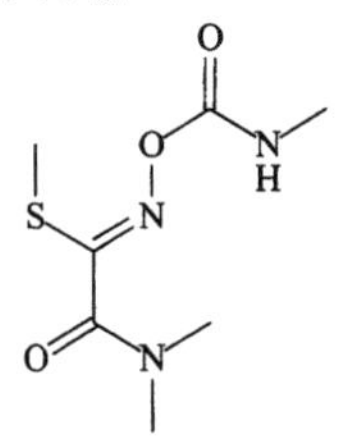	CAS No.	23135-22-0	fat	0.16	0.23
		MW, $g\,mol^{-1}$	219.26	brain	1.00	1.01
		Exp Kow @ pH 7.0	NA	rapid	0.94	0.94
		LogD @ pH 7.4	–0.47	kidney	0.95	0.98
		LogP	–0.47	liver	0.87	0.95
		p*K*a, MA	10.48	slow	0.91	0.94
		WS, $g\,L^{-1}$ (14.15)	87.52	skin	0.80	0.90
		Kp, $cm\,hr^{-1}$	NA			
		Log Kp	NA			
2	des-*N*-Oxamyl Ethanimidothioic acid, 2-(methylamino)-*N*[[(methylamino) carbonyl] oxy]-2-oxo-, methyl ester, (*Z* isomer shown, no CAS No. for isomer)					
	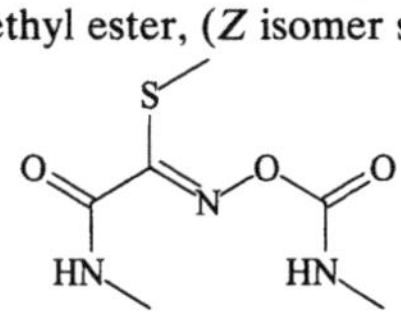	CAS No.	50917-40-3	fat	0.27	0.34
		MW, $g\,mol^{-1}$	205.24	brain	1.14	1.18
		Exp Kow @ pH 7.0	NA	rapid	0.98	0.98
		LogD @ pH 7.4	0.40	kidney	0.99	1.03
		LogP	0.40	liver	0.93	1.05
		p*K*a, MA	9.28	slow	0.94	1.00
		WS, $g\,L^{-1}$ (5.18)	19.10	skin	0.87	0.97
3	Oxamyl oxime Ethanimidothioic acid, 2-(dimethylamino)-*N*-hydroxy-2-oxo-, methyl ester; CAS No. for *Z* isomer, 66344-33-0 is shown; CAS No. for *E* isomer, 66344-32-9.					
	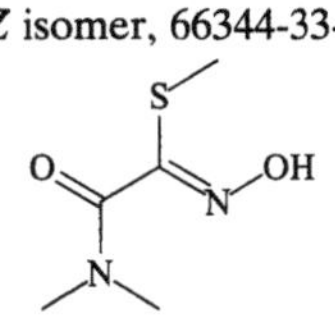	CAS No.	66344-33-0	fat	0.22	0.30
		MW, $g\,mol^{-1}$	162.21	brain	1.09	1.11
		Exp Kow @ pH 7.0	NA	rapid	0.97	0.96
		LogD @ pH 7.4	0.22	kidney	0.98	1.01
		LogP	0.34	liver	0.91	1.01
		p*K*a, MA	8.47	slow	0.93	0.98
		WS, g/L (23.97)	43.38	skin	0.84	0.94
4	Oxamyl nitrile, DMCF Carbonocyanidic amide, dimethyl-					
	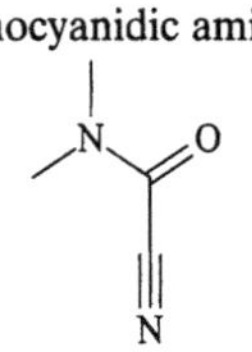	CAS No.	16703-51-8	fat	0.15	0.23
		MW, $g\,mol^{-1}$	98.10	brain	0.98	0.99
		Exp Kow @ pH 7.0	NA	rapid	0.94	0.94
		LogD @ pH 7.4	–0.93	kidney	0.94	0.97
		LogP	–0.93	liver	0.87	0.94
		p*K*a, MB	–2.15	slow	0.91	0.93
		WS, $g\,L^{-1}$ (147.30)	737.41	skin	0.79	0.89

Table 20. (*cont.*)

No.[a]	Pesticide, pesticide metabolite, and chemical structure[b]	Physical and chemical properties[c,d]		Partition coefficient[e]		
				Tissue	Rat	Human
5	des-*N*-Oxamyl oxime Ethanimidothioic acid, *N*-hydroxy-2-(methylamino)-2-oxo-, methyl ester					
	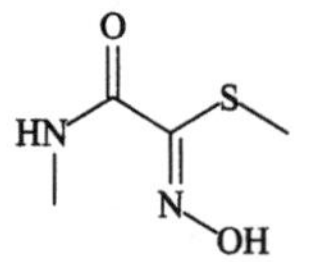	CAS No.	66157-67-3	fat	0.26	0.33
		MW, $g\,mol^{-1}$	148.18	brain	1.12	1.16
		Exp Kow @ pH 7.0	NA	rapid	0.98	0.97
		LogD @ pH 7.4	0.35	kidney	0.99	1.03
		LogP	0.38	liver	0.92	1.04
		p*K*a, MA	9.63	slow	0.94	0.99
		WS, $g\,L^{-1}$ (15.42)	38.83	skin	0.86	0.96
6	Oxamyl oxime glucuronic acid 6-({[(1*Z*)-2-(Dimethylamino)-1-(methylthio)-2-oxoethylidene]amino}oxy)-3,4,5-trihydroxytetrahydro-2*H*-pyran_2-carboxylic acid					
	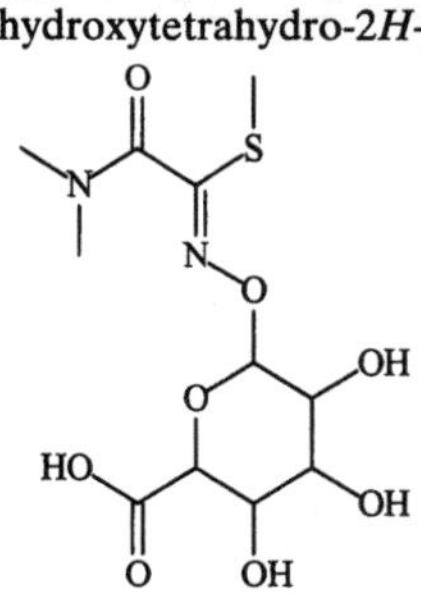	CAS No.	NA	fat	0.14	0.22
		MW, $g\,mol^{-1}$	338.33	brain	0.98	0.99
		Exp Kow @ pH 7.0	NA	rapid	0.94	0.94
		LogD @ pH 7.4	−3.03	kidney	0.94	0.97
		LogP	0.70	liver	0.86	0.94
		p*K*a, MA	2.68	slow	0.91	0.93
		WS, $g\,L^{-1}$ (1000.00)	2817.66	skin	0.79	0.88
7	Oxamyl acid (*N*,*N*-dimethyloxamic acid) Acetic acid, (dimethylamino) oxo-					
	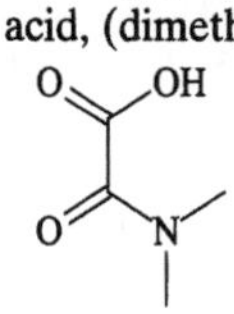	CAS No.	32833-96-8	fat	0.14	0.22
		MW, $g\,mol^{-1}$	117.10	brain	0.98	0.99
		Exp Kow @ pH 7.0	NA	rapid	0.94	0.94
		LogD @ pH 7.4	−4.98	kidney	0.94	0.97
		LogP	−1.28	liver	0.86	0.94
		p*K*a, MA	3.13	slow	0.91	0.93
		WS, $g\,L^{-1}$ (1000.0)	1840664.00	skin	0.79	0.88
8	des-*N*-Oxamyl oxime glucuronic acid 3,4,5-Trihydroxy-6-({[(1*Z*)-2-(methylamino)-1-(methylthio)-2-oxoethylidene] amino}oxy)tetrahydro-2*H*-pyran-2-carboxylic acid					
	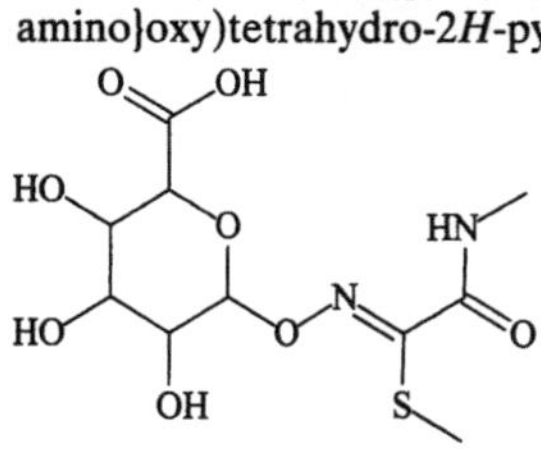	CAS No.	NA	fat	0.14	0.22
		MW, $g\,mol^{-1}$	324.31	brain	0.98	0.99
		Exp Kow @ pH 7.0	NA	rapid	0.94	0.94
		LogD @ pH 7.4	−2.99	kidney	0.94	0.97
		LogP	0.73	liver	0.86	0.94
		p*K*a, MA	2.68	slow	0.91	0.93
		WS, $g\,L^{-1}$ (1000.0)	3096.82	skin	0.79	0.88

Table 20. (*cont.*)

No.[a]	Pesticide, pesticide metabolite, and chemical structure[b]	Physical and chemical properties[c,d]		Partition coefficient[e] Tissue	Rat	Human
9	Glucuronide of *N,N*-dimethyloxamic acid 6-[2-(Dimethylamino)(oxo)acetoxy]-3,4,5-trihydroxytetrahydro-2*H*-pyran-2-carboxylic acid					
	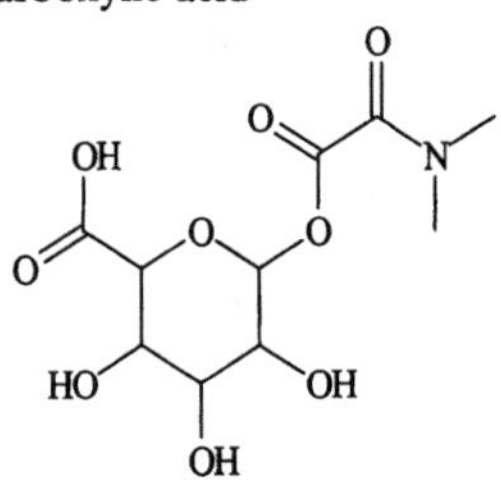	CAS No.	NA	fat	0.14	0.22
		MW, g mol^{-1}	293.23	brain	0.98	0.99
		Exp Kow @ pH 7.0	NA	rapid	0.94	0.94
		LogD @ pH 7.4	–4.65	kidney	0.94	0.97
		LogP	–0.93	liver	0.86	0.94
		p*K*a, MA	2.64	slow	0.91	0.93
		WS, g L^{-1} (1000.0)	125778.30	skin	0.79	0.88
10	des-*N*-Oxamyl acid (*N*-methyloxamic acid) Acetic acid, (methylamino) oxo-					
	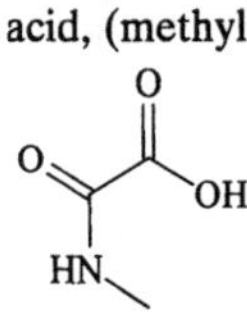	CAS No.	29262-58-6	fat	0.14	0.22
		MW, g mol^{-1}	103.08	brain	0.98	0.99
		Exp Kow @ pH 7.0	NA	rapid	0.94	0.94
		LogD @ pH 7.4	–4.99	kidney	0.94	0.97
		LogP	–1.24	liver	0.86	0.94
		p*K*a, MA	2.37	slow	0.91	0.93
		WS, g L^{-1} (1000.0)	2090361.00	skin	0.79	0.88
11	Glucuronide of *N*-methyloxamic acid 3,4,5-Trihydroxy-6-[2-(methylamino)(oxo)acetoxy]tetrahydro-2*H*-pyran-2-carboxylic acid					
	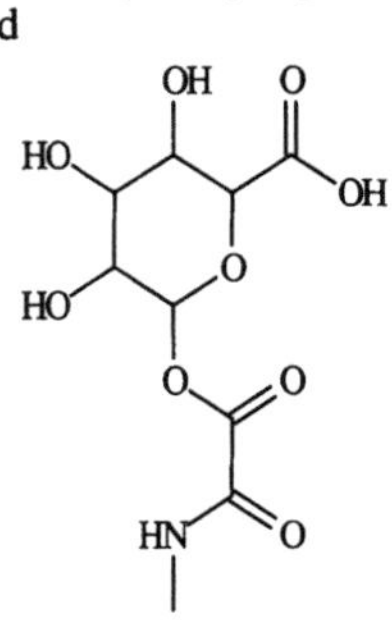	CAS No.	NA	fat	0.14	0.22
		MW, g mol^{-1}	279.20	brain	0.98	0.99
		Exp Kow @ pH 7.0	NA	rapid	0.94	0.94
		LogD @ pH 7.4	–4.61	kidney	0.94	0.97
		LogP	–0.89	liver	0.86	0.94
		p*K*a, MA	2.63	slow	0.91	0.93
		WS, g L^{-1} (1000.0)	140047.00	skin	0.79	0.88

[a]Parent chemical or metabolite number for cross-reference to Table 30 (Appendix B).
[b–e]See footnotes in Table 14 (Appendix A).

Table 21. Chemical Structure, Physical and Chemical Properties, and Tissue Partition Coefficients for Pirimicarb and the Resulting Metabolites.

No.[a]	Pesticide, pesticide metabolite, and chemical structure[b]	Physical and chemical properties[c,d]		Partition coefficient[e]		
				Tissue	Rat	Human
1	Pirimicarb Carbamic acid, dimethyl-, 2-(dimethylamino)-5,6-dimethyl-4-pyrimidinyl ester					
	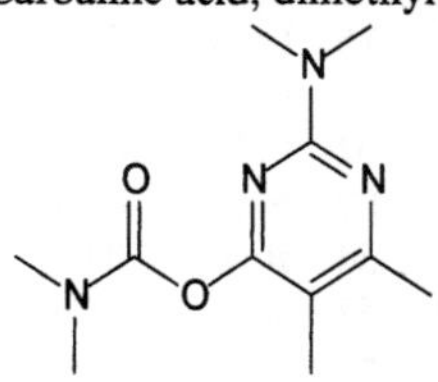	CAS No.	23103-98-2	fat	3.68	3.54
		MW, g mol^{-1}	238.29	brain	3.83	4.17
		Exp Kow @ pH 7.0	NA	rapid	1.78	1.58
		LogD @ pH 7.4	1.70	kidney	1.96	2.05
		LogP	1.70	liver	2.00	2.86
		p*K*a, MB	4.45	slow	1.49	2.04
		WS, g L^{-1} (11.24)	0.96	skin	2.28	2.29
		Kp, cm hr^{-1}	NA			
		Log Kp	NA			
2	Hydroxymethyl pirimicarb 2-(Dimethylamino)-5,6-dimethylpyrimidin-4-yl (hydroxymethyl)methylcarbamate					
	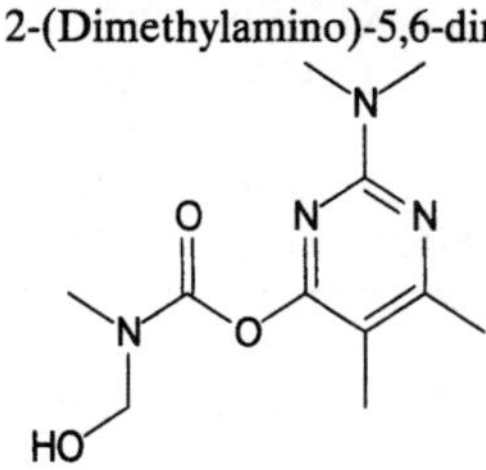	CAS No.	NA	fat	0.61	0.66
		MW, g mol^{-1}	254.29	brain	1.48	1.60
		Exp Kow @ pH 7.0	NA	rapid	1.09	1.06
		LogD @ pH 7.4	0.91	kidney	1.12	1.18
		LogP	0.91	liver	1.07	1.31
		p*K*a, MA	13.17	slow	1.01	1.15
		WS, g L^{-1} (65.65)	3.74	skin	1.05	1.15
3	Pirimicarb phenol 4(1*H*)-Pyrimidinone, 2-(dimethylamino), 5,6-dimethyl-					
	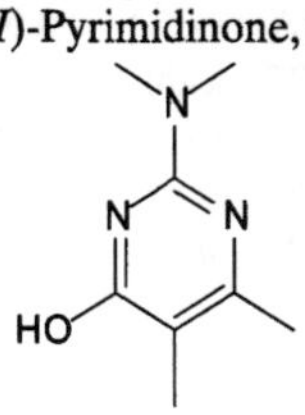	CAS No.	40778-16-3	fat	0.15	0.22
		MW, g mol^{-1}	167.21	brain	0.98	0.99
		Exp Kow @ pH 7.0	NA	rapid	0.94	0.94
		LogD @ pH 7.4	−0.99	kidney	0.94	0.97
		LogP	1.20	liver	0.87	0.94
		p*K*a, MA	7.58	slow	0.91	0.93
		WS, g L^{-1} (2.8)	444.05	skin	0.79	0.89
4	Demethyl pirimicarb phenol 4(1*H*)-Pyrimidinone, 5,6-dimethyl-2-(methylamino)-					
	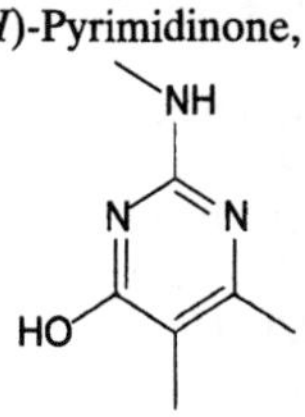	CAS No.	78195-30-9	fat	0.15	0.22
		MW, g mol^{-1}	153.18	brain	0.98	0.99
		Exp Kow @ pH 7.0	NA	rapid	0.94	0.94
		LogD @ pH 7.4	−1.42	kidney	0.94	0.97
		LogP	0.40	liver	0.86	0.94
		p*K*a, MA	7.66	slow	0.91	0.93
		WS, g L^{-1} (8.74)	1198.70	skin	0.79	0.88

Table 21. (*cont.*)

No.[a]	Pesticide, pesticide metabolite, and chemical structure[b]	Physical and chemical properties[c,d]		Partition coefficient[e] Tissue	Rat	Human
5	Amino pirimicarb phenol 4(1*H*)-Pyrimidinone, 2-amino-5,6-dimethyl-					
	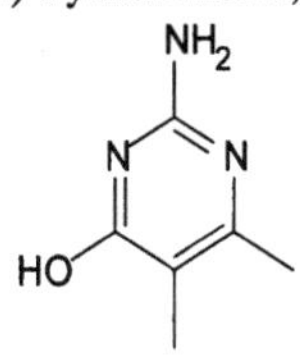	CAS No.	3977-23-9	fat	0.14	0.22
		MW, g mol^{-1}	139.16	brain	0.98	0.99
		Exp Kow @ pH 7.0	NA	rapid	0.94	0.94
		LogD @ pH 7.4	−2.02	kidney	0.94	0.97
		LogP	−0.09	liver	0.86	0.94
		p*K*a, MA	7.54	slow	0.91	0.93
		WS, g L^{-1} (16.72)	4482.52	skin	0.79	0.88
6	Hydroxymethyl demethyl pirimicarb phenol 4 (1*H*)-Pyrimidinone, 6-(hydroxymethyl)-5-methyl-2-(methylamino)-					
	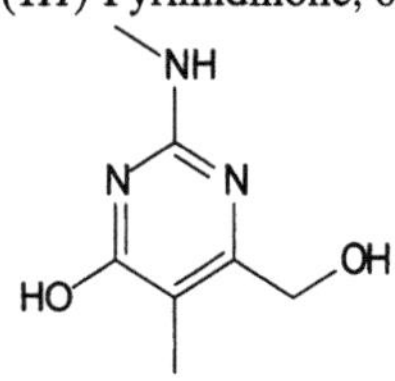	CAS No.	78195-34-3	fat	0.14	0.22
		MW, g mol^{-1}	169.18	brain	0.98	0.99
		Exp Kow @ pH 7.0	NA	rapid	0.94	0.94
		LogD @ pH 7.4	−2.31	kidney	0.94	0.97
		LogP	−1.24	liver	0.86	0.94
		p*K*a, MA	7.33	slow	0.91	0.93
		WS, g L^{-1} (46.15)	5826.93	skin	0.79	0.88
7	Pirimicarb phenyl glucuronic acid 6-{[2-(Dimethylamino)-5,6-dimethylpyrimidin-4-yl]oxy}-3,4,5-trihydroxytetrahydro-2*H*-pyran-2-carboxylic acid					
	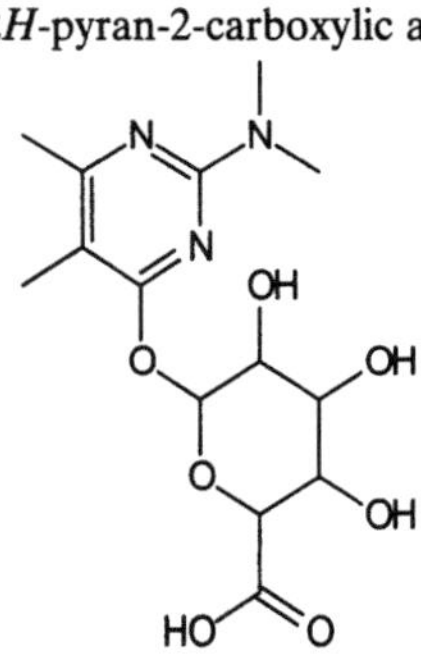	CAS No.	NA	fat	0.14	0.22
		MW, g mol^{-1}	343.34	brain	0.98	0.99
		Exp Kow @ pH 7.0	NA	rapid	0.94	0.94
		LogD @ pH 7.4	−3.77	kidney	0.94	0.97
		LogP	−0.15	liver	0.86	0.94
		p*K*a, MA	2.65	slow	0.91	0.93
		WS, g L^{-1} (16.60)	11268.07	skin	0.79	0.88
8	Hydroxymethyl pirimicarb phenol 4 (1*H*)-Pyrimidinone, 2-(dimethylamino)-6-(hydroxymethyl)-5-methyl-					
	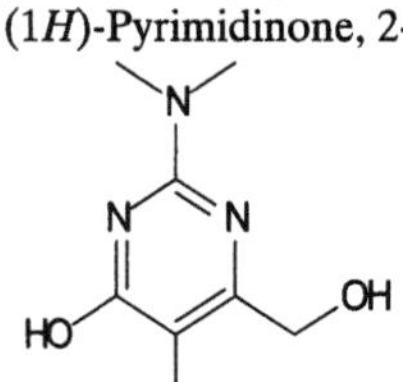	CAS No.	244247-04-9	fat	0.14	0.22
		MW, g mol^{-1}	183.21	brain	0.98	0.99
		Exp Kow @ pH 7.0	NA	rapid	0.94	0.94
		LogD @ pH 7.4	−2.0	kidney	0.94	0.97
		LogP	−0.44	liver	0.86	0.94
		p*K*a, MA	7.15	slow	0.91	0.93
		WS, g L^{-1} (14.1)	2711.32	skin	0.79	0.88

[a]Parent chemical or metabolite number for cross-reference to Table 31 (Appendix B).
[b–e]See footnotes in Table 14 (Appendix A).

Table 22. Chemical Structure, Physical and Chemical Properties, and Tissue Partition Coefficients for Propoxur and the Resulting Metabolites.

No.[a]	Pesticide, pesticide metabolite, and chemical structure[b]	Physical and chemical properties[c,d]		Partition coefficient[e]		
				Tissue	Rat	Human
1	Propoxur Phenol, 2-(1-methylethoxy)-, methylcarbamate					
	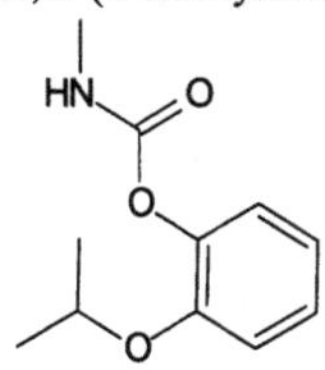	CAS No.	114-26-1	fat	2.88	2.80
		MW, g mol^{-1}	209.24	brain	3.29	3.61
		Exp Kow @ pH 7.0	NA	rapid	1.60	1.47
		LogD @ pH 7.4	1.60	kidney	1.77	1.86
		LogP	1.60	liver	1.79	2.52
		p*K*a, MA	12.28	slow	1.38	1.85
		WS, g L^{-1} (0.71)	1.69	skin	2.00	2.04
		Kp, cm hr^{-1}	NA			
		Log Kp	NA			
2	2-Isopropoxy phenol Phenol, 2-(1-methylethoxy)-					
	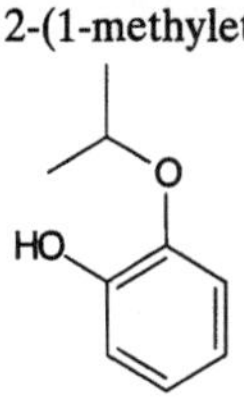	CAS No.	4812-20-8	fat	9.17	8.60
		MW, g mol^{-1}	152.19	brain	6.87	6.94
		Exp Kow @ pH 7.0	NA	rapid	2.08	2.13
		LogD @ pH 7.4	2.06	kidney	3.04	3.00
		LogP	2.07	liver	3.21	4.53
		p*K*a, MA	9.99	slow	2.12	3.01
		WS, g L^{-1} (2.77)	1.29	skin	3.86	3.51
3	2-OH Phenyl methylcarbamate 2-Hydroxy phenyl methylcarbamate					
	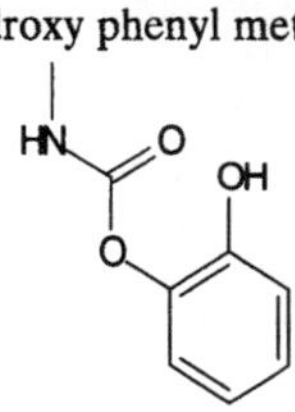	CAS No.	10309-97-4	fat	0.14	0.22
		MW, g mol^{-1}	167.16	brain	0.98	0.99
		Exp Kow @ pH 7.0	NA	rapid	0.94	0.94
		LogD @ pH 7.4	0.19	kidney	0.94	0.97
		LogP	0.22	liver	0.86	0.94
		p*K*a, MA	8.69	slow	0.91	0.93
		WS, g L^{-1} (10.69)	43.6	skin	0.79	0.88
4	5-OH Propoxur 1,3-Benzenediol, 4-(1-methylethoxy)-, 3-(methylcarbamate)					
	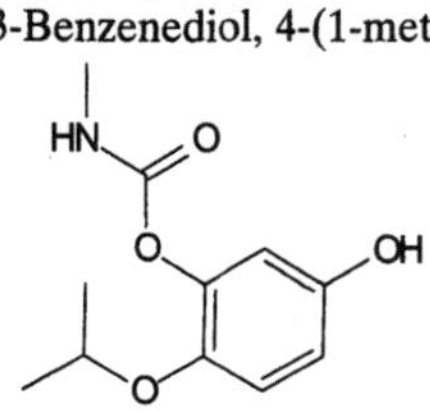	CAS No.	13200-88-9	fat	0.66	0.71
		MW, g mol^{-1}	225.24	brain	1.54	1.65
		Exp Kow @ pH 7.0	NA	rapid	1.10	1.07
		LogD @ pH 7.4	0.95	kidney	1.14	1.20
		LogP	0.95	liver	1.09	1.34
		p*K*a, MA	9.93	slow	1.02	1.17
		WS, g L^{-1} (2.02)	4.99	skin	1.08	1.18

Table 22. (*cont.*)

No.[a]	Pesticide, pesticide metabolite, and chemical structure[b]	Physical and chemical properties[c,d]		Partition coefficient[e] Tissue	Rat	Human
5	2-Isopropoxy 5-OH phenol 4-Isopropoxybenzene-1,3-diol					
	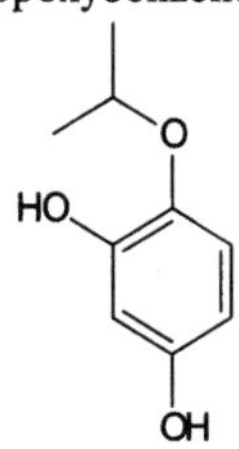	CAS No.	NA	fat	1.26	1.28
		MW, $g mol^{-1}$	168.19	brain	2.06	2.77
		Exp Kow @ pH 7.0	NA	rapid	1.26	1.20
		LogD @ pH 7.4	1.25	kidney	1.33	1.40
		LogP	1.26	liver	1.30	1.71
		p*K*a, MA	9.69	slow	1.13	1.38
		WS, $g L^{-1}$ (9.23)	5.37	skin	1.35	1.45
6	4-OH Propoxur Carbamic acid, methyl-, 4-hydroxy-2-isopropoxyphenyl ester					
	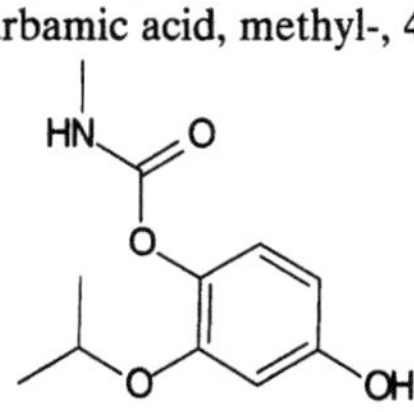	CAS No.	17595-59-4	fat	0.59	0.64
		MW, $g mol^{-1}$	225.24	brain	1.46	1.57
		Exp Kow @ pH 7.0	NA	rapid	1.08	1.05
		LogD @ pH 7.4	0.89	kidney	1.11	1.17
		LogP	0.89	liver	1.06	1.29
		p*K*a, MA	9.40	slow	1.01	1.14
		WS, $g L^{-1}$ (2.14)	5.61	skin	1.04	1.14
7	2-Isopropoxy 4-OH phenol Hydroquinone, isopropoxy-					
	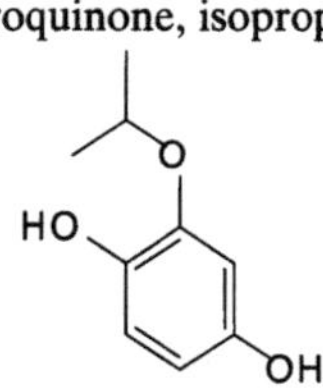	CAS No.	20042-92-6	fat	1.60	1.59
		MW, $g mol^{-1}$	168.19	brain	2.33	2.57
		Exp Kow @ pH 7.0	NA	rapid	1.34	1.26
		LogD @ pH 7.4	1.35	kidney	1.42	1.51
		LogP	1.35	liver	1.40	1.89
		p*K*a, MA	10.47	slow	1.18	1.49
		WS, $g L^{-1}$ (8.51)	4.41	skin	1.49	1.58
8	N-Hydroxymethyl propoxur Carbamic acid, (hydroxymethyl)-, 2-(1-methylethoxy) phenyl ester					
	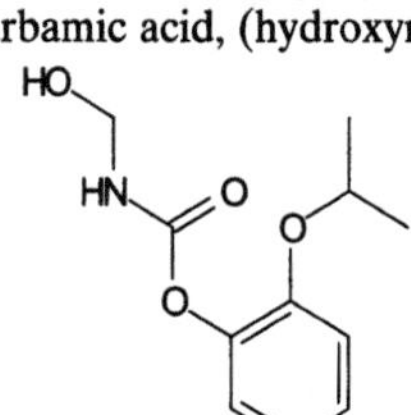	CAS No.	10310-16-4	fat	0.50	0.56
		MW, $g mol^{-1}$	225.24	brain	1.38	1.46
		Exp Kow @ pH 7.0	NA	rapid	1.05	1.03
		LogD @ pH 7.4	0.80	kidney	1.08	1.13
		LogP	0.80	liver	1.02	1.23
		p*K*a, A	13.59	slow	0.99	1.10
		WS, $g L^{-1}$ (4.22)	6.70	skin	0.99	1.10

Table 22. (*cont.*)

No.[a]	Pesticide, pesticide metabolite, and chemical structure[b]	Physical and chemical properties[c,d]		Partition coefficient[e] Tissue	Rat	Human
9	2-Isopropoxy phenyl sulfuric acid Phenol, 2-(1-methylethoxy)-, hydrogen sulfate					
	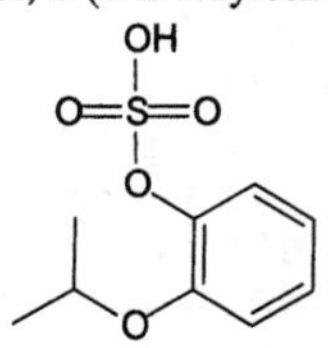	CAS No.	152242-55-2	fat	0.14	0.22
		MW, $g\,mol^{-1}$	232.25	brain	0.98	0.99
		Exp Kow @ pH 7.0	NA	rapid	0.94	0.94
		LogD @ pH 7.4	−1.75	kidney	0.94	0.97
		LogP	1.75	liver	0.86	0.94
		p*K*a, MA	−3.97	slow	0.91	0.93
		WS, $g\,L^{-1}$ (1000.0)	923.43	skin	0.79	0.88
10	2-Isopropoxy phenyl glucuronic acid β-D-glucopyranosiduronic acid, 2-(1-methylethoxy) phenyl					
	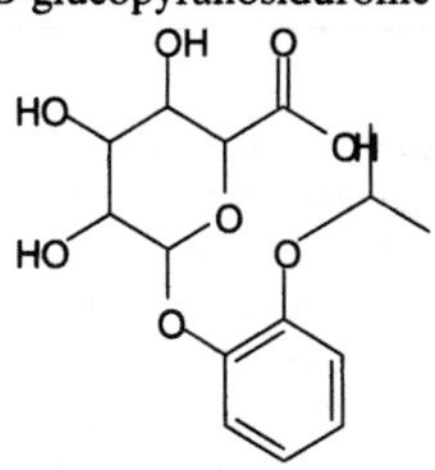	CAS No.	152242-54-1	fat	0.14	0.22
		MW, $g\,mol^{-1}$	328.31	brain	0.98	1.09
		Exp Kow @ pH 7.0	NA	rapid	0.94	0.96
		LogD @ pH 7.4	−4.10	kidney	0.94	1.00
		LogP	−0.39	liver	0.86	1.00
		p*K*a, MA	2.79	slow	0.91	0.97
		WS, $g\,L^{-1}$ (1000.0)	26524.48	skin	0.79	0.93
11	2-Hydroxyphenyl methylcarbamate sulfuric acid 2-(Sulfooxy) phenyl methylcarbamate					
	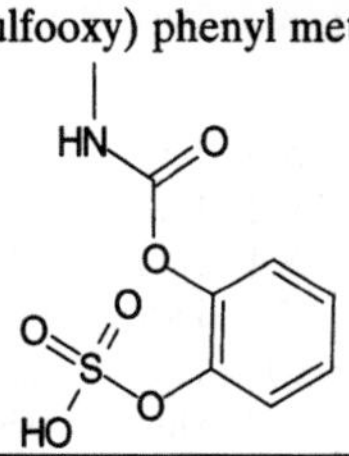	CAS No.	NA	fat	0.14	0.22
		MW, $g\,mol^{-1}$	247.23	brain	0.98	0.99
		Exp Kow @ pH 7.0	NA	rapid	0.94	0.94
		LogD @ pH 7.4	−3.6	kidney	0.94	0.97
		LogP	−0.1	liver	0.86	0.94
		p*K*a, MA	−4.25	slow	0.91	0.93
		WS, $g\,L^{-1}$ (1000.0)	29084.9	skin	0.79	0.88
12	2-Hydroxyphenyl methylcarbamate glucuronic acid 3,4,5-Trihydroxy-6-{2-[(methylcarbamoyl)oxy][phenoxy} tetrahydro-2*H*-pyran-2-carboxylic acid					
	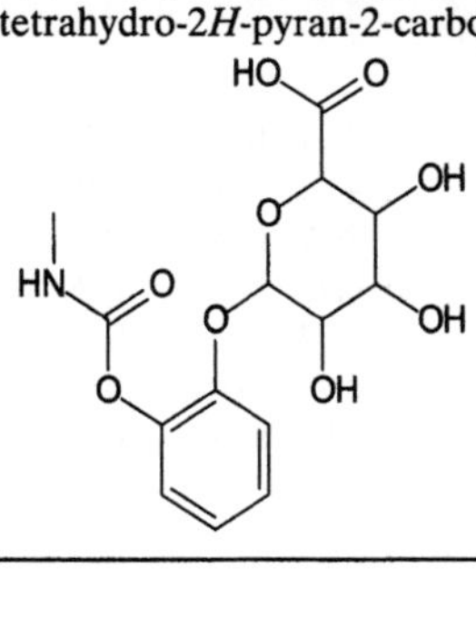	CAS No.	NA	fat	0.14	0.22
		MW, $g\,mol^{-1}$	343.29	brain	0.98	0.99
		Exp Kow @ pH 7.0	NA	rapid	0.94	0.94
		LogD @ pH 7.4	−5.95	kidney	0.94	0.97
		LogP	−2.23	liver	0.86	0.94
		p*K*a, MA	2.77	slow	0.91	0.93
		WS, $g\,L^{-1}$ (1000.0)	819992.6	skin	0.79	0.88

Table 22. (*cont.*)

No.[a]	Pesticide, pesticide metabolite, and chemical structure[b]	Physical and chemical properties[c,d]		Partition coefficient[e] Tissue	Rat	Human
13	5-OH Propoxur sulfuric acid					
	2-Isopropoxy-5-(sulfooxy) phenyl methylcarbamate					
	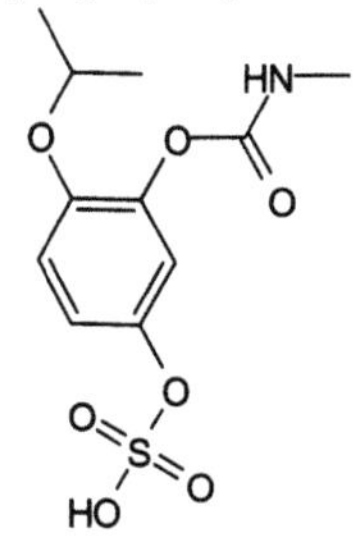	CAS No.	NA	fat	0.14	0.22
		MW, g mol^{-1}	305.30	brain	0.98	0.99
		Exp Kow @ pH 7.0	NA	rapid	0.94	0.94
		LogD @ pH 7.4	−2.73	kidney	0.94	0.97
		LogP	0.77	liver	0.86	0.94
		p*K*a, MA	−4.09	slow	0.91	0.93
		WS, g L^{-1} (1000.0)	2452.24	skin	0.79	0.88
14	5-OH Propoxur glucuronic acid					
	4-Isopropoxy-3-[(methylcarbamoyl)oxy]phenyl hexopyranosiduronic acid					
	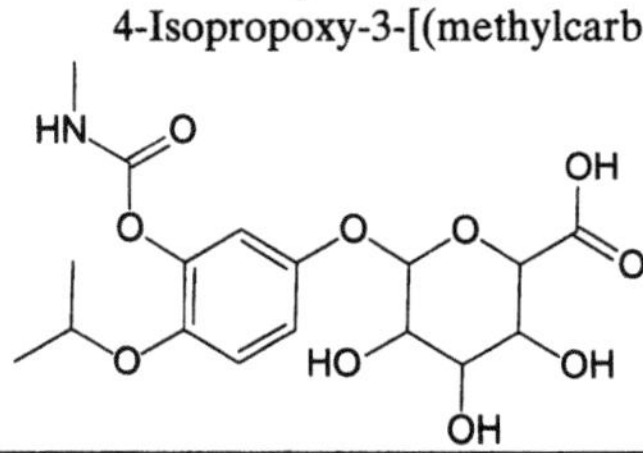	CAS No.	NA	fat	0.14	0.22
		MW, g mol^{-1}	401.37	brain	0.98	0.99
		Exp Kow @ pH 7.0	NA	rapid	0.94	0.94
		LogD @ pH 7.4	−5.08	kidney	0.94	0.97
		LogP	−1.37	liver	0.86	0.94
		p*K*a, MA	2.77	slow	0.91	0.93
		WS, g L^{-1} (1000.0)	65419.22	skin	0.79	0.88
15	2-Isopropoxy 5-OH phenyl sulfuric acid					
	5-Hydroxy-2-isopropoxyphenyl hydrogen sulfate					
	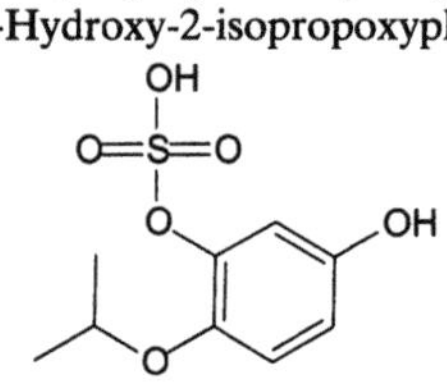	CAS No.	NA	fat	0.14	0.22
		MW, g mol^{-1}	248.25	brain	0.98	0.99
		Exp Kow @ pH 7.0	NA	rapid	0.94	0.94
		LogD @ pH 7.4	−2.4	kidney	0.94	0.97
		LogP	1.10	liver	0.86	0.94
		p*K*a, MA	−4.21	slow	0.91	0.93
		WS, g L^{-1} (1000.0)	2709.80	skin	0.79	0.88
16	2-Isopropoxy 5-OH phenyl glucuronic acid					
	3,5,5-Trihydroxy-6-(5-hydroxy-2-isopropoxyphenoxy) tetrahydro-2*H*-pyran-2-carboxylic acid					
	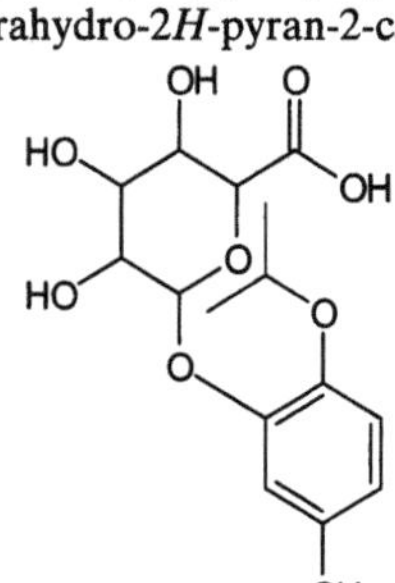	CAS No.	NA	fat	0.14	0.22
		MW, g mol^{-1}	344.32	brain	0.98	0.99
		Exp Kow @ pH 7.0	NA	rapid	0.94	0.94
		LogD @ pH 7.4	−4.75	kidney	0.94	0.97
		LogP	−1.03	liver	0.86	0.94
		p*K*a, MA	2.76	slow	0.91	0.93
		WS, g L^{-1} (1000.0)	76371.80	skin	0.79	0.88

Table 22. (*cont.*)

No.[a]	Pesticide, pesticide metabolite, and chemical structure[b]	Physical and chemical properties[c,d]		Partition coefficient[e]		
				Tissue	Rat	Human
17	4-OH Propoxur sulfuric acid 2-Isopropoxy-4-(sulfooxy) phenyl methylcarbamate					
	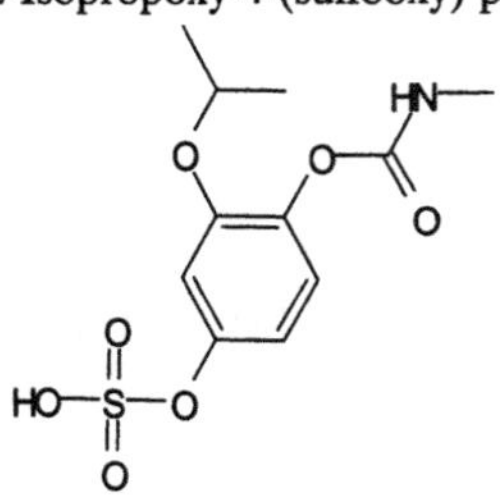	CAS No.	NA	fat	0.14	0.22
		MW, g mol^{-1}	305.31	brain	0.98	0.99
		Exp Kow @ pH 7.0	NA	rapid	0.94	0.94
		LogD @ pH 7.4	−2.73	kidney	0.94	0.97
		LogP	0.77	liver	0.86	0.94
		p*K*a, MA	−4.24	slow	0.91	0.93
		WS, g L^{-1} (1000.0)	2452.24	skin	0.79	0.88
18	4-OH propoxur glucuronic acid 2,3,4-Trihydroxy-5-{3-isopropoxy-4-[(methylcarbamoyl)oxy]phenoxy} cyclohexanecarboxylic acid					
	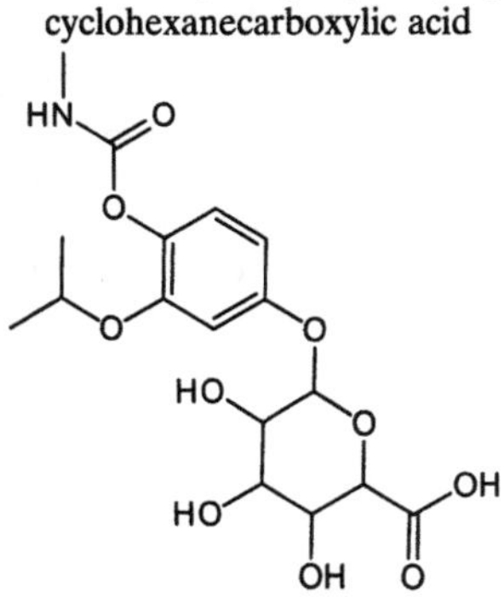	CAS No.	NA	fat	0.14	0.22
		MW, g mol^{-1}	401.37	brain	0.98	0.99
		Exp Kow @ pH 7.0	NA	rapid	0.94	0.94
		LogD @ pH 7.4	−5.08	kidney	0.94	0.97
		LogP	−1.37	liver	0.86	0.94
		p*K*a, MA	2.77	slow	0.91	0.93
		WS, g L^{-1} (1000.0)	65419.22	skin	0.79	0.88
19	2-Isopropoxy 4-OH phenyl sulfuric acid 4-Hydroxy-2-isopropoxyphenyl hydrogen sulfate					
	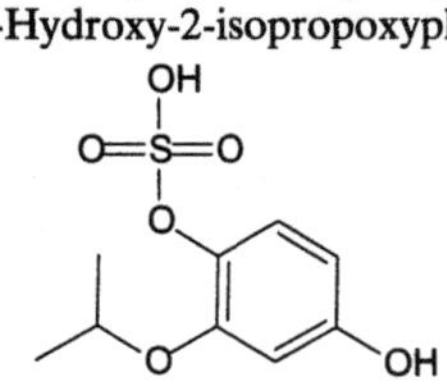	CAS No.	NA	fat	0.14	0.22
		MW, g mol^{-1}	248.25	brain	0.98	0.99
		Exp Kow @ pH 7.0	NA	rapid	0.94	0.94
		LogD @ pH 7.4	−2.47	kidney	0.94	0.97
		LogP	1.04	liver	0.86	0.94
		p*K*a, MA	−4.24	slow	0.91	0.93
		WS, g L^{-1} (1000.0)	3109.80	skin	0.79	0.88

Table 22. (*cont.*)

No.[a]	Pesticide, pesticide metabolite, and chemical structure[b]	Physical and chemical properties[c,d]		Partition coefficient[e] Tissue	Rat	Human
20	2-Isopropoxy 4-OH phenyl glucuronic acid 3,4,5-Trihydroxy-6-(4-hydroxy-2-isopropoxyphenoxy) tetrahydro-2*H*-pyran-2-carboxylic acid					
	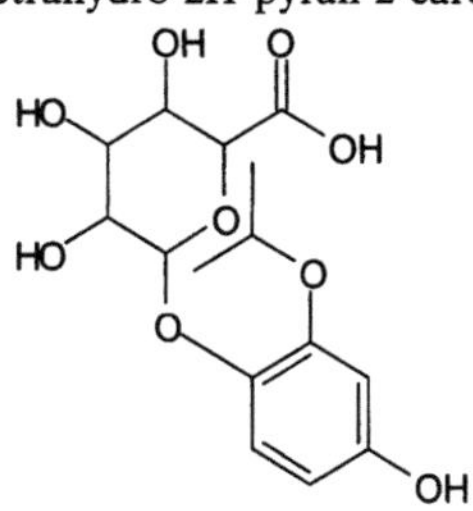	CAS No.	NA	fat	0.14	0.22
		MW, $g\,mol^{-1}$	344.32	brain	0.98	0.99
		Exp Kow @ pH 7.0	NA	rapid	0.94	0.94
		LogD @ pH 7.4	−4.81	kidney	0.94	0.97
		LogP	−1.09	liver	0.86	0.94
		p*K*a, MA	2.78	slow	0.91	0.93
		WS, $g\,L^{-1}$ (1000.0)	85935.58	skin	0.79	0.88
21	N-Hydroxymethyl propoxur sulfuric acid 2-Isopropoxyphenyl [(sulfooxy)methyl]carbamate					
	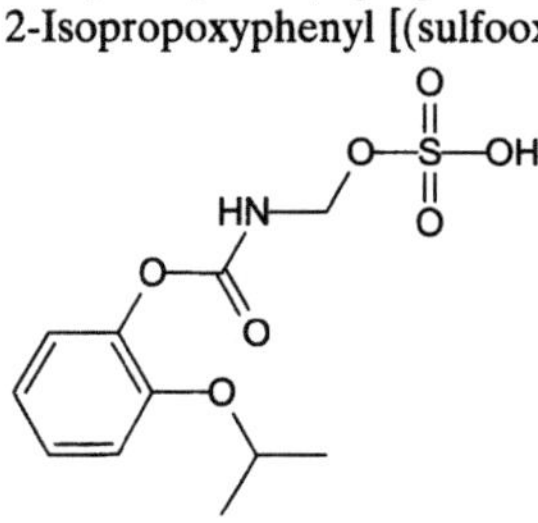	CAS No.	NA	fat	0.14	0.22
		MW, $g\,mol^{-1}$	305.31	brain	0.98	0.99
		Exp Kow @ pH 7.0	NA	rapid	0.94	0.94
		LogD @ pH 7.4	−2.43	kidney	0.94	0.97
		LogP	1.07	liver	0.86	0.94
		p*K*a, MA	−3.88	slow	0.91	0.93
		WS, $g\,L^{-1}$ (1000.0)	1359.5	skin	0.79	0.88
22	N-Hydroxymethyl propoxur glucuronic acid {[(2-Isopropoxyphenoxy)carbonyl]amino}methyl hexopyranosiduronic acid					
	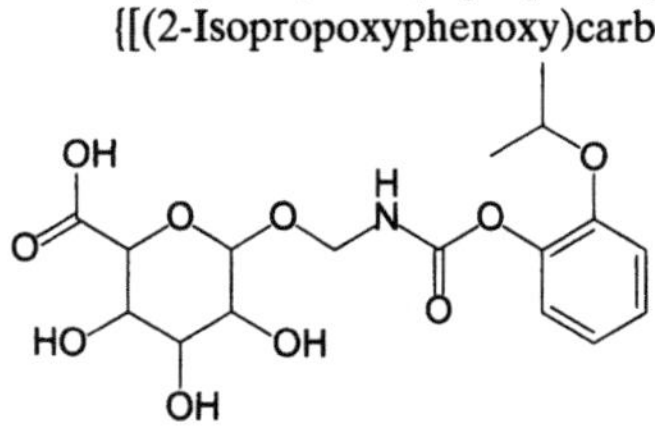	CAS No.	NA	fat	0.14	0.22
		MW, $g\,mol^{-1}$	401.37	brain	0.98	0.99
		Exp Kow @ pH 7.0	NA	rapid	0.94	0.94
		LogD @ pH 7.4	−4.93	kidney	0.94	0.97
		LogP	−1.22	liver	0.86	0.94
		p*K*a, MA	2.84	slow	0.91	0.93
		WS, $g\,L^{-1}$ (1000.0)	48729.4	skin	0.79	0.88

[a]Parent chemical or metabolite number for cross-reference to Table 32 (Appendix B).
[b–e]See footnotes in Table 14 (Appendix A).

Table 23. Chemical Structure, Physical and Chemical Properties, and Tissue Partition Coefficients for Thiodicarb and the Resulting Metabolites.

No.[a]	Pesticide, pesticide metabolite, and chemical structure[b]	Physical and chemical properties[c,d]		Partition coefficient[e]		
				Tissue	Rat	Human
1	Thiodicarb Ethanimidothioic acid, *N-N'*-[thiobis[(methylimino) carbonyloxy]] bis-, dimethyl ester					
	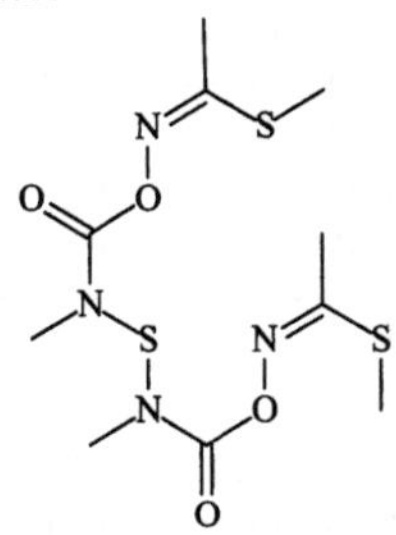	CAS No.	59669-26-0	fat	2.38	2.33
		MW, $g\,mol^{-1}$	354.47	brain	2.93	3.23
		Exp Kow @ pH 7.0	NA	rapid	1.51	1.39
		LogD @ pH 7.4	1.52	kidney	1.64	1.73
		LogP	1.52	liver	1.64	2.29
		p*K*a, MB	−2.09	slow	1.31	1.72
		WS, $g\,L^{-1}$ (0.19)	0.29	skin	1.81	1.88
		Kp, $cm\,hr^{-1}$	NA			
		Log Kp	NA			
2	*syn*-(*Z*)-Methomyl Ethanimidothioic acid, *N*-[[(methylamino)carbonyl]oxy]- methyl ester, (*Z*)-					
	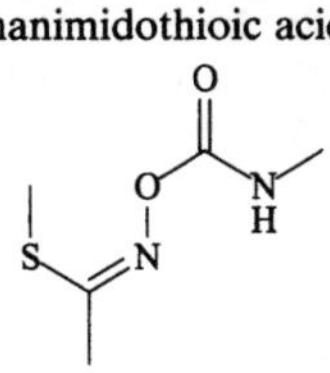	CAS No.	19928-37-1	fat	0.36	0.42
		MW, $g\,mol^{-1}$	162.21	brain	1.23	1.29
		Exp Kow @ pH 7.0	NA	rapid	1.01	1.00
		LogD @ pH 7.4	0.60	kidney	1.03	1.07
		LogP	0.60	liver	0.96	1.12
		p*K*a, MA	13.27	slow	0.96	1.04
		WS, $g\,L^{-1}$ (11.38)	20.55	skin	0.92	1.02
3	*anti*-(*E*)-Methomyl Ethanimidothioic acid, *N*-[[(methylamino)carbonyl]oxy]- methyl ester, (*E*)-					
	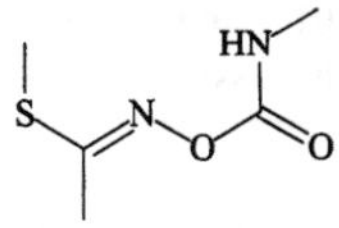	CAS No.	19928-35-9	fat	0.36	0.42
		MW, $g\,mol^{-1}$	162.21	brain	1.23	1.29
		Exp Kow @ pH 7.0	NA	rapid	1.01	1.00
		LogD @ pH 7.4	0.60	kidney	1.03	1.07
		LogP	0.60	liver	0.96	1.12
		p*K*a, MA	13.27	slow	0.96	1.04
		WS, $g\,L^{-1}$ (11.38)	20.55	skin	0.92	1.02
4	*syn*-(*Z*)-Methomyl oxime Ethanimidothioic acid, *N*-hydroxy-, methyl ester, (*Z*)-					
	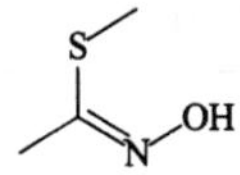	CAS No.	19125-12-3	fat	0.95	0.98
		MW, $g\,mol^{-1}$	105.16	brain	1.79	1.96
		Exp Kow @ pH 7.0	NA	rapid	1.18	1.13
		LogD @ pH 7.4	1.12	kidney	1.23	1.30
		LogP	1.12	liver	1.19	1.52
		p*K*a, MA	11.26	slow	1.07	1.27
		WS, $g\,L^{-1}$ (16.69)	12.47	skin	1.21	1.31

Table 23. (*cont.*)

No.[a]	Pesticide, pesticide metabolite, and chemical structure[b]	Physical and chemical properties[c,d]		Partition coefficient[e] Tissue	Rat	Human
5	*anti*-(*E*)-Methomyl oxime Ethanimidothioic acid, N-hydroxy-, methyl ester, (E)-					
	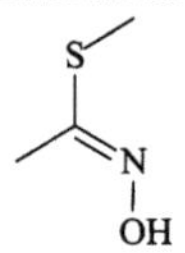	CAS No.	19145-16-5	fat	0.95	0.98
		MW, $g\,mol^{-1}$	105.16	brain	1.79	1.96
		Exp Kow @ pH 7.0	NA	rapid	1.18	1.13
		LogD @ pH 7.4	1.12	kidney	1.23	1.30
		LogP	1.12	liver	1.19	1.52
		p*K*a, MA	11.26	slow	1.07	1.27
		WS, $g\,L^{-1}$ (16.69)	12.47	skin	1.21	1.31
6	*S*-Methyl, *N*-methyl carbamic acid Carbamothioic acid, methyl-, *S*-methyl ester					
	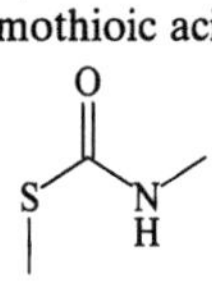	CAS No.	22013-97-4	fat	0.24	0.31
		MW, $g\,mol^{-1}$	105.16	brain	1.11	1.14
		Exp Kow @ pH 7.0	NA	rapid	0.97	0.97
		LogD @ pH 7.4	0.30	kidney	0.99	1.02
		LogP	0.30	liver	0.91	1.03
		p*K*a, MA	12.77	slow	0.93	0.99
		WS, $g\,L^{-1}$ (41.17)	62.53	skin	0.85	0.95
7	Ethylium Ethylium, 1-[(methylthio) imino]-					
	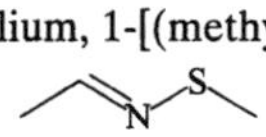	CAS No.	61599-22-2	fat	1.13	1.15
		MW, $g\,mol^{-1}$	89.16	brain	1.95	2.14
		Exp Kow @ pH 7.0	NA	rapid	1.22	1.17
		LogD @ pH 7.4	1.20	kidney	1.29	1.36
		LogP	1.20	liver	1.25	1.63
		p*K*a, MB	4.01	slow	1.11	1.33
		WS, $g\,L^{-1}$ (9.58)	11.81	skin	1.29	1.39
8	Carbon dioxide					
	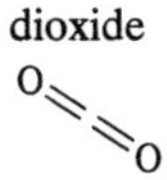	CAS No.	124-38-9	fat	0.53	0.58
		MW, $g\,mol^{-1}$	44.01	brain	1.40	1.50
		Exp Kow @ pH 7.0	NA	rapid	1.06	1.04
		LogD @ pH 7.4	0.83	kidney	1.09	1.14
		LogP	0.83	liver	1.03	1.25
		p*K*a	NA	slow	0.99	1.11
		WS, $g\,L^{-1}$ (358.82)	25.72	skin	1.01	1.11

Table 23. (*cont.*)

No.[a]	Pesticide, pesticide metabolite, and chemical structure[b]	Physical and chemical properties[c,d]		Partition coefficient[e]		
				Tissue	Rat	Human
9	Acetonitrile					
	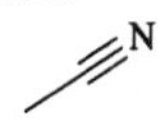	CAS No.	75-05-8	fat	0.16	0.23
		MW, $g\,mol^{-1}$	41.05	brain	1.00	1.01
		Exp Kow @ pH 7.0	NA	rapid	0.94	0.94
		LogD @ pH 7.4	–0.45	kidney	0.95	0.98
		LogP	–0.45	liver	0.87	0.95
		p*K*a	NA	slow	0.91	0.94
		WS, $g\,L^{-1}$ (122.06)	312.45	skin	0.80	0.90
10	Hydrogen cyanide					
	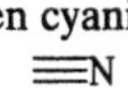	CAS No.	74-90-8	fat	0.15	0.23
		MW, $g\,mol^{-1}$	27.03	brain	1.02	1.03
		Exp Kow @ pH 7.0	NA	rapid	0.95	0.95
		LogD @ pH 7.4	–0.25	kidney	0.95	0.98
		LogP	–0.25	liver	0.88	0.96
		p*K*a	NA	slow	0.91	0.95
		WS, $g\,L^{-1}$ (89.68)	175.60	skin	0.81	0.90
11	Glutathione conjugate of acetonitrile 2-Amino-5-({2-[(carboxylmethyl)amino]-1-[(ethanimidoylthio)methyl]-2-oxoethyl}amino)-5-oxopenanoic acid					
	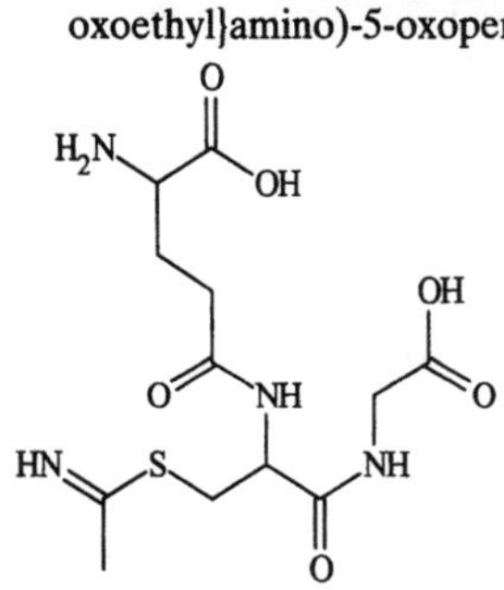	CAS No.	NA	fat	0.14	0.22
		MW, $g\,mol^{-1}$	348.38	brain	0.98	0.99
		Exp Kow @ pH 7.0	NA	rapid	0.94	0.94
		LogD @ pH 7.4	–3.98	kidney	0.94	0.97
		LogP	–1.11	liver	0.86	0.94
		p*K*a, MA, A	3.57	slow	0.91	0.93
			3.94			
		WS, $g\,L^{-1}$ (3.56)	15877.38	skin	0.80	0.88
12	Glutamylcysteine conjugate of acetonitrile 2-Amino-5-{[1-carboxy-2-(ethanimidoylthio)ethyl}-5-oxopentanoic acid					
	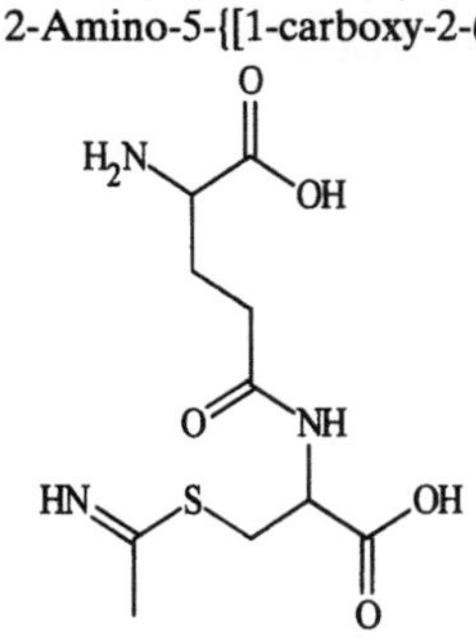	CAS No.	NA	fat	0.14	0.22
		MW, $g\,mol^{-1}$	291.33	brain	0.98	0.99
		Exp Kow @ pH 7.0	NA	rapid	0.94	0.94
		LogD @ pH 7.4	–3.67	kidney	0.94	0.97
		LogP	–0.92	liver	0.86	0.94
		p*K*a, MA, A	3.15	slow	0.91	0.93
			3.94			
		WS, $g\,L^{-1}$ (4.3)	18780.45	skin	0.79	0.88

Table 23. (*cont.*)

No.[a]	Pesticide, pesticide metabolite, and chemical structure[b]	Physical and chemical properties[c,d]		Partition coefficient[e] Tissue	Rat	Human
13	Cysteinylglycine conjugate of acetonitrile {[2-Amino-3-(ethanimidoylthio)propanoyl]amino}acetic acid					
	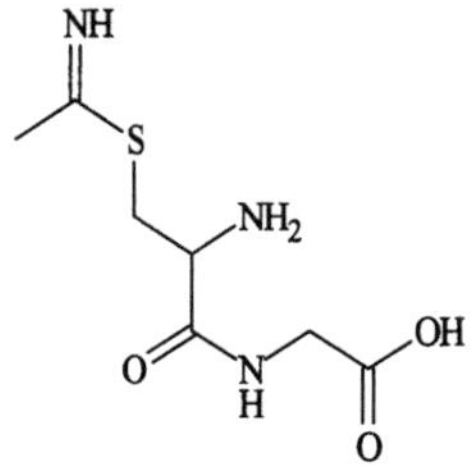	CAS No.	NA	fat	0.14	0.22
		MW, $g mol^{-1}$	219.26	brain	0.98	0.99
		Exp Kow @ pH 7.0	NA	rapid	0.94	0.94
		LogD @ pH 7.4	−3.62	kidney	0.94	0.97
		LogP	−0.85	liver	0.86	0.94
		p*K*a, MA	3.61	slow	0.91	0.93
		WS, $g L^{-1}$ (22.28)	42875.65	skin	0.80	0.88
14	Cysteine conjugate of acetonitrile 2-Amino-3-(ethanimidoylthio)propanoic acid					
	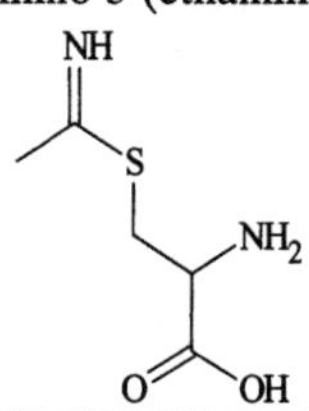	CAS No.	NA	fat	0.14	0.22
		MW, $g mol^{-1}$	162.21	brain	0.98	0.99
		Exp Kow @ pH 7.0	NA	rapid	0.94	0.94
		LogD @ pH 7.4	−2.67	kidney	0.94	0.97
		LogP	−0.025	liver	0.86	0.94
		p*K*a, MA	3.54	slow	0.91	0.93
		WS, $g L^{-1}$ (25.95)	12745.72	skin	0.80	0.88
15	*N*-Sulfate-cysteine conjugate of acetonitrile 3-Mercapto-2-{[(1*Z*)-*N*-(sulfinooxy)ethanimidoyl]amino}propanoic acid					
	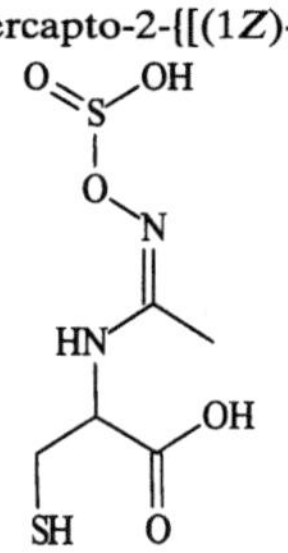	CAS No.	NA	fat	0.14	0.22
		MW, $g mol^{-1}$	242.27	brain	0.98	0.99
		Exp Kow @ pH 7.0	NA	rapid	0.94	0.94
		LogD @ pH 7.4	−4.73	kidney	0.94	0.97
		LogP	−0.083	liver	0.86	0.94
		p*K*a, MA, A	2.15 3.24	slow	0.91	0.93
		WS, $g L^{-1}$ (1000.0)	285646.80	skin	0.80	0.88

[a]Parent chemical or metabolite number for cross-reference to Table 33 (Appendix B).
[b–e]See footnotes in Table 14 (Appendix A).

Appendix B. Metabolic Pathways and Preliminary Metabolic Rate Constants (V_{max}, K_m) for the Metabolism of Parent Carbamates and Metabolites.

Table 24. Biotransformation and Elimination Paths of Aldicarb and the Resulting Metabolites and Preliminary Liver V_{max} and K_m Values.[a]

No.[b]	Initiating pesticide/metabolite[c]	Resulting metabolite
1	Aldicarb	
		Aldicarb oxime
		Aldicarb sulfoxide
		Aldicarb nitrile
		Aldicarb eliminated
2	Aldicarb sulfoxide	
		Aldicarb oxime sulfoxide
		Aldicarb sulfone
		Aldicarb nitrile sulfoxide
		Aldicarb sulfoxide eliminated
3	Aldicarb sulfone	
		Aldicarb oxime sulfone
		Aldicarb nitrile sulfone
		Aldicarb sulfone eliminated
4	Aldicarb oxime	
		Aldicarb aldehyde
		Aldicarb oxime eliminated
5	Aldicarb oxime sulfoxide	
		Aldicarb aldehyde sulfoxide
		Aldicarb oxime sulfoxide eliminated
6	Aldicarb oxime sulfone	
		Aldicarb aldehyde sulfone
		Aldicarb oxime sulfone eliminated
7	Aldicarb nitrile	
		Aldicarb nitrile eliminated
8	Aldicarb nitrile sulfoxide	
		Aldicarb nitrile sulfoxide eliminated
9	Aldicarb nitrile sulfone	
		Aldicarb nitrile sulfone eliminated

Table 24. (*cont.*)

No.[b]	Initiating pesticide/metabolite[c]	Resulting metabolite
10	Aldicarb aldehyde	
		Aldicarb alcohol
		Aldicarb acid
11	Aldicarb aldehyde sulfoxide	
		Aldicarb alcohol sulfoxide
		Aldicarb acid sulfoxide
12	Aldicarb aldehyde sulfone	
		Aldicarb alcohol sulfone
		Aldicarb acid sulfone
13	Aldicarb alcohol	
		Aldicarb alcohol eliminated
14	Aldicarb alcohol sulfoxide	
		Aldicarb alcohol sulfoxide eliminated
15	Aldicarb alcohol sulfone	
		Aldicarb alcohol sulfone eliminated
16	Aldicarb acid	
		Aldicarb acid eliminated
17	Aldicarb acid sulfoxide	
		Aldicarb acid sulfoxide eliminated
18	Aldicarb acid sulfone	
		Aldicarb acid sulfone eliminated

[a]Preliminary estimates of liver V_{max} (μmol $hr^{-1}kg^{-1}$ bw, unscaled) and K_m (μM) are equal to 10.0.

[b]Parent chemical or metabolite number for cross-reference in Table 14 (Appendix A).

[c]Refer to Table 14 for structure based on No.

Table 25. Biotransformation and Elimination Paths of Carbaryl and the Resulting Metabolites and Preliminary Liver V_{max} and K_m Values[a].

No.[b]	Initiating pesticide/metabolite[c]	Resulting metabolite
1	Carbaryl	
		Naphthol
		4-OH Carbaryl
		5-OH Carbaryl
		3,4-Dihydrodihydroxy carbaryl
		5,6-Dihydro-dihydroxy carbaryl
		Hydroxymethyl carbaryl
2	4-OH carbaryl	
		4-OH Naphthol
		4-OH Carbaryl sulfuric acid
		4-OH Carbaryl glucuronic acid
		4-OH Carbaryl eliminated
3	5-OH carbaryl	
		5-OH Naphthol
		5-OH Carbaryl sulfuric acid
		5-OH Carbaryl glucuronic acid
		5-OH Carbaryl eliminated
4	3,4-Dihydrodihydroxy carbaryl	
		3,4-Dihydrodihydroxy naphthol
		3,4-Dihydrodihydroxy carbaryl sulfuric acid
		3,4-Dihydrodihydroxy carbaryl glucuronic acid
		3,4-Dihydrodihydroxy carbaryl eliminated
5	5,6-Dihydrodihydroxy carbaryl	
		5,6-Dihydrodihydroxy naphthol
		5,6-Dihydrodihydroxy carbaryl sulfuric acid
		5,6-Dihydrodihydroxy carbaryl glucuronic acid
		5,6-Dihydrodihydroxy carbaryl eliminated
6	Hydroxymethyl carbaryl	
		1-Naphthol
		Hydroxymethyl carbaryl sulfuric acid
		Hydroxymethyl carbaryl glucuronic acid
		Hydroxymethyl carbaryl eliminated

Table 25. (*cont.*)

No.[b]	Initiating pesticide/metabolite[c]	Resulting metabolite
7	Naphthol	
		2-OH Naphthol
		4-OH Naphthol
		5-OH Naphthol
		3,4-Dihydrodihydroxy naphthol
		5,6-Dihydrodihydroxy naphthol
		Naphthyl sulfuric acid
		Naphthyl glucuronic acid
		Naphthol eliminated
8	1-Hydroxy-2-naphthol	
		2-OH Naphthyl sulfuric acid
		2-OH Naphthyl glucuronic acid
		2-OH Naphthol eliminated
9	4-OH naphthol	
		4-OH Naphthyl sulfuric acid
		4-OH Naphthyl glucuronic acid
		4-OH Naphthol eliminated
10	5-OH naphthol	
		5-OH Naphthyl sulfuric acid
		5-OH Naphthyl glucuronic acid
		5-OH Naphthol eliminated
11	3,4-Dihydrodihydroxy naphthol	
		3,4-Dihydrodihydroxy naphthyl sulfuric acid
		3,4-Dihydrodihydroxy naphthyl glucuronic acid
		3,4-Dihydrodihydroxy naphthol eliminated
12	5,6-Dihydrodihydroxy naphthol	
		5,6-Dihydrodihydroxy naphthyl sulfuric acid
		5,6-Dihydrodihydroxy naphthyl glucuronic acid
		5,6-Dihydrodihydroxy naphthol eliminated
13	Naphthyl sulfuric acid	
		Naphthyl sulfuric acid eliminated
14	Naphthyl glucuronic acid	
		Naphthyl glucuronic acid eliminated

Table 25. (*cont.*)

No.[b]	Initiating pesticide/metabolite[c]	Resulting metabolite
15	2-OH naphthyl sulfuric acid	2-OH naphthyl sulfuric acid eliminated
16	2-OH naphthyl glucuronic acid	2-OH naphthyl glucuronic acid eliminated
17	4-OH naphthyl sulfuric acid	4-OH naphthyl sulfuric acid eliminated
18	4-OH naphthyl glucuronic acid	4-OH naphthyl glucuronic acid eliminated
19	5-OH naphthyl sulfuric acid	5-OH naphthyl sulfuric acid eliminated
20	5-OH naphthyl glucuronic acid	5-OH naphthyl glucuronic acid eliminated
21	3,4-Dihydrodihydroxy naphthyl sulfuric acid	3,4-Dihydrodihydroxy naphthyl sulfuric acid eliminated
22	3,4-Dihydrodihydroxy naphthyl glucuronic acid	3,4-Dihydrodihydroxy naphthyl glucuronic acid eliminated
23	5,6-Dihydrodihydroxy naphthyl sulfuric acid	5,6-Dihydrodihydroxy naphthyl sulfuric acid eliminated
24	5,6-Dihydrodihydroxy naphthyl glucuronic acid	5,6-Dihyrodihydroxy naphthyl glucuronic acid eliminated
25	4-OH carbaryl sulfuric acid	4-OH carbaryl sulfuric acid eliminated
26	4-OH carbaryl glucuronic acid	4-OH carbaryl glucuronic acid eliminated

Table 25. (*cont.*)

No.[b]	Initiating pesticide/metabolite[c]	Resulting metabolite
27	5-OH carbaryl sulfuric acid	5-OH carbaryl sulfuric acid eliminated
28	5-OH carbaryl glucuronic acid	5-OH carbaryl glucuronic acid eliminated
29	3,4-Dihydrodihydroxy carbaryl sulfuric acid	3,4-Dihydrodihydroxy carbaryl sulfuric acid eliminated
30	3,4-Dihydrodihydroxy carbaryl glucuronic acid	3,4-Dihydrodihydroxy carbaryl glucuronic acid eliminated
31	5,6-Dihydrodihydroxy carbaryl sulfuric acid	5,6-Dihydrodihydroxy carbaryl sulfuric acid eliminated
32	5,6-Dihydrodihydroxy carbaryl glucuronic acid	5,6-Dihydrodihydroxy carbaryl glucuronic acid eliminated
33	Hydroxymethyl carbaryl sulfuric acid	Hydroxymethyl carbaryl sulfuric acid eliminated
34	Hydroxymethyl carbaryl glucuronic acid	Hydroxymethyl carbaryl glucuronic acid eliminated

[a]Preliminary estimates of liver V_{max} (µmol $hr^{-1}kg^{-1}$ bw, unscaled) and K_m (µM) are equal to 10.0. Additional published values can be found in Table 34 (Appendix B).

[b]Parent chemical or metabolite number for cross-reference in Table 15 (Appendix A).

[c]Refer to Table 15 for structure based on No.

Table 26. Biotransformation and Elimination Paths of Carbofuran and the Resulting Metabolites and Preliminary Liver V_{max} and K_m Values[a].

No.[b]	Initiating pesticide/metabolite[c]	Resulting metabolite
1	Carbofuran	
		Carbofuran-7-phenol
		3-OH carbofuran
		Hydroxymethyl carbofuran
		Carbofuran eliminated
2	Carbofuran 7-phenol	
		3,7-Diol
		7-Phenyl sulfuric acid
		7-Phenyl glucuronic acid
		7-Phenol eliminated
3	3-OH carbofuran	
		3,7-Diol
		3-Keto carbofuran
		3-OH carbofuran sulfuric acid
		3-OH carbofuran glucuronic acid
		3-OH carbofuran eliminated
4	Hydroxymethyl carbofuran	
		Hydroxymethyl carbofuran sulfuric acid
		Hydroxymethyl carbofuran glucuronic acid
		Hydroxymethyl carbofuran eliminated
5	Carbofuran 3,7-diol	
		3-Keto-7-phenol
		3,7-Diol sulfuric acid
		3,7-Diol glucuronic acid
		3,7-Diol eliminated
6	3-Keto carbofuran	
		3-Keto-7-phenol
		3-Keto carbofuran eliminated
7	3-Keto-7-phenol	
		3-Keto-7-phenyl sulfuric acid
		3-Keto-7-phenyl glucuronic acid
		3-Keto-7-phenol eliminated
8	7-Phenyl sulfuric acid	
		7-Phenyl sulfuric acid eliminated
9	7-Phenyl glucuronic acid	
		7-Phenyl glucuronic acid eliminated

Table 26. (*cont.*)

No.[b]	Initiating pesticide/metabolite[c]	Resulting metabolite
10	3-OH carbofuran sulfuric acid	3-OH carbofuran sulfuric acid eliminated
11	3-OH carbofuran glucuronic acid	3-OH carbofuran glucuronic acid eliminated
12	Hydroxymethyl carbofuran sulfuric acid	Hydroxymethyl carbofuran sulfuric acid eliminated
13	Hydroxymethyl carbofuran glucuronic acid	Hydroxymethyl carbofuran glucuronic acid eliminated
14	3,7-Diol sulfuric acid	3,7-Diol sulfuric acid eliminated
15	3,7 Diol glucuronic acid	3,7-Diol glucuronic acid eliminated
16	3-Keto-7-phenyl sulfuric acid	3-Keto-7-phenyl sulfuric acid eliminated
17	3-Keto-7-phenyl glucuronic acid	3-Keto-7-phenyl glucuronic acid eliminated

[a]Preliminary estimates of liver V_{max} (μmol hr^{-1}kg^{-1} bw, unscaled) and K_m (μM) are equal to 10.0.

Additional published values can be found in Table 34 (Appendix B).

[b]Parent chemical or metabolite number for cross-reference in Table 16 (Appendix A).

[c]Refer to Table 16 for structure based on No.

Table 27. Biotransformation and Elimination Paths of Formetanate and the Resulting Metabolites and Preliminary Liver V_{max} and K_m Values[a].

No.[b]	Initiating pesticide/metabolite[c]	Resulting metabolite
1	Formetanate	
		Demethylformetanate
		3-Formaminophenyl *N*-methylcarbamate
2	Demethylformetanate	
		3-Formaminophenol
		Demethylformetanate eliminated
3	3-Formaminophenyl N-methylcarbamate	
		3-Formaminophenol
4	3-Formaminophenol	
		3-Aminophenol
		3-Formaminophenyl glucuronic acid
		3-Formaminophenyl sulfuric acid
5	3-Aminophenol	
		3-Acetamidophenol
		3-Aminophenyl glucuronic acid
		3-Aminophenyl sulfuric acid
6	3-Acetamidophenol	
		3-Acetamidophenyl glucuronic acid
		3-Acetamidophenyl sulfuric acid
7	3-Formaminophenyl glucuronic acid	
		3-Formaminophenyl glucuronic acid eliminated
8	3-Formaminophenyl sulfuric acid	
		3-Formaminophenyl sulfuric acid eliminated
9	3-Aminophenyl glucuronic acid	
		3-Aminophenyl glucuronic acid eliminated
10	3-Aminophenyl sulfuric acid	
		3-Aminophenyl sulfuric acid eliminated
11	3-Acetamidophenyl glucuronic acid	
		3-Acetamidophenyl glucuronic acid eliminated
12	3-Acetamidophenyl sulfuric acid	
		3-Acetamidophenyl sulfuric acid eliminated

[a]Preliminary estimates of liver V_{max} ($\mu mol\,hr^{-1}kg^{-1}$ bw, unscaled) and K_m (μM) are equal to 10.0.

[b]Parent chemical or metabolite number for cross-reference in Table 17 (Appendix A).

[c]Refer to Table 17 for structure based on No.

Table 28. Biotransformation and Elimination Paths of Methiocarb and the Resulting Metabolites and Preliminary Liver V_{max} and K_m Values[a].

No.[b]	Initiating pesticide/metabolite[c]	Resulting metabolite
1	Methiocarb	
		Methiocarb sulfoxide
		Methiocarb phenol
		Hydroxymethyl methiocarb
2	Methiocarb sulfoxide	
		Methiocarb sulfone
		Methiocarb sulfoxide phenol
3	Methiocarb sulfone	
		Methiocarb sulfone phenol
4	Hydroxymethyl methiocarb	
		Hydroxymethyl methiocarb glucuronic acid
		Hydroxymethyl methiocarb sulfuric acid
		Hydroxymethyl methiocarb eliminated
5	Methiocarb phenol	
		Methiocarb phenyl glucuronic acid
		Methiocarb phenyl sulfuric acid
6	Methiocarb phenol sulfoxide	
		Methiocarb sulfoxide phenyl glucuronic acid
		Methiocarb sulfoxide phenyl sulfuric acid
7	Methiocarb phenol sulfone	
		Methiocarb sulfone phenyl glucuronic acid
		Methiocarb sulfone phenyl sulfuric acid
8	Hydroxymethyl methiocarb glucuronic acid	
		Hydroxymethyl methiocarb glucuronic acid eliminated
9	Hydroxymethyl methiocarb sulfuric acid	
		Hydroxymethyl methiocarb sulfuric acid eliminated
10	Methiocarb phenyl glucuronic acid	
		Methiocarb phenyl glucuronic acid eliminated

Table 28. (*cont.*)

No.[b]	Initiating pesticide/metabolite[c]	Resulting metabolite
11	Methiocarb phenyl sulfuric acid	
		Methiocarb phenyl sulfuric acid eliminated
12	Methiocarb sulfoxide phenyl glucuronic acid	
		Methiocarb sulfoxide phenyl glucuronic acid eliminated
13	Methiocarb sulfoxide phenyl sulfuric acid	
		Methiocarb sulfoxide phenyl sulfuric acid eliminated
14	Methiocarb sulfone phenyl glucuronic acid	
		Methiocarb sulfone phenyl glucuronic acid eliminated
15	Methiocarb sulfone phenyl sulfuric acid	
		Methiocarb sulfone phenyl sulfuric acid eliminated

[a]Preliminary estimates of liver V_{max} ($\mu mol\,hr^{-1}kg^{-1}$ bw, unscaled) and K_m (μM) are equal to 10.0.

[b]Parent chemical or metabolite number for cross-reference in Table 18 (Appendix A).

[c]Refer to Table 18 for structure based on No.

Table 29. Biotransformation and Elimination Paths of Methomyl and the Resulting Metabolites and Preliminary Liver V_{max} and K_m Values[a].

No.[b]	Initiating pesticide/metabolite[c]	Resulting metabolite
1	*syn*-(*Z*)-methomyl	
		anti-(*E*)-methomyl
		syn-(*Z*)-methomyl oxime
2	*anti*-(*E*)-methomyl	
		anti-(*E*)-methomyl oxime
3	*syn*-(*Z*)-methomyl oxime	
		S-Methyl, *N*-methyl carbamic acid
4	*anti*-(*E*)-methomyl oxime	
		Ethylium

Table 29. (*cont.*)

No.[b]	Initiating pesticide/metabolite[c]	Resulting metabolite
5	*S*-Methyl, *N*-methyl carbamic acid	
		Carbon dioxide
6	Ethylium	
		Acetonitrile
7	Carbon dioxide	
		Carbon dioxide eliminated
8	Acetonitrile	
		Hydrogen cyanide
		Glutathione conjugate of acetonitrile
		Acetonitrile eliminated
9	Hydrogen cyanide	
		Hydrogen cyanide eliminated
10	Glutathione conjugate of acetonitrile	
		Glutamylcysteine conjugate of acetonitrile
		Cysteinylglycine conjugate of acetonitrile
		Glutathione conjugate of acetonitrile eliminated
11	Glutamylcysteine conjugate of acetonitrile	
		Cysteine conjugate of acetonitrile
		Glutamylcysteine conjugate of acetonitrile eliminated
12	Cysteinylglycine conjugate of acetonitrile	
		Cysteine conjugate of acetonitrile
		Cysteinylglycine conjugate of acetonitrile eliminated
13	Cysteine conjugate of acetonitrile	
		Cysteine-sulfate conjugate of acetonitrile
		Cysteine conjugate of acetonitrile eliminated
14	*N*-Sulfate-cysteine conjugate of acetonitrile	
		Cysteine-sulfate conjugate of acetonitrile eliminated

[a]Preliminary estimates of liver V_{max} (μmol $hr^{-1}kg^{-1}$ bw, unscaled) and K_m (μM) are equal to 10.0.

[b]Parent chemical or metabolite number for cross-reference in Table 19 (Appendix A).

[c]Refer to Table 19 for structure based on No.

Table 30. Biotransformation and Elimination Paths of Oxamyl and the Resulting Metabolites and Preliminary Liver V_{max} and K_m Values[a].

No.[b]	Initiating pesticide/metabolite[c]	Resulting metabolite
1	Oxamyl	des *N* oxamyl Oxamyl oxime Oxamyl nitrile
2	des *N* oxamyl	des *N* Oxamyl oxime
3	Oxamyl oxime	Oxamyl oxime glucuronic acid des *N* oxamyl oxime
4	Oxamyl nitrile, DMCF	Oxamyl acid
5	des *N* oxamyl oxime	des *N* oxamyl oxime glucuronic acid
6	Oxamyl oxime glucuronic acid	Oxamyl oxime glucuronic acid eliminated
7	Oxamyl acid (*N,N*-dimethyloxamic acid)	des *N* oxamyl acid Glucuronide of Oxamyl acid
8	des *N* oxamyl oxime glucuronic acid	des *N* oxamyl oxime glucuronic acid eliminated
9	Glucuronide of *N,N*-dimethyloxamic acid	Glucuronide of *N,N*-dimethyloxamic acid eliminated
10	des *N*-oxamyl acid (*N*-methyloxamic acid)	Glucuronide of *N*-methyloxamic acid
11	Glucuronide of *N*-methyloxamic acid	Glucuronide of *N*-methyloxamic acid eliminated

[a]Preliminary estimates of liver V_{max} (μmol $hr^{-1}kg^{-1}$ bw, unscaled) and K_m (μM) are equal to 10.0.

[b]Parent chemical or metabolite number for cross-reference in Table 20 (Appendix A).

[c]Refer to Table 20 for structure based on No.

Table 31. Biotransformation and Elimination Paths of Pirimicarb and the Resulting Metabolites and Preliminary Liver V_{max} and K_m Values[a].

No.[b]	Initiating pesticide/metabolite[c]	Resulting metabolite
1	Pirimicarb	
		Pirimicarb phenol
		Hydroxymethyl pirimicarb
2	Hydroxymethyl pirimicarb	
		Pirimicarb phenol
		Demethyl pirimicarb phenol
		Amino pirimicarb phenol
3	Pirimicarb phenol	
		Demethyl pirimicarb phenol
		Hydroxymethyl pirimicarb phenol
		Pirimicarb phenyl glucuronic acid
		Pirimicarb phenol eliminated
4	Demethyl pirimicarb phenol	
		Amino pirimicarb phenol
		Hydroxymethyl demethyl pirimicarb phenol
		Demethyl pirimicarb phenol eliminated
5	Amino pirimicarb phenol	
		Amino pirimicarb phenol eliminated
6	Hydroxymethyl demethyl pirimicarb phenol	
		Hydroxymethyl demethyl pirimicarb phenol eliminated
7	Pirimicarb phenyl glucuronic acid	
		Pirimicarb phenyl glucuronic acid eliminated
8	Hydroxymethyl pirimicarb phenol	
		Hydroxymethyl pirimicarb phenol eliminated

[a]Preliminary estimates of liver V_{max} ($\mu mol\,hr^{-1}kg^{-1}$ bw, unscaled) and K_m (μM) are equal to 10.0.

[b]Parent chemical or metabolite number for cross-reference in Table 21 (Appendix A).

[c]Refer to Table 21 for structure based on No.

Table 32. Biotransformation and Elimination Paths of Propoxur and the Resulting Metabolites and Preliminary Liver V_{max} and K_m Values[a].

No.[b]	Initiating pesticide/metabolite[c]	Resulting metabolite
1	Propoxur	
		2-Isopropoxy phenol
		5-OH propoxur
		4-OH propoxur
		Hydroxymethyl propoxur
		Propoxur eliminated
2	2-Isopropoxy phenol	
		2-Isopropoxy phenyl sulfuric acid
		2-Isopropoxy phenyl glucuronic acid
		2-Isopropoxy phenol eliminated
3	2-OH phenyl methylcarbamate	
		2-OH phenyl methylcarbamate sulfuric acid
		2-OH phenyl methylcarbamate glucuronic acid
		2-OH phenyl methylcarbamate eliminated
4	5-OH propoxur	
		2-Isopropoxy 5-OH phenol
		5-OH propoxur sulfuric acid
		5-OH propoxur glucuronic acid
		5-OH propoxur eliminated
5	2-Isopropoxy 5-OH phenol	
		2-Isopropoxy 5-OH phenyl sulfuric acid
		2-Isopropoxy 5-OH phenyl glucuronic acid
		2-Isopropoxy 5-OH phenol eliminated
6	4-OH propoxur	
		2-Isopropoxy 4-OH phenol
		4-OH propoxur sulfuric acid
		4-OH propoxur glucuronic acid
		4-OH propoxur eliminated
7	2-Isopropoxy 4-OH phenol	
		2-Isopropoxy phenyl sulfuric acid
		2-Isopropoxy phenyl glucuronic acid
		2-Isopropoxy 4-OH phenol eliminated
8	N-Hydroxymethyl propoxur	
		Hydroxymethyl propoxur sulfuric acid
		Hydroxymethyl propoxur glucuronic acid
		Hydroxymethyl propoxur eliminated

Table 32. (*cont.*)

No.[b]	Initiating pesticide/metabolite[c]	Resulting metabolite
9	2-Isopropoxy phenyl sulfuric acid	2-Isopropoxy phenyl sulfuric acid eliminated
10	2-Isopropoxy phenyl glucuronic acid	2-Isopropoxy phenyl glucuronic acid eliminated
11	2-OH phenyl methylcarbamate sulfuric acid	2-OH phenyl methylcarbamate sulfuric acid eliminated
12	2-OH phenyl methylcarbamate glucuronic acid	2-OH phenyl methylcarbamate glucuronic acid eliminated
13	5-OH propoxur sulfuric acid	5-OH propoxur sulfuric acid eliminated
14	5-OH propoxur glucuronic acid	5-OH propoxur glucuronic acid eliminated
15	2-Isopropoxy 5-OH phenyl sulfuric acid	2-Isopropoxy 5-OH phenyl sulfuric acid eliminated
16	2-Isopropoxy 5-OH phenyl glucuronic acid	2-Isopropoxy 5-OH phenyl glucuronic acid eliminated
17	4-OH propoxur sulfuric acid	4-OH propoxur sulfuric acid eliminated
18	4-OH propoxur glucuronic acid	4-OH propoxur glucuronic acid eliminated
19	2-Isopropoxy 4-OH phenyl sulfuric acid	2-Isopropoxy 4-OH phenyl sulfuric acid eliminated

Table 32. (*cont.*)

No.[b]	Initiating pesticide/metabolite[c]	Resulting metabolite
20	2-Isopropoxy 4-OH phenyl glucuronic acid	
		2-Isopropoxy 4-OH phenyl glucuronic acid eliminated
21	*N*-Hydroxymethyl propoxur sulfuric acid	
		Hydroxymethyl propoxur sulfuric acid eliminated
22	*N*-Hydroxymethyl propoxur glucuronic acid	
		Hydroxymethyl propoxur glucuronic acid eliminated

[a]Preliminary estimates of liver V_{max} ($\mu mol\,hr^{-1}kg^{-1}$ bw, unscaled) and K_m (μM) are equal to 10.0.
[b]Parent chemical or metabolite number for cross-reference in Table 22 (Appendix A).
[c]Refer to Table 22 for structure based on No.

Table 33. Biotransformation and Elimination Paths of Thiodicarb and the Resulting Metabolites and Preliminary Liver V_{max} and K_m Values[a].

No.[b]	Initiating pesticide/metabolite[c]	Resulting metabolite
1	Thiodicarb	
		syn-(*Z*)-methomyl
		anti-(*E*)-methomyl
2	*syn*-(*Z*)-methomyl	
		ani-(*E*)-methomyl
		syn-(*Z*)-methomyl oxime
3	*anti*-(*E*)-methomyl	
		anti-(*E*)-methomyl oxime
4	*syn*-(*Z*)-methomyl oxime	
		S-Methyl, *N*-methyl carbamic acid
5	*anti*-(*E*)-methomyl oxime	
		Ethylium
6	*S*-Methyl, *N*-methyl carbamic acid	
		Carbon dioxide
7	Ethylium	
		Acetonitrile

Table 33. (*cont.*)

No.[b]	Initiating pesticide/metabolite[c]	Resulting metabolite
8	Carbon dioxide	
		Carbon dioxide eliminated
9	Acetonitrile	
		Hydrogen cyanide
		Glutathione conjugate of acetonitrile
		Acetonitrile eliminated
10	Hydrogen cyanide	
		Hydrogen cyanide eliminated
11	Glutathione conjugate of acetonitrile	
		Glutamylcysteine conjugate of acetonitrile
		Cysteinylglycine conjugate of acetonitrile
		Glutathione conjugate of acetonitrile eliminated
12	Glutamylcysteine conjugate of acetonitrile	
		Cysteine conjugate of acetonitrile
		Glutamylcysteine conjugate of acetonitrile eliminated
13	Cysteinylglycine conjugate of acetonitrile	
		Cysteine conjugate of acetonitrile
		Cysteinylglycine conjugate of acetonitrile eliminated
14	Cysteine conjugate of acetonitrile	
		Cysteine-sulfate conjugate of acetonitrile
		Cysteine-sulfate conjugate of acetonitrile eliminated
15	N-Sulfate-cysteine conjugate of acetonitrile	
		Cysteine-sulfate conjugate of acetonitrile eliminated

[a]Preliminary estimates of liver V_{max} (μmol hr^{-1}kg^{-1} bw, unscaled) and K_m (μM) are equal to 10.0.

[b]Parent chemical or metabolite number for cross-reference in Table 23 (Appendix A).

[c]Refer to Table 23 for structure based on No.

Table 34. Published Liver Microsomal V_{max} and K_m Values for Carbaryl and Carbofuran.

	Biotransformation and elimination paths		Liver microsomes		
Parent chemical	Initiating pesticide/ metabolite (no.[a])	Resulting metabolite	V_{max}[b] $nmol\,min^{-1}\,mg^{-1}$ protein	K_m $\mu mol\,L^{-1}$	Source
Carbaryl	Carbaryl, HLM[c]	4-OH Carbaryl	0.87	349	Tang et al. 2002
	Carbaryl, HLM	5-OH Carbaryl	0.04	349	Tang et al. 2002
	Carbaryl, HLM	Hydroxymethyl carbaryl	0.57	81.0	Tang et al. 2002
Carbofuran	Carbofuran (rat output)[d]	Carbofuran-7-phenol	0.012	25.0	Zhang et al. 2006
	Carbofuran (rat output)[d]	3-OH carbofuran	0.024	12.5	Zhang et al. 2006
	Carbofuran, HLM	3-OH carbofuran	3.30	1974	Usmani et al. 2004
	Carbofuran, RLM[c]	3-OH carbofuran	5.50	207	Usmani et al. 2004

[a]From metabolism pathways in Tables 25 and 26.

[b]HLM V_{max} values ($nmol\,min^{-1}\,mg^{-1}$ protein) and may be expressed as *in vivo* values ($\mu mol\,hr^{-1}\,kg^{-1}$ bw); this is accomplished by multiplying the *in vitro* values by $60\,min\,hr^{-1}$; 30 mg microsomal protein/g liver; and by 27 g liver/kg bw to give the *in vivo* value in $nmol\,hr^{-1}kg^{-1}$ bw and dividing the resultant by 1000.

[c]HLM, human liver microsomes; RLM, rat liver microsomes.

[d]ERDEM V_{max} (mmol/L/H) was converted to nmol min^{-1} mg^{-1} protein by using 30 mg protein/g liver.

Table 35. Published CYP V_{max} and K_m Values for Carbaryl and Carbofuran.

	Biotransformation and elimination paths		Liver		
Parent	Initiating pesticide and CYP (no.[a])	Resulting metabolite	V_{max}[b] nmol min^{-1} $nmol^{-1}$ isoform	K_m µmol L^{-1}	Source
Carbaryl	Carbaryl, CYP1A1	4-OH carbaryl	3.89	20	Tang et al. 2002
	Carbaryl, CYP1A2		1.72	58	
	Carbaryl, CYP2B6		0.80	11	
	Carbaryl, CYP2C19		2.21	44	
	Carbaryl, CYP3A4		5.81	235	
	Carbaryl, CYP1A1	5-OH carbaryl	4.81	15	Tang et al. 2002
	Carbaryl, CYP1A2		2.62	89	
	Carbaryl, CYP2B6		0.29	110	
	Carbaryl, CYP2C19		0.99	62	
	Carbaryl, CYP3A4		2.34	281	
	Carbaryl, CYP1A1	Hydroxymethyl carbaryl	1.19	51	Tang et al. 2002
	Carbaryl, CYP1A2		4.80	36	
	Carbaryl, CYP2B6		15.54	45	
	Carbaryl, CYP2C19		3.46	15	
	Carbaryl, CYP3A4		1.47	156	
Carbofuran	Carbofuran, CYP3A4	3-OH carbofuran	40.3	742	Usmani et al. 2004
	Carbofuran, CYP1A2	3-OH carbofuran	22.0	238	Usmani et al. 2004
	Carbofuran, CYP2C19	3-OH carbofuran	18.0	199	Usmani et al. 2004

[a]From metabolism pathways in Tables 25 and 26. [b]These values may be converted to *in vivo* values (nmol hr^{-1} kg^{-1} bw) by multiplying the values by CYP content in nmol mg^{-1} microsomal protein, 60 min hr^{-1}; 30 mg microsomal protein/g liver; and by 27 g liver/kg bw to give the *in vivo* value.

Appendix C. Physiological model for aldicarb depicting its metabolic pathway.

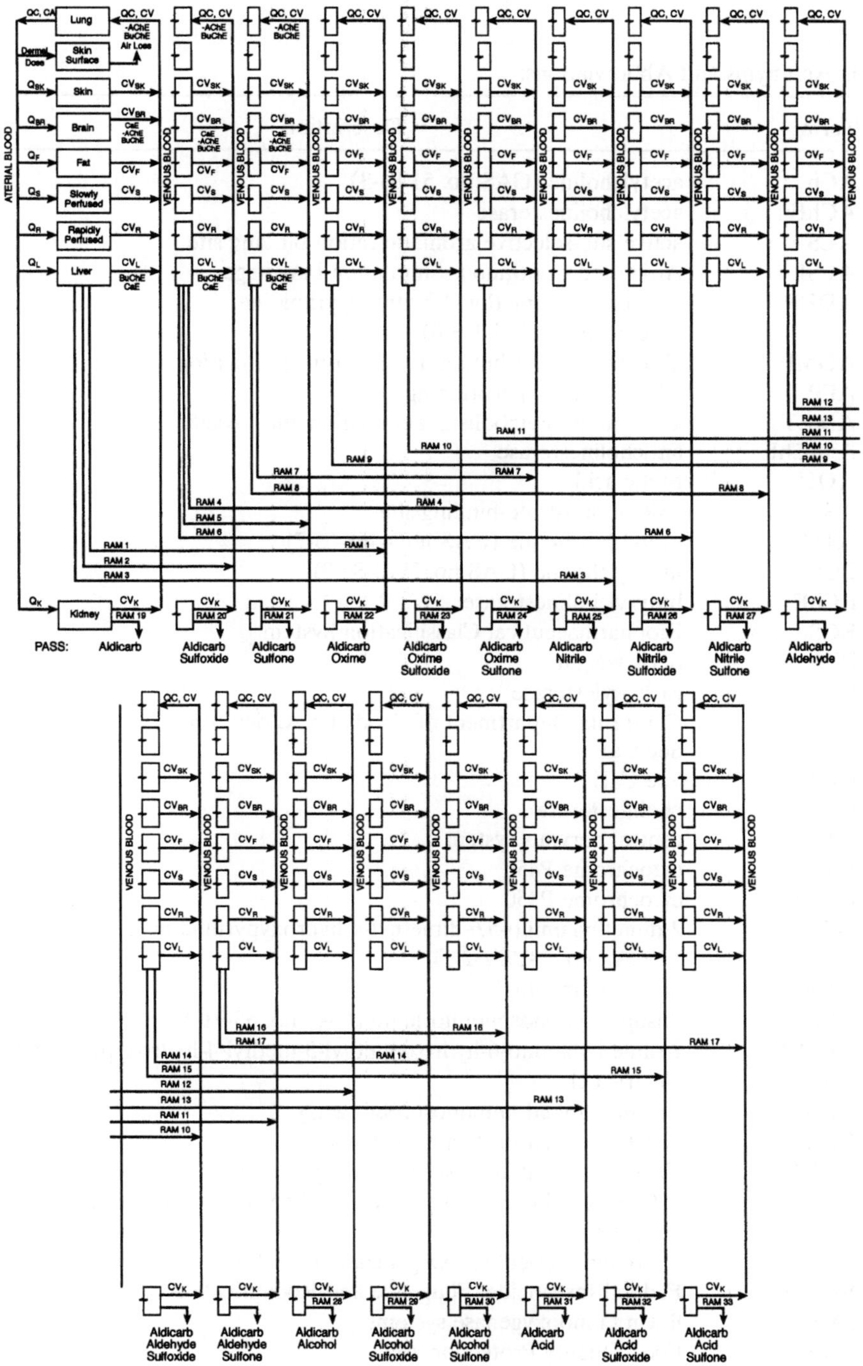

Fig. 9. Physiologically based pharmacokinetic/pharmacodynamic model for oral and dermal exposure to aldicarb. The model includes the inhibition and recovery of blood AChE and BChE, brain AChE, BChE, and CaE, and liver BChE and CaE.

Appendix D. Nomenclature.

i. Acronyms and Abbreviations:

Term	Description
ACh	acetylcholine (CAS no. 51-84-3)
AChE	acetylcholinesterase
ACS	active site-selective aromatic cation-binding site
ACSL	Advanced Continuous Simulation Language
ADHP	2-amino-5,6-dimethyl-4-hydroxypyrimidine (CAS no. 78195-313-0)
ADME	absorption, distribution, metabolism, elimination
AFP	adaptive fuzzy partitioning
AMET	absorption, metabolism, elimination and toxicity
AntiChE	anticholinesterase
AOH	acetic acid
AS	anionic substrate-binding site
ATCh	acetylthiocholine (CAS no. 4468-05-7)
BCh	butyrylcholine (CAS no. 3922-86-9)
BChE	butyrylcholinesterase
BCS	Biopharmaceutical Classification System
BW	body weight
CaE	carboxylesterase
CDFA	California Department of Food and Agriculture
CE	See CaE
CES	See CaE
ChE	cholinesterase
CNS	central nervous system
CYP	cytochrome P450
CYP450	cytochrome P450
DDHP	2-dimethylamino-5,6-dimethyl-4-hydroxypyrimidine (CAS no. 78195-32-1)
DEAE	diethylaminoethyl
DFP	diisopropyl phosphofluoridate (CAS no. 55-91-4)
DHHP	2-dimethylamino-6-hydroxymethyl-5-methyl-4-hydroxypyrimidine
EE	electric eel
ELISA	Enzyme-linked immunosorbent assay
EPA	Environmental Protection Agency
ERDEM	Exposure-Related Dose Estimating Model (see web site: http://www.epa.gov/heasdweb/erdem/erdem.htm)
ES	esteratic site
FAO	Food and Agriculture Organization, WHO
FIFRA	Federal Insecticide, Fungicide and Rodenticide Act
FMO	flavin monooxygenase systems
FQPA	Food Quality Protection Act
GA/SW	genetic algorithm concepts and a stepwise technique

i. Acronyms and Abbreviations: (*cont.*)

Term	Description
HB	hydrogen bonding
hCE-1, hCE-2	human liver isozymes
HLM	human liver microsome
HPLC	high performance liquid chromatography
HuAChE	human AChE
IP	intraperitoneally
IPP	2-isopropoxyphenol (CAS no. 4812-20-8)
iso-OMPA	tetraisopropylpyrophosphoramide (CAS no. 513-00-8)
LD_{50}	lethal dose for 50% of the population
LOAEL	lowest observed adverse effect level
MDHP	2-methylamino-5,6-dimethyl-4-hydroxypyrimidine (CAS no. 78195-33-2)
MEPQ	7-(methylethoxyphosphinyloxy)-1-methylquinolinium iodide (CAS no. 95230-44-7)
MLM	mouse liver microsome
MR	molar refractivity
MRP2	multidrug-resistant protein
NADPH	nicotinomide adenine dinucleotide phosphate
NCEA	National Center for Environmental Assessment
NOAEL	no observable adverse effect level
OATP	organic anion-transporting peptides
OP	organophosphorus
ORMUCS	**or**dered **mu**lticategorical **c**lassification method using the **s**implex technique
PAPS	3′-phosphoadenosinne-5′-phosphosulfate
PAS	peripheral anionic binding site or sites
PBPK/PD	physiologically based pharmacokinetic/pharmacodynamic
PCB	polychlorinated biphenyl
QSAR	quantitative structure–analysis relationship
RBC	red blood cells
RfD	reference dose
RLM	rat liver microsome
RPF	relative potency factor
SDS-PAGE	sodium dodecyl sulfate-polyacrylamide gel electrophoresis
TCh	thiocholine
TLC	thin-layer chromatography
UDP	umbelliferyl diethyl phosphate (CAS no. 299-49-6)
UDPGA	uridine diphosphate glucuronic acid
UGT	UDP-glucuronosyltransferases
WHO	World Health Organization

ii. Chemical and Mathematical Expressions:

Expression	Description
α	molecular polarizability
[E]	enzyme concentration
$[E]_t$; $[E]_T$	total enzyme
[EH], EH	concentration of the active enzyme
[ES]	enzyme–substrate complex
[IB], IB	concentration of carbamate in the incubation mixture
[S]	substrate concentration
AC_t	total AChE
A_{iACHEB}	amount of inhibited blood AChE
A_{rsal}	amount of chemical in reference saline
A_{ssal}	amount of chemical in the sample saline after filtration
b	coefficient from nonlinear equation fitting Eq. 32; reflects efficiency of hydrolysis of ternary complex SES relative to ES
BH	leaving group
CB	carbamic acid concentration or concentration of pesticide in skin
C_{ACHEB}	concentration of free AChE in blood
Ca^{max}	maximum H-bond acceptor descriptor in a molecule
CB	carbamate inhibitor
$C_{Carbamate}$	concentration of carbamate in blood
C_{fil}	concentration of the chemical in the ultrafiltrate from reference saline
CHG	indicator variable for charged substituents
C_j	concentration of chemical in tissue j
C_{sal}	concentration of the chemical in the reference saline before ultrafiltration
C_{ss}	concentration of chemical in the sample saline after filtration
$D^*_{vo:w}$	vegetable oil:buffer, nonionized and ionized species at pH 7.4; or distribution between vegetable oil and buffer at pH 7.4 (Eq. 15)
$D_{vo:w}$	distribution between vegetable oil:buffer, neutral compound
$D_{vo:w,\ pH7.4}$	distribution between *n*-octanol and buffer at pH 7.4 (ACD values)
E	enzyme
E_0	initial enzyme
EA	acylated enzyme
EC	carbamylated enzyme
EHAB	enzyme inhibitor or Michaelis–Menten complex
EHCB	enzyme–carbamate complex
EH_t	EH at time *t*
EI	enzyme inhibitor complex
Es	Taft steric parameter
fu_b	binding to protein in blood
fu_t	binding to protein in tissue
HB	leaving group
$\mathfrak{I}$	inductive parameter
I_0	initial inhibitor
I_{50}	molar concentrations resulting in 50% inhibition

ii. Chemical and Mathematical Expressions: (*cont.*)

Expression	Description
K	K_m, substrate concentration that results in half-maximum velocity
k_{+1}	equilibrium reaction, forward velocity
k_{-1}	equilibrium reaction, backward velocity
k_{+2}	carbamylation reaction, same as k_2
k_{+3}	decarbamylation rate
k_2	carbamylation rate
K_{ss}	substrate inhibition constant
K_a	binding affinity constant
k_{cat}	hydrolysis rate constant, hr^{-1}
K_d	equilibrium or affinity constant
k_i	bimolecular inhibtion rate constant
K_{iACHEB}	AChE bimolecular inhibition rate constant
K_m	substrate concentration that results in half-maximum velocity
K_{ow}	octanol:water partition coefficient
k_p	turnover rate
LD_{50}	lethal dose for 50% of the population
Log D	distribution coefficient for partially dissociated compounds
Log $D_{pH7.4}$	Log D calculated with pH of 7.4
Log P	log of octanol:water partition coefficient, neutral compounds
nl	neutral lipids
p	plasma
P	partition coefficient
$PC_{b:adipose}$	blood:adipose tissue partition coefficient
$PC_{b:liver}$	blood:liver partition coefficient
PC_t	tissue:blood partition coefficient
ph	phospholipds
pI	plasma
P_j	partition coefficient for tissue j
pK_a	acid–base ionization constant
$P_{o:w}$	*n*-octanol:buffer partition coefficient, nonionized species at pH 7.4
$P_{t:b}$	tissue:blood partition coefficient
$P_{t:p\ adipose}$	partition coefficient between adipose tissue and plasma
$P_{t:p\ nonadipose}$	partition coefficient between nonadipose tissue and plasma
q^{+max}	maximum positive charge
R_f	recovered fraction
RGMR	MR of certain parts of ring substituents
ΣCa	sum of H-bond factor values for all acceptor substructures in a molecule
ΣCd	sum of H-bond factor values for all donor atoms in a molecule
SES	termary enzyme complex
ΣQ+	sum of all positive partial atomic charges for all atoms in the molecule
t	tissue
t	inhibition time
$t_{1/2}$	half-life

ii. Chemical and Mathematical Expressions: (*cont.*)

Expression	Description
V	fractional tissue volume content
V_B	volume of blood
V_i	initial velocity
V_j	volume of tissue j
V_{max}	half-maximum velocity
V_{nlb}	fractional volume of neutral nonpolar lipids in blood
V_{nlt}	fractional volume of neutral nonpolar lipids in tissue
V_{phlb}	fractional volume of phospholipids in blood
V_{phlt}	fractional volume of phospholipids in tissue
V_{ss}	volume of saline in the test vial
V_{wb}	fractional volume of water in blood
V_{wt}	fractional volume of water in tissue
w	water
WS	water solubility

Rev Environ Contam Toxicol 193:213–285

Persistent Organic Pollutants in Vietnam: Environmental Contamination and Human Exposure

Tu Binh Minh, Hisato Iwata, Shin Takahashi, Pham Hung Viet, Bui Cach Tuyen, and Shinsuke Tanabe

Contents

I. Introduction

Global contamination and toxic effects of persistent organic pollutants (POPs) have been an emerging environmental issue and have received considerable attention during the past four decades. Although the extent of contamination by POPs has been dominant in industrialized nations, an

---

Communicated by G.W. Ware.

T.B. Minh, H. Iwata, S. Takahashi, S. Tanabe (✉)
Center for Marine Environmental Studies (CMES), Ehime University, Bunkyo-cho 2-5, Matsuyama 790-8577, Japan.

P.H. Viet
Center for Environmental Technology and Sustainable Development (CETASD), Hanoi National University, 334 Nguyen Trai Street, Thanh Xuan, Hanoi, Vietnam.

B.C. Tuyen
Nong Lam University, Linh Trung, Thu Duc District, Hochiminh City, Vietnam.

increasing number of recent investigations have highlighted the role of the Asia-Pacific region as a potential source of emission for these chemicals, particularly to pristine areas such as the Arctic and the Antarctic (Tanabe et al. 1994; Tanabe 2000, 2002; Tanabe and Subramanian 2006).

In view of environmental contamination, Vietnam has been well known as a land of extensive spraying of Agent Orange during the Vietnam War. The high degree of dioxin contamination in some military bases and Agent Orange-sprayed areas in South Vietnam has received particular attention during the past 30 years. In addition to the dioxin contamination caused by Agent Orange, the rapid agricultural and industrial growth in this country lends credence to the widespread contamination of POPs. Vietnam is located at the center of the Southeast Asian region (Fig. 1); it has more than 300 km of coastal area and two major agricultural production areas: the Red River Delta in the north and the Mekong River Delta in the south. These two deltas are inhabited by more than 30 million people and are two of the most densely populated areas in the world. The Mekong River Delta has recently become one of the most productive agricultural regions of Southeast Asia. Such a strategic geographical position and the rapid agricultural development of Vietnam made this country an important region where extensive studies on environmental pollution have been carried out during the last two decades.

This chapter provides a comprehensive review of the studies dealing with POPs in Vietnam. Available data on POP contamination in Vietnam are compiled on the basis of various investigations in the framework of the

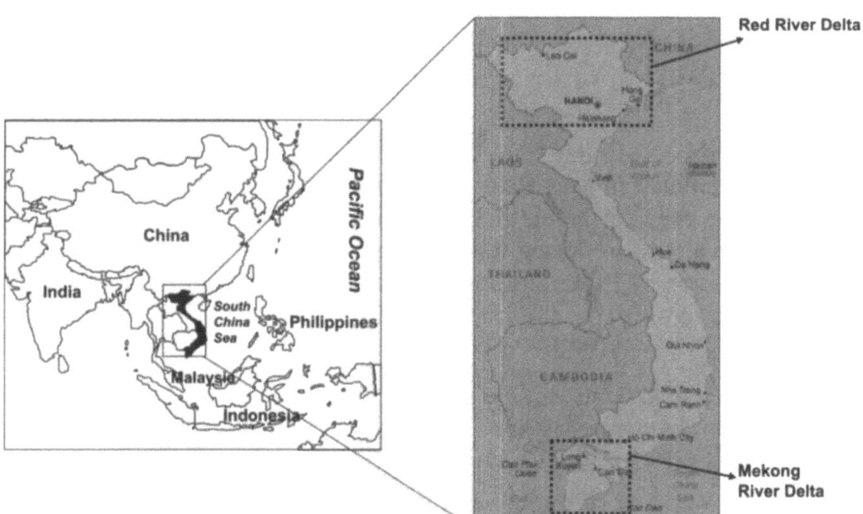

Fig. 1. Map of Vietnam. Vietnam is located at the center of the Southeast Asian region: it has two the largest deltas, Red River Delta and Mekong River Delta.

Asia-Pacific Mussels Watch Program, the 21st Century Center of Excellence Program, and the Core University Program supported by the Japan Society for the Promotion of Science (JSPS), which were conducted in our laboratory during the past decade. Results of these comprehensive studies are reviewed, and various issues of POPs contamination in Vietnam are discussed in a comparative point of view with the other countries in the Asia-Pacific region. In addition, results from other laboratories are also reviewed to help improve insights into the distribution, transport, bioaccumulation, and possible toxic implications on environmental quality and human health. This review focuses on the organochlorine insecticides such as 1,1,1-trichloro-2,2-bis(*p*-chlorophenyl)ethane (DDT) and its metabolites (DDTs), hexachlorocyclohexane isomers (HCHs), chlordane compounds (CHLs), and hexachlorobenzene (HCB). Residue levels of industrially derived contaminants such as polychlorinated biphenyls (PCBs), polychlorinated dibenzo-*p*-dioxins and dibenzofurans (PCDD/Fs), and polybrominated biphenyl ethers (PBDEs) are also reviewed.

II. Production and Use

In general, information on the production and usage of POPs, particularly organochlorine (OC) insecticides and PCBs in Vietnam, as well as some other developing countries in the East and South Asian region, is still limited or obscure. Systematic inventory of toxic manmade chemicals is lacking in these countries because of their limited capacity to conduct comprehensive monitoring surveys. Recently, the United Nations Environment Program (UNEP) has initiated various monitoring programs for POPs at regional and global levels, and the results have been summarized at different workshops. According to these reports, the active ingredients for insecticides were not produced in Vietnam. In fact, before 1985, pesticides such as DDT and HCB were imported from the former Soviet Union and some socialistic countries up to a quantity of 6,500–9,000 t/yr (Sinh et al. 1999). The statistical data showed that the total quantity of DDT imported into Vietnam for malaria control from 1957 to 1990 was 24,042 t. During 1986–1990, approximately 800 t was used (Sinh et al. 1999). These amounts are still lower than those in some other countries in regions such as Malaysia, Indonesia, and India. DDT usage for malaria control ceased in Vietnam in 1995, and other insecticides such as pyrethroid compounds have been used as substitutes for DDT (Sinh et al. 1999).

The information on PCB usage in Vietnam is still obscure. Data indicate that about 27,000–30,000 t of oils contaminated by PCBs were imported from the former USSR, China, and Rumania (Sinh et al. 1999). In addition, electrical equipment containing PCBs, such as transformers, was also imported from Australia until the mid-1980s (Kannan et al. 1995). Yet another possible source of PCBs in Vietnam are the weapons used extensively during the Indochina War (Thao et al. 1993a,b). The major source of dioxins in Vietnam in the past was Agent Orange and other herbicides sprayed in South Vietnam during the American war. Recently, Stellman and coworkers (Stellman et al. 2003) provided revised estimates of the amounts of herbicides used in Vietnam. During 1961–1971, at least about 45 million L Agent Orange was sprayed (Stellman et al. 2003). 2,4,5-T, a constituent of Agent Orange, is known to be contaminated with 2,3,7,8-tetrachlorodibenzo-*p*-dioxin (TCDD). However, the combustion-derived sources of dioxins in Vietnam are unknown. Various kinds of combustion processes may facilitate the widespread contamination of dioxins and related compounds in Vietnam.

III. Contamination Status

A. Air, Water, Sediments, and Soils

Comprehensive monitoring surveys have been conducted to examine the distribution of POPs such as PCBs, DDTs, HCHs, and HCB in air, water, and sediments from estuarine environments from various countries in the Asia-Pacific comprising Japan, India, Vietnam, Thailand, Indonesia, Malaysia, the Philippines, and Australia (Iwata et al. 1994). These investigations reported the presence of higher residues of DDTs and HCHs in air and water from coastal and estuarine areas in the developing countries of tropical and subtropical regions (India, Thailand, and Vietnam), than in developed nations (Japan and Australia). A compilation of available data for Vietnam is given in Tables 1 and 2. The distribution in air, water, and sediments from north, middle, and south regions of Vietnam showed relatively higher DDT concentrations, supporting the concept of widespread contamination of this insecticide throughout the country. This result suggests extensive use of DDT for agricultural purposes in the past and for malaria control until very recently. Interestingly, in a survey conducted about 10 yr later than the survey by Iwata et al. (1994) (in 1998/1999) covering an extended area along Red River and Duong River, the two biggest rivers in northern Vietnam, elevated concentrations of DDTs, HCHs, and CHLs were reported (Hung and Thiemann 2002). The levels of DDTs, HCHs, and CHLs in Red River and Duong River were apparently higher than those reported in the early 1990s surveys. In addition, wastewater collected from extensive human activity areas such as canals of Tu Liem district, a suburb of Hanoi city (northern Vietnam), and Thi Nghe River, Hochiminh (southern Vietnam) contained elevated concentrations of DDTs (see Table 1). It is also interesting to note that levels found in a recent survey (in suburb Hanoi; Hung and Thiemann 2002) were higher than those reported a decade ago (Iwata et al. 1994). Although backgrounds of analytical methods and sampling locations are different among studies, these observations suggest that the use of DDT for malaria control was relatively extensive until very recently in both northern and southern Vietnam.

Table 1. Concentrations (ng L^{-1}) of Persistent Organochlorines in Water from Vietnam.

Location	Sample	Year	*n*	DDTs	HCHs	CHLs	PCBs	Reference	Remarks
Red River, Hanoi	River water	1990	1	0.68	3.2	0.045	0.84	Iwata et al. 1994	DDTs: sum of *p,p′*-DDE, *p,p′*-DDD and *p,p′*-DDT
Phu Loc Lake, Hue	Lake water	1990	1	0.29	18	0.21	1.2	Iwata et al. 1994	
Huong River, Hue	River water	1900	1	1.1	1.9	0.07	1.6	Iwata et al. 1994	HCHs: sum of α-, β-, and γ-HCH
Nha Be River, Hochiminh City	Estuarine water	1990	1	4.7	31	0.55	2.7	Iwata et al. 1994	CHLs: sum of *cis*-chlordane, *trans*-chlordane, *cis*-nonachlor, *trans*-nonachlor and oxychlordane
Gua canal, Cu Chi	River water	1990	1	0.6	9.5	0.16	1.9	Iwata et al. 1994	
Long Tau River, Hochiminh City	Estuarine water	1990	1	0.55	5.2	0.13	0.57	Iwata et al. 1994	PCBs: quantified by an equivalent mixture of Kanechlor preparations (KC-300, KC-400, KC-500 and KC-600)
Thi Nghe canal, Hochiminh City	Estuarine water	1990	1	25	19	1	8	Iwata et al. 1994	
Red River and Duong River, northern Vietnam	River water	1998/99						Hung and Thiemann 2002	DDTs: sum of *p,p′*-DDE, *p,p′*-DDD, *p,p′*-DDT, *o,p′*-DDE, *o,p′*-DDD, *o,p′*-DDT
Dry season			18	44 (0.55–320)[a]	17 (1.6–83)	21 (0.81–110)	—		
Rainy season			18	56 (8.4–230)	29 (3.1–97)	27 (1.6–130)	—		HCHs: sum of α-, β-, γ-, and δ-HCH CHLs: sum of heptachlor and heptachlorepoxide
Lakes in Hanoi: West Lake, Thuyen Quang Lak	Lake water	1998/99						Hung and Thiemann 2002	
Bay Mau Lake and Ba Mau Lake									
Dry season			4	2.1 (0.21–3.6)	0.69 (0.26–2.16)	2.9 (1.1–4.8)	—		
Rainy season			4	5.1 (0.65–15)	32 (0.69–120)	13 (2–51)	—		
Irrigation canal, Tu Liem, suburb Hanoi	River water	1998/99						Hung and Thiemann 2002	
Dry season			6	59 (8.2–130)	7.2 (1.6–1.8)	6.9 (6.0–7.5)	—		
Rainy season			6	50 (11–110)	17 (5.5–26)	15 (1.2–25)	—		
Wells, Gia Lam, suburb Hanoi	Groundwater	1998–1999						Hung and Thiemann 2002	
Dry season			2	0.17 (0.11–0.23)	0.21 (0.19–0.22)	0.17	—		
Rainy season			2	0.09	0.04	<0.5	—		

[a]Mean (range).

Table 2. Concentrations (ng g^{-1} dry wt) of Persistent Organochlorines in Sediments and Soils From Vietnam.

Location	Sample/sampling site characteristic	Year	*n*	DDTs
Phu Da, Hue	Sediment, paddy field	1990	1	0.52
A Luoi, Binh Tri Thien	Sediment, municipal sewage	1990	1	68
Long Giang Canal, Duyen Hai	Sediment, paddy field	1990	1	8
Lo Giang River, Duyen Hai	Sediment, paddy field	1990	1	1
Don Canal, Duyen Hai	Sediment, paddy field	1990	1	3.3
La Canal, Duyen Hai	Sediment, paddy field	1990	1	10
Ben Nghe Canal, Hochiminh City	Sediment, municipal sewage	1990	1	120
Tan Binh, Tay Ninh	Sediment, paddy field	1990	1	7.8
Gua Canal, Cu Chi	Sediment, paddy field	1990	1	0.37
Song Long Tau, Duyen Hai, Hochiminh City	Sediment, paddy field	1990	6	17 (2.1–47)
Thi Nghe Canal, Hochiminh City	Sediment, municipal sewage	1990	3	530 (360–790)
Thinh Liet, Thanh Tri, Hanoi	Soil, paddy field	1990	1	330
Vinh Quynh, Thanh Tri, Hanoi	Soil, paddy field	1990	2	76 (31–120)
Yen Duyen, Thanh Tri, Hanoi	Soil, paddy field	1990	2	47 (19–74)
Phu Da, Binh Tri, Thien	Soil, paddy field	1990	1	38
Hong Ha, A luoi, Binh Tri, Thien	Soil, paddy field	1990	1	23
Ta Bat, Aluoi, Binh Tri, Thien	Soil, paddy field	1990	1	11
Son Thuy, Aluoi, Binh Tri, Thien	Soil, paddy field	1990	1	1300
Hong Kim, Aluoi, Binh Tri, Thien	Soil, paddy field	1990	1	9.7
An Hoa, Hue	Soil, paddy field	1990	1	69
Phu Loc, Hue	Soil, upland	1990	4	1.9 (0.73–4.4)
Binh Khanh, Duyen Hai, Hochiminh City	Soil, paddy field	1990	2	27 (24–30)
Tan Thuan, Nha Be, Hochiminh City	Soil, paddy field	1990	1	64
My Hung, Cu Chi	Soil, paddy field	1990	1	7.9

HCHs	CHLs	PCBs	Reference	Remarks
0.43	0.072	0.65	Iwata et al. 1994	DDTs: sum of *p,p'*-DDE, *p,p'*-DDD and *p,p'*-DDT HCHs: sum of α-, β-, and γ-HCH CHLs: sum of *cis*-chlordane, *trans*-chlordane, *cis*-nonachlor, *trans*-nonachlor and oxychlordane PCBs: quantified by an equivalent mixture of Kanechlor preparations (KC-300, KC-400, KC-500 and KC-600)
2.4	0.79	0.18	Iwata et al. 1994	
1.1	0.42	3.7	Iwata et al. 1994	
0.45	0.15	2.2	Iwata et al. 1994	
0.94	0.21	2.1	Iwata et al. 1994	
2.3	0.58	9.7	Iwata et al. 1994	
5.2	8.8	140	Iwata et al. 1994	
0.7	0.31	—	Iwata et al. 1994	
0.63	0.24	0.22	Iwata et al. 1994	
0.82 (0.5–1.3)	0.26 (0.14–0.46)	6.2 (2.3–8.9)	Iwata et al. 1994	
8.5 (6.0–12)	20 (19–20)	570 (440–630)	Iwata et al. 1994	
18	—	39	Thao et al. 1993	DDTs: sum of *p,p'*-DDE, *p,p'*-DDD, *p,p'*-DDT, *o,p'*-DDT HCHs: sum of α-, β-, γ- and δ-HCH CHLs: sum of *cis*-chlordane, *trans*-chlordane, *cis*-nonachlor, *trans*-nonachlor and oxychlordane PCBs: quantified by an equivalent mixture of Kanechlor preparations (KC-300, KC-400, KC-500 and KC-600)
30 (4–55)	—	11 (5.5–17)	Thao et al. 1993	
1.8 (1.3–2.3)	—	21 (13–28)	Thao et al. 1993	
2.8	—	8.6	Thao et al. 1993	
4.7	—	5.2	Thao et al. 1993	
1.4	—	2.6	Thao et al. 1993	
5.7	—	12	Thao et al. 1993	
0.31	—	0.61	Thao et al. 1993	
5.3	—	5	Thao et al. 1993	
0.15 (0.09–0.21)	—	2.1 (1.4–4.0)	Thao et al. 1993	
1.5 (1.4–1.5)	—	3.9 (1.6–6.1)	Thao et al. 1993	
4	—	12	Thao et al. 1993	
0.23	—	1.5	Thao et al. 1993	

Table 2. (*cont.*)

Location	Sample/sampling site characteristic	Year	n	DDTs
Ba Diep, Hooc Mon	Soil, upland	1990	1	280
An Phu, Cu Chi	Soil, upland	1990	3	14 (1.9–37)
Tan Binh, Tan Bien, Tay Ninh	Soil, paddy field	1990	1	5.5
Thanh Phu, Tan Bien, Tay Ninh	Soil, paddy field	1990	1	350
Duong Minh Chau, Tay Ninh	Soil, paddy field	1991	1	1.9
Duong Minh Chau, Tay Ninh	Soil, green bean field	1991	1	2.4
Tan Bien, Tay Ninh	Soil, paddy field	1991	1	6.6
Hoa Thanh, Tay Ninh	Soil, sugarcane and peanut	1991	1	5.1
Tan Chau, Tay Ninh	Soil, rubber plantation	1991	1	1.6
Tan Chau, Tay Ninh	Soil, sugarcane	1991	1	290
Tan Chau, Tay Ninh	Soil, harvested crop	1991	5	88 (1.0–270)
Tan Bien, Tay Ninh	Soil, paddy field	1991	1	5.7
Go Dau, Tay Ninh	Soil, harvested crop	1991	1	130
Trang Bang, Tay Ninh	Soil, harvested crop	1991	1	150
Duong Minh Chau, Tay Ninh	Non-cultivated soil	1991	1	1.5
Tan Bien, Tay Ninh	Non-cultivated soil, Former US air base ground	1991	1	13
Tan Chau, Tay Ninh	Non-cultivated soil, US bombed site	1991	1	0.46
Tan Bien, Tay Ninh	Non-cultivated soil	1991	6	11 (0.25–38)
Tan Uyen, southern Vietnam	Non-cultivated soil	1991	2	14 (2.0–26)
Saigon River, Hochiminh City	River sediment	1996	11	80 (1.8–250)
				DS/RS[a]
Hanoi	Sediment, irrigation canal, northwest Hanoi	1995/96	1	6.9/13

HCHs	CHLs	PCBs	Reference	Remarks
0.79	—	130	Thao et al. 1993	
2.3 (1.9–2.5)	—	3.7 (3.0–4.2)	Thao et al. 1993	
0.15	—	1.6	Thao et al. 1993	
3.4	—	320	Thao et al. 1993	
2	—	2.3	Thao et al. 1993	
0.57	—	1.8	Thao et al. 1993	
0.78	—	2.3	Thao et al. 1993	
1.3	—	1.6	Thao et al. 1993	
1.4	—	1	Thao et al. 1993	
0.27	—	0.45	Thao et al. 1993	
0.92 (0.09–2.3)	—	34 (0.95–150)	Thao et al. 1993	
0.42	—	1	Thao et al. 1993	
0.46	—	32	Thao et al. 1993	
0.47	—	38	Thao et al. 1993	
0.8	—	5	Thao et al. 1993	
1.8	—	92	Thao et al. 1993	
0.95	—	25	Thao et al. 1993	
0.71 (0.09–1.9)	—	4.8 (0.23–13)	Thao et al. 1993	
1.3 (0.44–2.1)	—	8.0 (3.1–13)	Thao et al. 1993	
—	—	220	Phuong et al. 1998	DDTs: sum of *p,p′*-DDE, *p,p′*-DDD and *p,p′*-DDT PCBs: sum of 6 congeners
DS/RS 0.14/0.07	—	DS/RS 4.1/0.98	Nhan et al. 1998	DDTs: sum of *p,p′*-DDE, *p,p′*-DDD and *p,p′*-DDT

Table 2. (*cont.*)

Location	Sample/sampling site characteristic	Year	*n*	DDTs
Hanoi	Sediment, paddy field, southwest Hanoi	1995/96	1	7.5/14
Ba Lat estuary, Red River, northern Vietnam	Sediment	1995/96	1	7.1/3
Cua Lan estuary, Thai Binh coast lines, northern Vietnam	Sediment, intertidal mudflat areas	1995/96	1	5.8/4.9
Tra Ly estuary, Thai Binh coast lines, northern Vietnam	Sediment, intertidal mudflat areas	1995/96	1	7.3/5.1
Diem Dien estuary, Thai Binh coast lines, northern Vietnam	Sediment, intertidal mudflat areas	1995/96	1	6.2/4.6
Tra Co beach, Mong Cai, northern coast of Vietnam	Marine sediment, intertidal mudflat areas	1997	1 (pooled)	10
Mong Duong, northern coast of Vietnam	Marine sediment, intertidal mudflat areas	1997	1 (pooled)	8.1
Ha Long, northern coast of Vietnam	Marine sediment, intertidal mudflat areas	1997	1 (pooled)	7.2
Hai Phong, northern coast of Vietnam	Marine sediment, intertidal mudflat areas	1997	1 (pooled)	6.7
Ba Lat estuary, northern coast of Vietnam	Marine sediment, intertidal mudflat areas	1997	1 (pooled)	6.3
Cau Dien, Nhue River, suburb Hanoi	Sediment, canal, densely populated industrial area	1997	2	43 (15–71)
Nhue River, suburb Hanoi	Sediment, canal, rural area	1997	1	8.3
Nhue River, suburb Hanoi	Sediment, canal	1997	1	21
Hanoi downtown	Sediment, canal, densely populated industrial area	1997	1	13
ToLich River, suburb Hanoi	Sediment, canal, densely populated industrial area	1997	2	59 (36–81)
Thuong Tin, suburb Hanoi	Sediment, canal, southern Hanoi city	1997	2	37 (24–50)
Thuong Tin, suburb Hanoi	Sediment, canal, rural area, southern Hanoi city	1997	1	7.4

HCHs	CHLs	PCBs	Reference	Remarks
0.41/0.16	—	6.0/1.3	Nhan et al. 1998	PCBs: quantified by Arochlor 1260 HCHs: γ-HCH only
0.5/0.05	—	1.1/0.7	Nhan et al. 1998	
0.48/0.025	—	0.87/0.41	Nhan et al. 1998	
0.62/0.13	—	0.36/0.32	Nhan et al. 1998	
0.36/0.11	—	0.23/0.11	Nhan et al. 1998	
34	—	22	Nhan et al. 1999	DDTs: sum of *p,p'*-DDE, *p,p'*-DDD and *p,p'*-DDT
4.1	—	0.51	Nhan et al. 1999	HCHs: sum of α-, β, and γ-HCH PCBs: sum of 13
1.8	—	11	Nhan et al. 1999	congeners
1.7	—	18	Nhan et al. 1999	
1.2	—	0.33	Nhan et al. 1999	
0.34 (0.09–0.58)	—	1.7 (0.97–2.51)	Nhan et al. 2001	DDTs: sum of *p,p'*-DDE, *p,p'*-DDD, *p,p'*-DDT, *o,p'*-DDE, *o,p'*-DDD,
0.78	—	0.74	Nhan et al. 2001	*o,p'*-DDT and DDMU PCBs: sum 13 congeners
0.32	—	18	Nhan et al. 2001	HCHs: sum of α-, β, and γ-HCH
0.44	—	5.3	Nhan et al. 2001	CHLs: sum *cis*-chlordane, *trans*-chlordane & *trans*-nonacl
2.0 (0.85–3.12)	—	29 (24–34)	Nhan et al. 2001	
1.3 (0.59–1.98)	—	20 (16–24)	Nhan et al. 2001	
0.65	—	3.1	Nhan et al. 2001	

Table 2. (*cont.*)

Location	Sample/sampling site characteristic	Year	*n*	DDTs
Gia Lam, suburb Hanoi	Sediment, canal, rural area, eastern Hanoi city	1997	1	17
Dong Anh, suburb Hanoi	Sediment, canal, rural area, northern Hanoi city	1997	1	23
Ha Long Bay, northern Vietnam	estuary sediment	1998	—	28
Viet Tri, northern Vietnam	Industrial areas	1998	—	5.2
Hanoi	Soils, municipal dumping sites	1999–01	7	19 (1.9–52)[b]
Hanoi	Soils, reference areas relative to the municipal dumping sites	1999–01	3	3.2 (2.2–4.3)
Hochiminh City	Soils, municipal dumping sites	1999–01	6	23 (1.1–83)
Hochiminh City	Soils, reference areas relative to the municipal dumping sites	1999–01	3	3.8 (0.41–10)
Canal, Can Tho, Mekong River, southern Vietnam	Sediment, canals in Cantho city	2003–04	4	2.8 (1.8–4.3)
Hau River, Mekong River delta, southern Vietnam	Sediment, river	2003–04	7	0.96 (0.043–1.9)
Canals, Hochiminh city, Saigon River, southern Vietnam	Sediment, canals, densely populated areas	2004	5	37
Saigon River, Hochiminh City	Sediment, river	2004	5	6.9
Saigon River, downstream and coastal areas	Sediment, downstream and coastal area	2004	9	1.2

[a]Dry season/rainy season.
[b]Mean (range).

HCHs	CHLs	PCBs	Reference	Remarks
0.46	—	1.9	Nhan et al. 2001	
0.07	—	1.8	Nhan et al. 2001	
6.1	—	37	Viet et al. 2000	
0.68	—	2.3	Viet et al. 2000	
0.83 (0.29–2.2)	0.45 (<0.01–1.4)	12 (2.2–20)	Minh et al. 2004	DDTs: sum of *p,p′*-DDE, *p,p′*-DDD, and *p,p′*-DDT HCHs: sum of α-, β-, and γ-HCH CHLs: sum of *cis*-chlordane, *trans*-chlordane, *cis*-nonachlor, *trans*-nonachlor and oxychlodane PCBs: quantified by an equivalent mixture of Kanechlor preparations (KC-300, KC-400, KC-500 and KC-600)
0.14 (0.11–0.16)	0.15 (0.11–0.18)	0.73 (0.45–1.1)	Minh et al. 2004	
0.37 (0.21–0.87)	1.3 (0.15–2.5)	22 (4.2–40)	Minh et al. 2006a	
0.36 (0.32–0.43)	0.2 (0.16–0.25)	3.6 (2.5–4.4)	Minh et al. 2006a	
0.04 (<0.02–0.11)	0.2 (0.12–0.35)	1.8 (0.12–3.7)	Iwata et al. 2004	DDTs: sum of *p,p′*-DDE, *p,p′*-DDD, and *p,p′*-DDT HCHs: sum of α-, β-, and γ-HCH CHLs: sum of *cis*-chlordane, *trans*-chlordane, *cis*-nonachlor, and *trans*-nonachlor PCBs: quantified by an equivalent mixture of Kanechlor preparations (KC-300, KC-400, KC-500 and KC-600)
<0.02–0.044	0.06 (0.025–0.13)	0.21 (0.12–0.54)	Iwata et al. 2004	
—	—	81	Iwata et al. 2005	
—	—	8.5	Iwata et al. 2005	
—	—	0.92	Iwata et al. 2005	

Geographical distribution of OCs in sediments from Vietnam showed little variability compared to air and water (Fig. 2, Table 2). Consistent results were also observed in a wide area when considering distribution in the Asia-Pacific region (Iwata et al. 1994). Another survey also indicated rather uniform distribution of DDTs in sediments along the coasts of northern Vietnam, from the sites near the border of China down toward the Red River estuary (see Table 2; Nhan et al. 1998, 1999). Shorter residence time in the water phase and rapid volatilization of POPs from sediments as a consequence of the high temperatures in tropical region could be reasons for this uniform distribution (Iwata et al. 1994; Tanabe 2000). However, an interesting result was observed for HCHs, showing relatively higher concentrations in Mong Cai, a site near the border of China, than other locations toward the southern areas (see Fig. 2, Table 2). China has been known as one of the top global HCH users (Li 1999), and this may be the reason for such high concentrations in Mong Cai.

A recent study in the Sai Gon–Dong Nai River that examined residue concentrations of POPs in sediments collected from the canals of the metropolitan area in Hochiminh, the river, and downstream toward the coastal areas showed an interesting geographical distribution (Fig. 3) (Minh TB et al. 2005; Minh NH et al. 2007a). Concentrations of DDTs and PCBs were found to be highest in sediments from the city canals and showed an apparent decreasing trend toward the river and coastal areas. Residues in sediments collected from downstream and coastal areas were much lower than those in sediments from the river and city canals. This result clearly demonstrated the role of the urban areas in Hochiminh City with extensive industrial and human activities as the point sources of POPs pollution. The

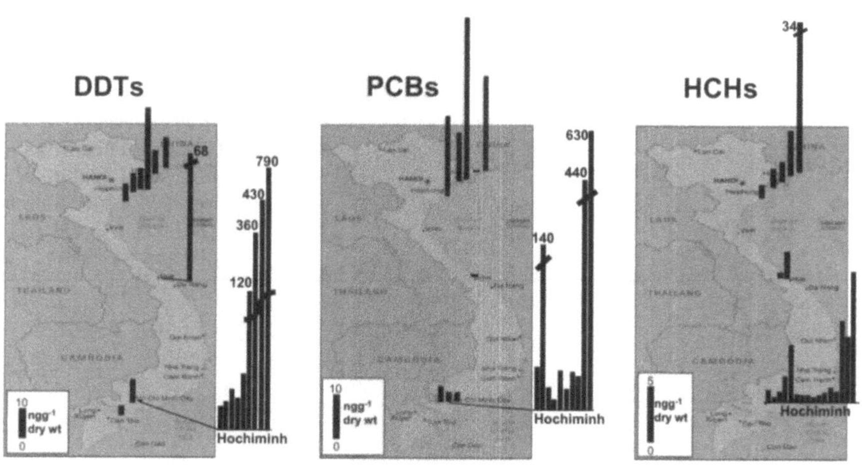

Fig. 2. Distribution of persistent organochlorines in sediment from various locations in Vietnam (Iwata et al. 1994; Nhan et al. 1998, 1999).

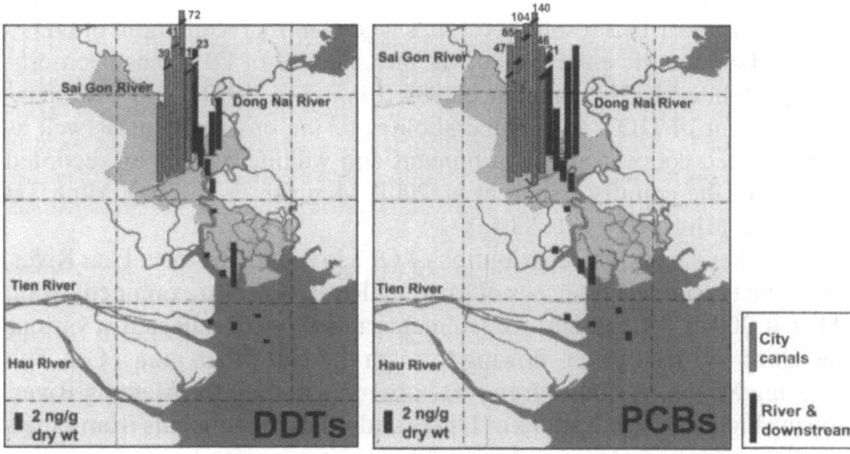

Fig. 3. Distribution of DDTs and PCBs in sediments from Sai Gon–Dong Nai River region, southern Vietnam (Minh TB et al. 2005; Minh NH et al. 2007a).

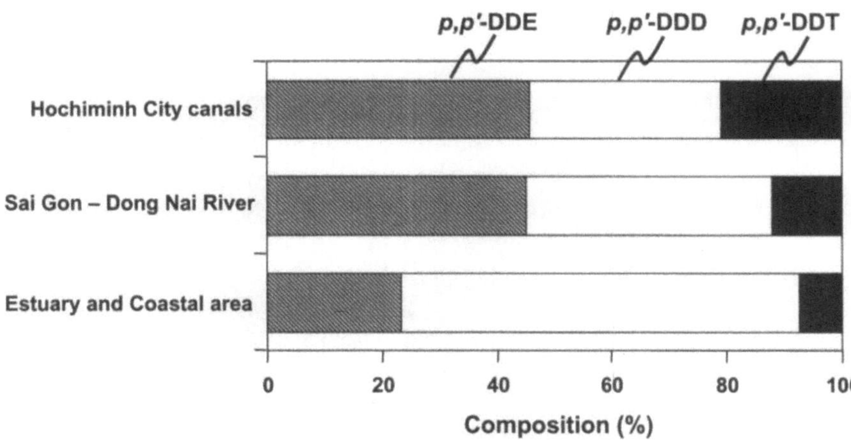

Fig. 4. Composition of DDT compounds in sediments from Sai Gon–Dong Nai River region, southern Vietnam (Minh TB et al. 2005; Minh NH et al. 2007a).

results on compositions of DDT compounds also further substantiate our findings (Fig. 4). Proportions of *p,p′*-DDT to total DDTs concentrations were found to be the highest in sediments collected from canals of metropolitan areas, followed by those in the sediments from the river and coastal areas. It is known that a higher proportion of *p,p′*-DDT indicates more recent residues of DDTs in the environment and biota (Strandberg et al. 1998).

This accumulation pattern is in good agreement with the results on the spatial distribution of total DDT concentrations, showing higher contamination in city canals and decreasing trends toward the downstream and coastal areas. This result therefore suggests more recent input of DDTs to the urban aquatic environment. In fact, the use of DDT has been officially prohibited in Vietnam since 1995 (Sinh et al. 1999). However, continuous input of DDT and its metabolites to the environment as well as their elevated concentrations in humans and wildlife have been recorded throughout the country (Nhan et al. 2001; Monirith et al. 2003; Minh TB et al. 2002; Minh NH et al. 2004a).

In another survey, the distribution of OCs in sediments from Hau River, one of the two largest branches of Mekong River, was also examined (Minh NH et al. 2004b, 2007b). Surface sediment samples were collected at various sites along the Hau River, flowing through the two urban areas, Can Tho and Long Xuyen. A similar trend was observed here in the Mekong River, showing higher concentrations of DDTs and PCBs in sediments from canals near the urban areas of Can Tho and Long Xuyen, with levels decreasing downstream (Fig. 5). In particular, DDT residues from the site near Can Tho and Long Xuyen were about 10–20 fold higher than those in the respective downstream sites. This result indicates that urban areas are

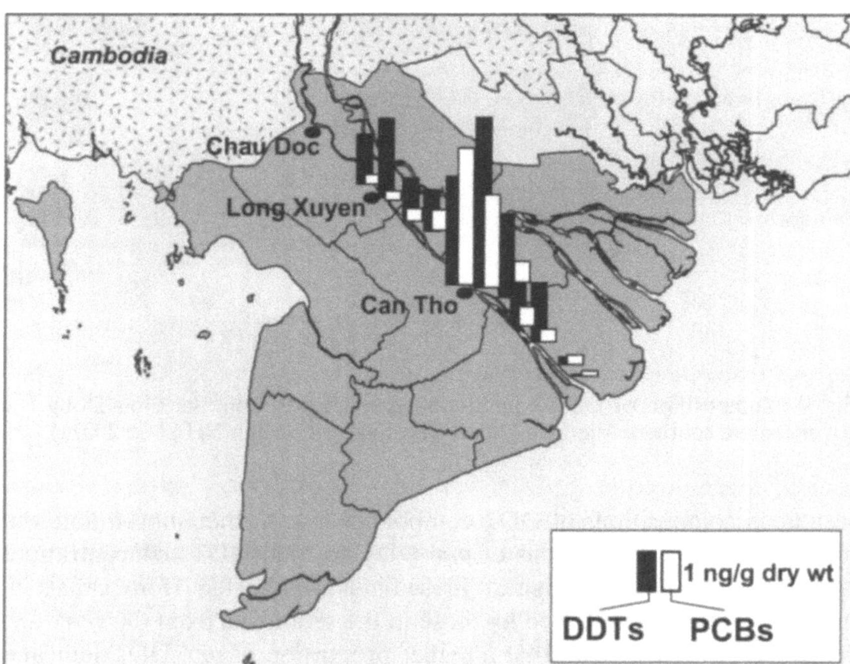

Fig. 5. Distribution of DDTs and PCBs in sediments from Hau River, the largest branch of the Mekong River, southern Vietnam (Minh NH et al. 2004b, 2007b).

major pollution sources of POPs. Inputs of DDT residues are probably derived from its application for hygienic purposes and malaria vector control in urban areas rather than DDT used for agricultural purposes.

As for soil samples, an extensive survey in northern and southern Vietnam indicated higher concentrations of DDTs from paddy field sites than from upland areas (see Table 2; Thao et al. 1993a,b). This result clearly reflects the status of DDT use as an insecticide in the past in Vietnam. In some specific sites in Tay Ninh, southern Vietnam, where a U.S. Army base was located, elevated PCB levels were recorded (Thao et al. 1993a,b). A recent survey in the open dumping sites for municipal wastes in Hanoi and Hochiminh City, the two largest cities in Vietnam, revealed that concentrations of DDTs and PCBs in soils collected from dumping sites were much higher than those in paddy fields far from the dumping sites (Fig. 6; Minh NH et al. 2006a). PCBs and OC insecticides likely originated from continuous loadings of municipal waste containing residues of these compounds. The open dumping sites therefore, may act as reservoir sources of PCBs and OC insecticides.

To understand the magnitude of contamination of POPs in Vietnam, concentrations in water and sediments were compared with those in other countries in Asia-Pacific (Fig. 7). Higher contamination of DDTs in the Vietnamese coastal environment was noticed, again indicating its extensive use of in Vietnam. Interestingly, elevated PCBs residues were also observed in water and sediments from the Mekong River estuary, southern Vietnam, and the levels were comparable to those reported in some locations in India, Japan, and Australia (Fig. 7). High PCB contamination in Vietnam observed

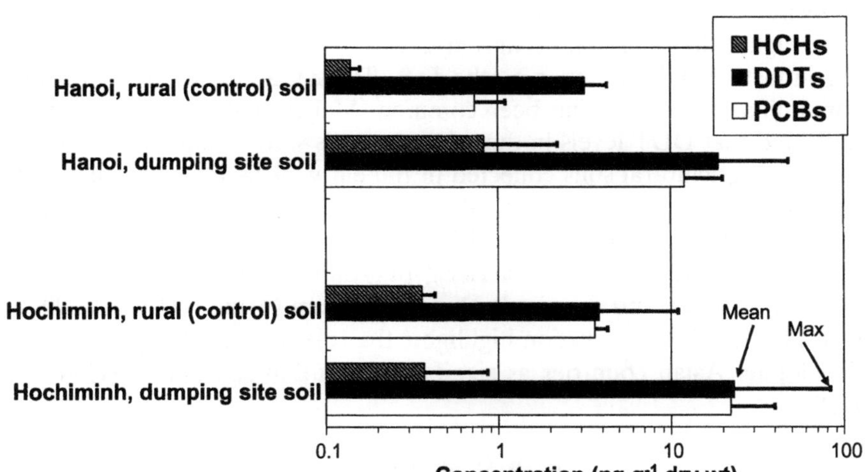

Fig. 6. Residue concentrations of persistent organochlorines in soils from open dumping sites and rural areas (control sites) in Hanoi and Hochiminh City, the two largest cities in Vietnam (Minh NH et al. 2006a).

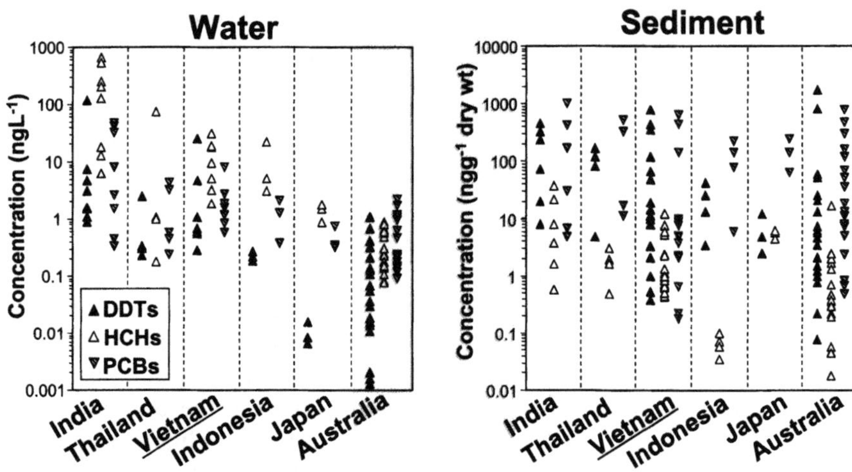

Fig. 7. Comparison of persistent organochlorine residues in surface water and sediment from different countries in Asia-Pacific (Iwata et al. 1994).

during our survey in the early 1990s could be derived from electrical equipment imported from industrialized nations, e.g., the former Soviet Union and Australia, and from leakage from army weapons extensively used in the Vietnam War during 1961–1971 (Thao et al. 1993a,b).

Available data on the worldwide comparison of PCBs and OC insecticide levels in open dumpsites are rather scarce. A comparison of PCBs and DDTs levels in dumping sites from Vietnam with those in soils from other countries is given in Fig. 8. In general, the levels of PCBs in dumping sites were higher than those in background soils from various countries, including industrialized nations, e.g., the U.S., Russia, and Italy, where heavy contamination of PCBs has been common (Meijer et al. 2003). Our survey indicated that DDT levels in dumping site soils were comparable to those found in agricultural soils collected in the early 1990s from various countries such as Russia (Iwata et al. 1995), Ireland (McGrath 1995), and Slovakia (Marta et al. 1997), and higher than in most of the urban soils collected in many countries such as Egypt (Ahmed et al. 1998), Korea (Kim and Smith 2001), and the rural areas of Vietnam (Minh NH et al. 2006a) (see Fig. 8). This observation highlights the role of open dumping sites in developing Asian countries as significant pollution sources of PCBs and DDTs.

B. Biological Samples

The suitability of using a wide range of biological samples as bioindicators for monitoring POPs contamination was extensively discussed by Tanabe and Subramanian (2006). Surveys examined POP residue levels in various

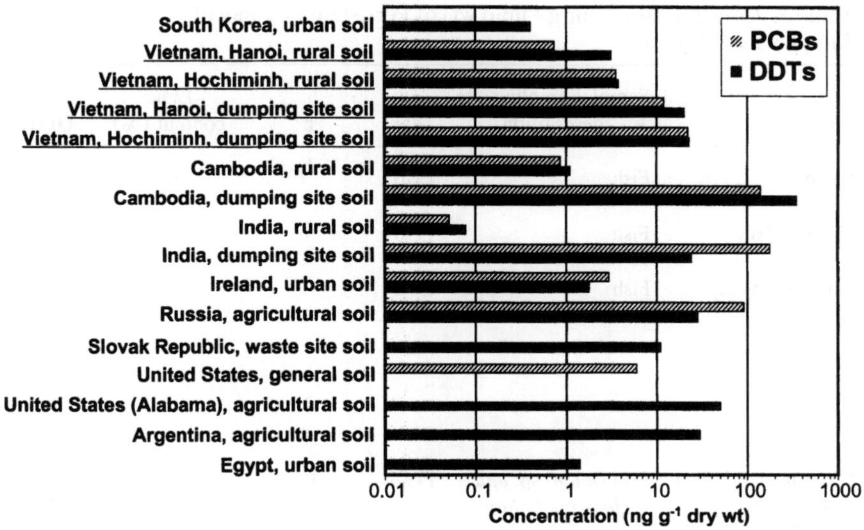

Fig. 8. Comparison of organochlorine residue levels in soils from different countries.

biological samples such as fish, bivalves, and birds from the Asia-Pacific region, including Vietnam, have made clear the status of contamination, distribution, and source allocation. An extensive study on OC contamination in fish from Asia and Oceania including Vietnam was carried out by Kannan et al. (1995). Similar to sediment samples, a relatively uniform distribution of OCs was observed in fish in various countries in the Asia-Pacific region. In Vietnam, residues of DDTs were relatively high in the surveys conducted in 1990 and 1997 (Table 3). Mussels collected from coastal areas in north and middle Vietnam contained elevated DDT concentrations (Table 3; Monirith et al. 2003). Subsequent surveys conducted by Nhan et al. (1998, 1999) examined OC distribution in clams from different sites along the northern coast, showing a very similar distribution in sediment samples. In particular, residue concentrations of DDTs, HCHs, and PCBs were relatively high in the sites near the border of China, and a decreasing trend was noticed toward the southern coastlines (Table 4). At the two estuary areas, Hai Phong harbor with extensive human and industrial activities, and Thai Binh province, one of the high rice production areas in Vietnam, higher concentrations were again observed. In general, the distribution and magnitude of contamination in sediment and biota (fish and bivalves) was very similar at both the local and regional scale, which can be attributed to the enhanced volatilization of semivolatile organic compounds because of the high temperatures prevailing in tropical ecosystems.

Polybrominated diphenyl ethers (PBDEs) are emerging contaminants that have received considerable attention in recent years. We conducted a

Table 3. Concentrations (ng g^{-1} lipid wt) of Persistent Organochlorines in Biological Samples from Vietnam.

Location	Sample/sampling site description	Year	n	Lipid content (%)	DDTs
Hanoi	Fish	1990	7	1.9	1,900 (680–4,000)[a]
Phu Da, Hue	Fish	1990	6	1.9	1,100 (210–2,700)
Hochiminh City	Fish	1990	6	1.9	1,100 (89–4,100)
Con Lu island, Red River estuary, northern Vietnam	Fish (*mugil* sp. *and chlorophthalmus* sp.)	1997	10	3.2	4,200 (330–8,500)
Con Lu island, Red River estuary	Shrimp	1997	20 (pooled as 1 sample)	2.8	160
Cat Ba island, Cai Hai province	Green Mussel, floating habitat	1997	38 (pooled as 1 sample)	1.1	530
Cat Hai province	Green Mussel, floating habitat	1997	34 (pooled as 1 sample)	0.9	300
Cat Hai province	Green Mussel, floating habitat	1997	8 (pooled as 1 sample)	0.7	2,500
Cat Hai province	Green Mussel, floating habitat	1997	12 (pooled as 1 sample)	2	420
Lach Truong, Thanh Hoa	Green Mussel, aquaculture	1997	33 (pooled as 1 sample)	1.2	610
Ron River estuary, Ky Anh	Green Mussel, fishing village	1997	50 (pooled as 1 sample)	0.6	470
Lang Co, Hue	Green Mussel	1997	143 (pooled as 1 sample)	0.9	34,000
Thi Nai, Binh Dinh	Green Mussel, urban, shipping traffic, aquaculture	1997	54 (pooled as 1 sample)	1.1	220
Phan Ri estuary, Phan Ri	Green Mussel, urban, fishing village	1997	30 (pooled as 1 sample)	1.1	240

HCHs	CHLs	PCBs	Reference	Remarks
120 (47–210)	7.9 (<0.5–17)	580 (270–950)	Kannan et al. 1995	DDTs: sum of *p,p′*-DDE, *p,p′*-DDD and
48 (32–74)	8.9 (>0.5–12)	630 (160–1,300)	Kannan et al. 1995	*p,p′*-DDT HCHs: sum of α-, β-, and
110 (33–200)	3.2 (<0.5–18)	950 (190–3,100)	Kannan et al. 1995	γ-HCH CHLs: sum of *cis*-chlordane, *trans*-chlordane, *cis*-nonachlor, *trans*-nonachlor and oxychlordane
350 (27–210)	110 (29–210)	110 (81–140)	Minh et al. 2002	PCBs: quantified by an equivalent mixture of Kanechlor perparations (KC-300, KC-400, KC-
150	3.9	100	Minh et al. 2002	500 and KC-600)
3.6	14	86	Monirith et al. 2003	DDTs: sum of *p,p′*-DDE, *p,p′*-DDD and
12	12	20	Monirith et al. 2003	*p,p′*-DDT HCHs: sum of α-, β-, and γ-HCH
5.7	24	450	Monirith et al. 2003	CHLs: sum of *cis*-chlordane, *trans*-chlordane,
3	5	110	Monirith et al. 2003	*cis*-nonachlor, *trans*-nonachlor and oxychlordane
3.3	13	65	Monirith et al. 2003	PCBs: quantified by an equivalent mixture of
5.5	20	190	Monirith et al. 2003	Kanechlor perparations (KC-300, KC-400, KC-500 and KC-600)
10	—	380	Monirith et al. 2003	
6.3	36	26	Monirith et al. 2003	
2.9	11	80	Monirith et al. 2003	

Table 3. (*cont.*)

Location	Sample/sampling site description	Year	n	Lipid content (%)	DDTs
Hanoi	Clam (*hyriopsis*), irrigation canal	1996	15 (pooled as 1 sample)	—	7,400
Hanoi	Carp (*cyprinus carpio*), rice field	1996	3 (pooled as 1 sample)	—	22,000
Balat estuary, Thai Binh	Shrimp (*metapenaeus*)	1996	35 (pooled as 1 sample)	—	1,000
Balat estuary, Thai Binh	Clam (*Meretrix meretrix*)	1996	43 (pooled as 1 sample)	—	830
Diem Dien estuary, Thaibinh	Clam (*Mactra quadrangularis*), intertidal mudflat coastal area	1996	13 (pooled as 1 sample)	—	1,000
Tra Co, Mong Cai, northern coast of Vietnam	Clam (*Meretrix meretrix*), intertidal mudflat areas	1997	1 pooled sample	—	1,200
Mong Duong, northern coast of Vietnam	Clam (*Meretrix meretrix*), intertidal mudflat areas	1997	1 pooled sample	—	590
Ha Long, Quang Ninh, northern coast of Vietnam	Clam (*Meretrix meretrix*), intertidal mudflat areas	1997	1 pooled sample	—	660
Hai Phong, northern coast of Vietnam	Clam (*Meretrix meretrix*). intertidal mudflat areas	1997	1 pooled sample	—	850
Ba Lat estuary, Thai Binh, northern coast of Vietnam	Clam (*Meretrix meretrix*), intertidal mudflat areas	1997	1 pooled sample	—	920
Cau Dien, Nhue River, Hanoi	Freshwater snails (*Angulyagra* sp.), densely populated industrial	1997	1 pooled sample	—	23,000
Cau Dien, Nhue River, Hanoi	Freshwater snails (*Angulyagra* sp.), densely populated industrial transformer production	1997	1 pooled sample	—	45,000
Nhue River, suburb Hanoi	Freshwater snails (*Angulyagra* sp.), rural area	1997	1 pooled sample	—	320

HCHs	CHLs	PCBs	Reference	Remarks
97	—	480	Nhan et al. 1998	DDTs: sum of *p,p′*-DDE, *p,p′*-DDD and *p,p′*-DDT HCHs: sum of α-, β-, and γ-HCH PCBs: sum 13 congeners
370	—	2,600	Nhan et al. 1998	
390	—	490	Nhan et al. 1998	
57	—	220	Nhan et al. 1998	
30	—	200	Nhan et al. 1998	
2,400	—	900	Nhan et al. 1999	DDTs: sum of *p,p′*-DDE and *p,p′*-DDD and *p,p′*-DDT HCHs: sum of α-, β-, and γ-HCH PCBs: sum 13 congeners
400	—	480	Nhan et al. 1999	
93	—	470	Nhan et al. 1999	
77	—	580	Nhan et al. 1999	
71	—	250	Nhan et al. 1999	
29	14	720	Nhan et al. 2001	DDTs sum of *p,p′*-DDE, *p,p′*-DDD, *p,p′*-DDT, *o,p′*-DDE, *p,p′*-DDD, *p,p′*-DDT and DDMU PCBs: sum 13 congeners HCHs: sum of α-, β-, and γ-HCH CHLs: sum of *cis*-chlordane, *trans*-chlordane, *trans*-nanachlor and heptachlor
21	27	2,200	Nhan et al. 2001	
9.8	—	330	Nhan et al. 2001	

Table 3. (*cont.*)

Location	Sample/sampling site description	Year	n	Lipid content (%)	DDTs
Hanoi downtown	Freshwater snails (*Angulyagra* sp.), densely populated industrial area	1997	1 pooled sample	—	2,700
To Lich River, suburb Hanoi	Freshwater snails (*Angulyagra* sp.), densely populated industrial area	1997	1 pooled sample	—	5,300
Red River, suburb Hanoi	Freshwater snails (*Angulyagra* sp.), rural area	1997	1 pooled sample	—	910
Gia Lam, suburb Hanoi	Freshwater snails (*Angulyagra* sp.), rural area	1997	1 pooled sample	—	1,000
Dong Anh, suburb Hanoi	Freshwater snails (*Angulyagra* sp.), rural area	1997	1 pooled sample	—	640
Mekong River, Can Tho	Catfish (*Pangasianodon hypophthalmus*), common aquaculture	2004	20	3.8 (0.6–7.2)	59 (7.9–150)
Can Tho	Catfish (*Clarias* sp.), from a pond near municipal dumping site	2004	5	3.6 (3.2–4.1)	390 (330–700)
Con Lu island, Red River estuary, northern Vietnam	Resident birds (including 7 species), wetland wintering grounds	1997	16	1.9–16	6,200(1,100–13,000)[b]
Con Lu island, Red River estuary, northern Vietnam	Migratory birds (including 17 species), wetland wintering grounds	1997	84	4.1–33	2,900 (750–6,800)

[a]Mean (range).

[b]Average concentrations of 7 resident and 17 migratory species.

HCHs	CHLs	PCBs	Reference	Remarks
13	1.8	1,000	Nhan et al. 2001	
19	—	3,000	Nhan et al. 2001	
8.2	2.2	480	Nhan et al. 2001	
15	—	340	Nhan et al. 2001	
11	1	280	Nhan et al. 2001	
0.47 (<0.03–1.5)	0.62 (<0.01–2.6)	7.2 (0.91–27)	Minh et al. 2005, 2006	DDTs: sum of *p,p′*-DDE, *p,p′*-DDD and *p,p′*-DDT HCHs: sum of α-, β-, and γ-HCH
2.2 (0.86–5.1)	5.7 (4.2–8.2)	50 (37–77)	Minh et al. 2005, 2006	
150 (23–310)	100 (5–550)	780 (250–2,400)	Minh et al. 2002	CHLs: sum of *cis*-chlordane, *trans*-chlordane, *cis*-nonachlor, *trans*-nonachlor and oxychlordane PCBs: quantified by an equivalent mixture of Kanechlor preparations (KC-300, KC-400, KC-500 and KC-600)
330 (20–1,700)	22 (5.3–130)	530 (82–1,600)	Minh et al. 2002	

Table 4. Concentrations of Persistent Organochlorines in Human Samples from Vietnam.

Locations	Year	Samples/sampling site despcription	Unit	*n*	DDTs
Me Tri and Tu Liem, Hanoi	2000	Breast milk, primiparous women near municipal dumping sites	$ng\,g^{-1}$ lipid		2,400
Me Tri and Tu Liem, Hanoi	2000	Breast milk, multiparous women near municipal dumping sites	$ng\,g^{-1}$ lipid		1,700
Me Tri and Tu Liem, Hanoi	2000	Breast milk, overall near municipal dumping sites	$ng\,g^{-1}$ lipid	42	2,100 (480–6,900)[a]
Vinh Loc and Dong Thanh, Hochiminh City	2001	Breast milk, primiparous women near municipal dumping sites	$ng\,g^{-1}$ lipid		3,020
Vinh Loc and Dong Thanh, Hochiminh City	2001	Breast milk, multiparous women near municipal dumping sites	$ng\,g^{-1}$ lipid		1,500
Vinh Loc and Dong Thanh, Hochiminh City	2001	Breast milk, overall near municipal dumping sites	$ng\,g^{-1}$ lipid	44	2,300 (440–17,000)
Tan Than village, suburb Hochiminh City	1985–87	Breast milk, rural area	$ng\,g^{-1}$ lipid	2	10,500
Song Be province, southern Vietnam	1985–87	Breast milk, rural area	$ng\,g^{-1}$ lipid	3	12,000
Hochiminh City	1985–87	Breast milk, urban area	$ng\,g^{-1}$ lipid	7	11,400
Hanoi, rural area	1994	Serum	$ng\,mL^{-1}$	30	12 (1.2–59)
Hanoi, urban area	1994	Serum	$ng\,mL^{-1}$	8	32 (12–68)
Hanoi, breast cancer cases	1994	Serum	$ng\,mL^{-1}$	21	16 (1.2–52)
Hanoi, control samples for breast cancer cohort survey	1994	Serum	$ng\,mL^{-1}$	21	21 (1.5–88)

[a]Mean (range).

HCHs	CHLs	HCB	PCBs	Reference	Remark
69	2.5	4.2	76	Minh et al. 2004	DDTs: sum of *p,p′*-DDE, *p,p′*-DDD and *p,p′*-DDT HCHs: β-HCH only
46	1.4	3.5	72	Minh et al. 2004	CHLs: sum of oxychlordane, *trans*-nonachlor and cis-nonachlor
58 (11–160)	2.0 (<0.72–13)	3.9 (0.62–9.5)	74 (26–210)	Minh et al. 2004	PCBs: quantified by an equivalent mixture of
14	7.8	2.8	88	Minh et al. 2004	Kanechlor preparations (KC-300, KC-400, KC-500 and KC-600)
13	6	2.1	70	Minh et al. 2004	
14 (4.1–35)	6.9 (1.3–2.6)	2.5 (1.3–10)	79 (29–200)	Minh et al. 2004	
25	<2.0	<2.0	49	Schecter et al. 1989a	DDTs: sum of *p,p′*-DDE and *p,p′*-DDT
37	<2.0	10	28	Schecter et al. 1989a	HCHs: sum of α-, β- and γ-HCH; CHLs: oxychlordane on
250	3	3	84	Schecter et al. 1989a	PCBs: sum of CB-138, CB-153 and CB-180
—	—	—	—	Schecter et al. 1997	DDTs: sum of *p,p′*-DDE and *p,p′*-DDT
—	—	—	—	Schecter et al. 1997	
—	—	—	—	Schecter et al. 1997	
—	—	—	—	Schecter et al. 1997	

preliminary survey to examine the levels of PBDEs in common aquaculture catfish from Can Tho, a city located on the Mekong River, in southern Vietnam (Minh NH et al. 2006b). Catfish from a pond located near a municipal waste site of Can Tho were also analyzed to evaluate the role of dumping site as a source of pollution. Interestingly, concentrations of PBDEs and persistent OCs such as PCBs, DDTs, CHLs, and HCHs in the dumping site fish were significantly higher than those in fish collected from a common aquaculture site (Fig. 9, Minh NH et al. 2006b), suggesting an additional exposure of the dumpsite catfish to PBDEs. In the dumping site, municipal wastes including household goods and small electric appliances, that may contain PBDEs as flame retardants, were dumped. Under ambient conditions, PBDEs may be emitted from such materials and consequently contaminate the dumping site soil. Therefore, it is anticipated that runoff and leaching water from the dumping site during flood and rains may have carried PBDEs to other vicinities, causing higher contamination in the catfish. Higher levels in fish from this study further highlighted the role of open dumping sites for municipal wastes as a conspicuous source of various toxic chemicals, including a new group of POPs such as PBDEs.

To our knowledge, there are very few studies on OC contamination in higher trophic animals from Vietnam. A survey conducted in 1997 determined OC concentrations in resident and migratory birds from North Vietnam (Minh TB et al. 2002). Resident birds contained higher concentrations of DDTs than those in migrants (Fig. 10); this indicates recent exposure to DDTs in resident birds from North Vietnam, where elevated DDT contamination is very common, as discussed earlier. Interestingly, accumulation of HCHs revealed a contrasting pattern, showing apparently greater concentration in migratory birds (Fig. 10),

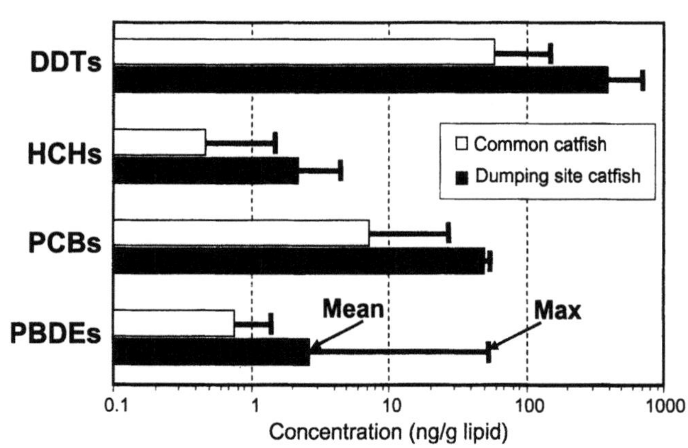

Fig. 9. Comparison of PBDE and persistent organochlorine residues in catfish from open dumping sites and in common aquaculture catfish from Can Tho City, Mekong River, southern Vietnam (Minh NH et al. 2006b).

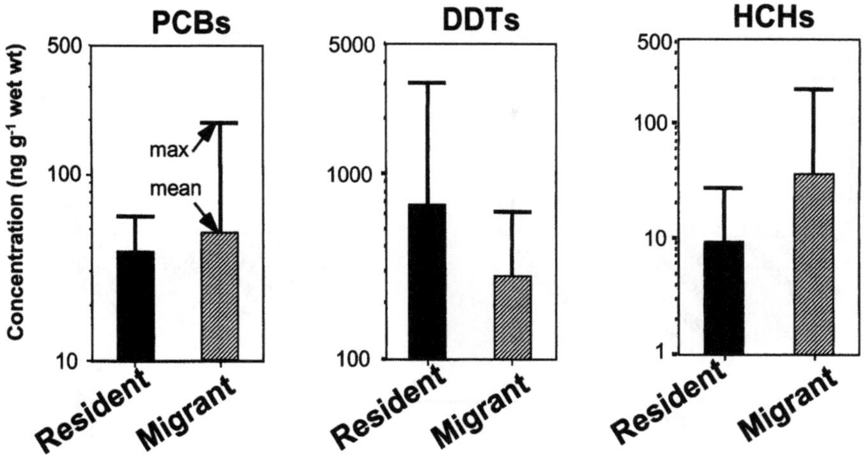

Fig. 10. Accumulation of persistent organochlorines in birds from Vietnam according to migratory behavior (Minh TB et al. 2002).

which could be the result of accumulation in stopover sites during migration in some polluted areas such as India, southern China, and Japan. The role of these countries as potential sources of HCH accumulation in wintering migratory birds breeding in Lake Baikal, Russia, has also been suggested in a recent study (Kunisue et al. 2002). Concentrations of PCBs were similar in residents and migratory species and levels were relatively low, indicating a smaller source of PCBs in North Vietnam in recent years. Thus, accumulation pattern of OCs in birds from North Vietnam according to their migratory behavior reflects the status of each contaminant. This phenomenon also suggests the suitability of using birds as bioindicators for monitoring POPs in the environment.

International comparison of OC residues in fish, mussels, and birds from Vietnam indicated relatively higher levels of DDTs in Vietnamese samples (Fig. 11). Results of recent surveys in the Asia-Pacific Mussel Watch Program showed that DDT levels in Vietnamese mussels were lower than in those in southern China and Hong Kong, but higher than those in samples collected from most of the other countries in the East Asian region (Fig. 11; Monirith et al. 2003). Kannan et al. (1995) reported mean concentrations of DDT in fish from Vietnam were the highest when compared with India, Thailand, Indonesia, and Australia. Contamination patterns were similar for birds, which showed that DDT levels in some resident species in Vietnam were among the highest values reported for the Asian countries surveyed (see Fig. 11; Minh TB et al. 2002). Although the recent amounts of DDT used in Vietnam were lower than those of other countries in the region, the extent of DDT contamination in environmental samples in Vietnam is higher. This observation suggests that the application

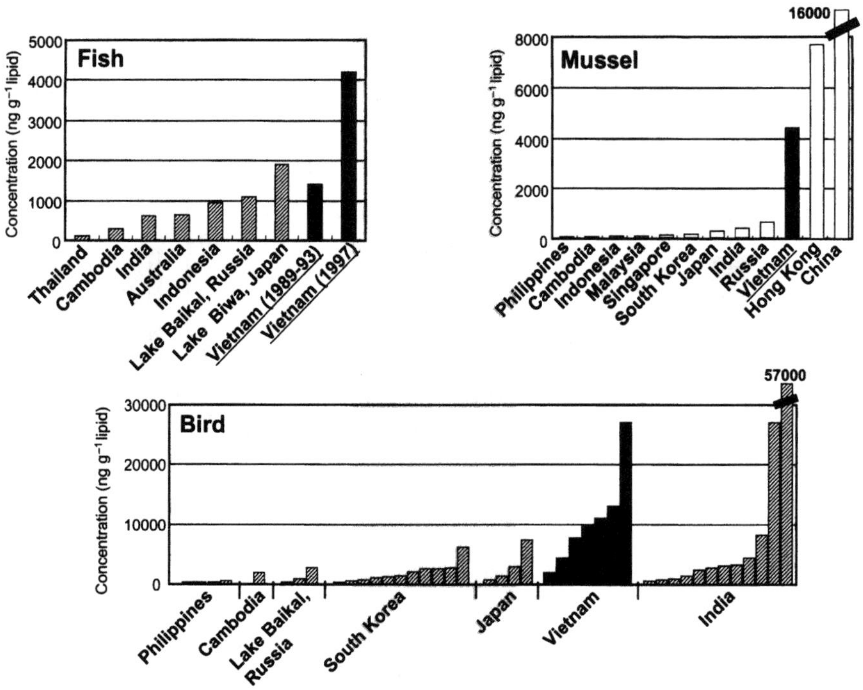

Fig. 11. Comparison of DDTs residues in fish, mussels, and birds in Asia-Pacific countries. [From: fish: Kannan et al. (1995), Nakata et al. (1995), Guruge et al. (1997), Monirith et al. (2000), Minh TB et al. (2002); mussels: Monirith et al. (2003); birds: Hoshi et al. (1998), Choi et al. 2001, Kunisue et al. (2002, 2003), Minh TB et al. (2002), Tanabe et al. (1998)].

of DDT in Vietnam has continued until very recently, resulting in elevated contamination of these compounds in different species occupying low to high trophic levels in the food chain.

C. Foodstuffs

The first examination of OC contamination in foodstuffs from Vietnam was conducted by Schecter et al. (1990a). Pork and chicken fat were analyzed, and relatively high levels of DDTs were found (190–5100 ng/g lipid). However, residues levels of other contaminants such as HCHs, CHLs, HCB, and PCBs were relatively low.

A subsequent investigation done by Kannan et al. (1992) in the samples collected during January 1990 and May 1991 examined various kinds of foodstuffs including rice, pulses, oil, butter, animal fat, meat, fish, and seafood from Hanoi, northern Vietnam, and Tay Ninh and Hochiminh City, southern Vietnam. The results demonstrated significant concentrations of PCBs in foods from both the northern and southern parts, which is consistent with the data reported for soils collected at the same locations (Thao et al. 1993a,b). DDT concentrations were prominent in all kinds of foodstuffs, particularly in oil and animal fat. The ongoing application of DDT at the time of this investigation was also evident. In particular, composition of DDT compounds showed that *p,p′*-DDT was the predominant isomer in rice, pulses, and oils, accounting for approximately 50% of the total DDT concentrations (Kannan et al. 1992). Very recently, Schecter et al. (2003) reported some alarmingly elevated concentrations of PCDD/Fs in foodstuffs from Bien Hoa City, southern Vietnam, where Agent Orange was sprayed during the Vietnam War. Several samples of duck, toad, chicken, and snakehead fish collected from the former U.S. Army base site contained elevated concentrations of DDTs. The high levels of DDTs found in meat of livestock and fish from these studies indicate the application of technical DDT to prevent malaria near cattle sheds, stagnant water bodies, and drainage channels, which are common breeding sites for mosquitoes (Kannan et al. 1992).

D. Human Exposure

Available data on human exposure to organochlorine insecticides and PCBs in Vietnamese residents are limited. A preliminary survey by Schecter et al. (1989a) reported very high concentrations of DDTs and HCHs in human breast milk from some locations in and around Hochiminh. DDT levels were higher in both rural and urban areas, ranging from 10,500 to 12,000 ng/g lipid. PCB levels, however, were relatively low compared to other countries. A subsequent study conducted in 2000–2001 provided more extensive data on human exposure and insights into the accumulation kinetics of PCBs and OC insecticides such as DDTs, HCHs, CHLs, HCB, and tris(4-chlorophenyl)methane (TCPMe), a newly detected environmental contaminant that exhibits weak endocrine-disrupting properties. Breast milk from 96 nursing women living near the dumping sites of municipal wastes in Hanoi and Hochiminh, the two largest metropolitan cities in Vietnam, was analyzed for these contaminants (see Table 4). In general, similar degrees of exposure to DDTs, CHLs, HCB, and PCBs were observed among samples from Hanoi and Hochiminh. Interestingly, HCH residues in breast milk of women from Hanoi were significantly higher than those in Hochiminh, suggesting recent high background levels of HCHs in the northern compared to the southern region. Consistent results where also observed in the survey on sediment and bivalves from various locations along the northern coast of Vietnam, which demonstrated relatively higher residues in sites near the China border (Nhan et al. 1998, 1999). Earlier reports also pointed out similar spatial distribution in different kinds of environmental samples, showing higher levels of HCHs in Hanoi compared to Hochiminh (Thao et al. 1993a; Iwata et al. 1994; Kannan et al. 1995). In addition to the influence of the possible transport from China, the world leading HCH user, differences in climate between Hanoi and Hochiminh could be an alternative explanation. The Mekong River delta in southern Vietnam is characterized by the typical tropical climate with high temperatures and heavy rainfall. Rapid volatilization of highly volatile HCH isomers may therefore be enhanced in southern Vietnam, resulting in lower residues in various environmental and human samples.

Similar to those observed in environmental samples, human exposures to DDTs in Vietnam were among the highest in the developing as well as developed nations. As discussed earlier, high DDT contamination in Vietnam has been apparent in many environmental samples. This is also the case for humans, which raises concern over the possible impacts on human health. These results suggest that Vietnam is a potential source of DDTs in the South Asian region.

E. Dioxin Contamination

Southern Vietnam has long been considered as a well-known region where Agent Orange was extensively sprayed during 1961–1971 in the Vietnam War, resulting in severe dioxin contamination of various environmental media and the food chain, including humans. During the past three decades, Schecter and coworkers have been conducting investigations on dioxin contamination in southern Vietnam, including sediments, foodstuffs, and, particularly, samples from humans living near the "hot spot" of dioxin contamination (Table 5). In general, dioxin contamination as a result of Agent Orange can be characterized by the predominant of 2,3,7,8-

Table 5. Concentration of PCDD/Fs and Dioxin-Like PCBs in Different Environmental Samples from Vietnam.

Location	Year	Sample/sampling site description	Unit	*n*	PCDDs
Soil, sediment					
Saigon River	1984/85	Sediment, industrial area	$pg\,g^{-1}$ dry wt	1	5,800
Dong Nai River	1984/85	Sediment, near Orange Agent spraying during Vietnam War	$pg\,g^{-1}$ dry wt	5	1,200 (830–1,400)
Red River	1984/85	Sediment, river	$pg\,g^{-1}$ dry wt	2	240 (220–250)
Bien Hoa	2000	Soil, near the Air Base, former Agen Orange Storage	$ng\,g^{-1}$ dry wy	4	590 (0.039–1,645)

tetrachlorodibenzo-*p*-dioxin (2,3,7,8-TCDD), the major contaminants of the herbicide 2,4,5-T, a constituent of Agent Orange. 2,3,7,8-TCDD is one of the most toxic congeners and has received considerable attention. Research has shown that this congener can caused large number of adverse health effects in animals (ATSDR 1998). As seen in Table 5, 2,3,7,8-TCDD was the major component, and its concentrations were higher in foods and human samples including blood, adipose tissues, and breast milk.

Dwernychuk et al. (2002) surveyed soils, sediments, and foods from Aluoi valley, a "hot spot" of Agent Orange spraying and former U.S. Army base sites. The highest concentrations of PCDD/Fs in soil were 2200 pg/g dry wt, which is much higher than those in background levels in many industrialized countries. Of greater concern, however, are high levels of PCDD/Fs in breast milk, a major dietary intake for breast-fed children. A survey throughout the valley indicates that breast-fed infants of primipara groups had intake values 27 fold exceeding the Tolerable Daily Intake (TDI) proposed by WHO. This fact highlights a new environmental issue: that the first child is exposed to high levels of contaminants and will always be at higher risk. Breast-feeding has been highly recommended by various organizations and scientists because of the benefits for children, such as passing antibodies to the infants and reducing the risk of allergic reactions. However, for communities with high exposure to toxic contaminants such as people in underdeveloped countries such as Vietnam, breast-feeding unfortunately has become an effective route of transfer of toxic contaminants to the next generation. Scientific and social efforts are therefore urgently needed to mitigate exposure to reduce body burdens of dioxins in both adults and children.

PCDFs	Total PCDD/Fs	Dioxin-like PCBs	Reference	Remark
1,000	6,800	—	Schecter et al. 1989b	
54 (5–140)	1,300 (840–1,500)	—	Schecter et al. 1989b	
4.2 (3.4–4.9)	240 (220–255)	—	Schecter et al. 1989b	
93 (39–147)	680 (39–1,790)	—	Schecter et al. 2001	

Table 5. (*cont.*)

Location	Year	Sample/sampling site description	Unit	*n*	PCDDs
Bien Hung Lake, Bien Hoa	2000	Sediment, former Air Base	$pg\,g^{-1}$ dry wt	3	340 (199–532)
Bien Hung Lake, Bien Hoa	2000	Sediment, close to the former Air Base, Agent Orange Storage	$pg\,g^{-1}$ dry wt	3	1,640 (1,410–1,970)
Bien Hung Lake, Bien Hoa	2000	Sediment, downstream and upstream	$pg\,g^{-1}$ dry wt	2	600 (500–700)
Dong Nai River, southern Vietnam	2000	Sediment, reservoir of Bien Hung Lake	$pg\,g^{-1}$ dry wt	2	630 (540–715)
Hanoi	2000	Sediment, control samples for samples near former Air Base	$pg\,g^{-1}$ dry wt	1	400
Aso, Aluoi valey, Hue	1996	Soil, former US Army Force Base	$pg\,g^{-1}$ dry wt	1	650
Aso, Aluoi valey, Hue	1996	Sediment, pond, former US Army Force base	$pg\,g^{-1}$ dry wt	1	500
Aso, Aluoi valey, Hue	1997	Soil, former US Army Force Base	$pg\,g^{-1}$ dry wt	1	1,600
Aso, Aluoi valey, Hue	1997	Soil, former US airstrip	$pg\,g^{-1}$ dry wt	1	880
Aso, Aluoi valey, Hue	1997	Soil, manioc fields	$pg\,g^{-1}$ dry wt	1	170
Aso, Aluoi valey, Hue	1997	Sediment, pond, former US Army Force base	$pg\,g^{-1}$ dry wt	4	150 (32–250)
Aso, Aluoi valey, Hue	1999	Soil, former US Army Force Base	$pg\,g^{-1}$ dry wt	10	360 (98–730)
Tabat, Aluoi valey, Hue	1999	Soil, former US Army Force Base	$pg\,g^{-1}$ dry wt	9	740 (430–1,100)
Aluoi, Aluoi valey, Hue	1999	Soil, former US Army Force Base	$pg\,g^{-1}$ dry wt	9	270 (110–430)
Adot, Aluoi valey, Hue	1999	Soil, areas sprayed with Agent Orange	$pg\,g^{-1}$ dry wt	3	350 (48–550)

PCDFs	Total PCDD/Fs	Dioxin-like PCBs	Reference	Remark
10 (9–11)	1,050 (208–542)	—	Schecter et al. 2001	
107 (89–134)	1,720 (1,510–2,100)	—	Schecter et al. 2001	
24 (19–29)	625 (720–530)	—	Schecter et al. 2001	
4.4 (2.0–6.8)	630 (544–720)	—	Schecter et al. 2001	
70	470	—	Schecter et al. 2001	
92	740	—	Dwernychuk et al. 2002	
5.6	510	—	Dwernychuk et al. 2002	
98	1,700	—	Dwernychuk et al. 2002	
80	960	—	Dwernychuk et al. 2002	
7.1	180	—	Dwernychuk et al. 2002	
2.7 (1.1–3.7)	150 (33–250)	—	Dwernychuk et al. 2002	
33 (5.6–75)	390 (100–810)	—	Dwernychuk et al. 2002	
14 (5.2–29)	750 (440–1,100)	—	Dwernychuk et al. 2002	
8.9 (5.0–19)	280 (120–440)	—	Dwernychuk et al. 2002	
2.7 (2–3.1)	350 (50–550)	—	Dwernychuk et al. 2002	

Table 5. (*cont.*)

Location	Year	Sample/sampling site description	Unit	*n*	PCDDs
Huonglam, Aluoi valey, Hue	1999	Soil, areas sprayed with Agent Orange	$pg\,g^{-1}$ dry wt	2	150 (72–220)
Huongphong, Aluoi valey, Hue	1999	Soil, areas sprayed with Agent Orange	$pg\,g^{-1}$ dry wt	2	260 (240–280)
Phuvinh, Aluoi valey, Hue	1999	Soil, areas sprayed with Agent Orange	$pg\,g^{-1}$ dry wt	2	600 (560–630)
Hongthuong, Aluoi valey, Hue	1999	Soil, areas sprayed with Agent Orange	$pg\,g^{-1}$ dry wt	1	2,200
Bodot market, Aluoi valey, Hue	1999	Soil, areas sprayed with Agent Orange	$pg\,g^{-1}$ dry wt	1	1,300
Sonthuy, Aluoi valey, Hue	1999	Soil, areas sprayed with Agent Orange	$pg\,g^{-1}$ dry wt	2	1,700 (1,200–2,100)
Hongquang, Aluoi valey, Hue	1999	Soil, areas sprayed with Agent Orange	$pg\,g^{-1}$ dry wt	1	88
Aluoi, Aluoi valey, Hue	1999	Soil, areas sprayed with Agent Orange	$pg\,g^{-1}$ dry wt	1	46
Aluoi market, Aluoi valey, Hue	1999	Soil, areas sprayed with Agent Orange	$pg\,g^{-1}$ dry wt	1	830
Hongkim, Aluoi valey, Hue	1999	Soil, areas sprayed with Agent Orange	$pg\,g^{-1}$ dry wt	1	87
Hongvan, Aluoi valey, Hue	1999	Soil, areas sprayed with Agent Orange	$pg\,g^{-1}$ dry wt	2	110 (70–140)
Taymo, Hanoi	2000	Soil, open dumping site for	$pg\,g^{-1}$ dry wt	9	3,100
		municipal wastes	TEQ ($pg\,g^{-1}$ dry wt)		31
Taymo, Hanoi	2000	Soil, paddy field, control sites for	$pg\,g^{-1}$ dry wt	1	360
		Taymo open dumping site	TEQ ($pg\,g^{-1}$ dry wt)		0.94
Dongthanh, Hochiminh City	2001	Soil, open dumping site for	$pg\,g^{-1}$ dry wt	8	320
		municipal wastes	TEQ ($pg\,g^{-1}$ dry wt)		1.4

PCDFs	Total PCDD/Fs	Dioxin-like PCBs	Reference	Remark
8.7 (1.3–16)	160 (73–240)	—	Dwernychuk et al. 2002	
6.1 (3.9–8.2)	270 (240–290)	—	Dwernychuk et al. 2002	
6.4 (4.3–8.5)	610 (560–640)	—	Dwernychuk et al. 2002	
5.5	2,210	—	Dwernychuk et al. 2002	
110	1,400	—	Dwernychuk et al. 2002	
59 (7.1–110)	1,800 (1,200–2,200)	—	Dwernychuk et al. 2002	
6	94	—	Dwernychuk et al. 2002	
3.3	49	—	Dwernychuk et al. 2002	
14	840	—	Dwernychuk et al. 2002	
2.1	89	—	Dwernychuk et al. 2002	
2.6 (2.3–2.8)	110 (72–140)	—	Dwernychuk et al. 2002	
2,900	6,000 (125–50,500)	2,600 (59–7,100)	Minh et al. 2003	
64	95 (0.4–850)	7.3 (0.22–59)	Minh et al. 2003	
11	370	130	Minh et al. 2003	
0.08	1	0.097	Minh et al. 2003	
47	370 (21–880)	910 (65–2,000)	Minh et al. 2003	
0.8	2.2 (0.02–4.4)	0.49 (0.01–1.0)	Minh et al. 2003	

Table 5. (*cont.*)

Location	Year	Sample/sampling site description	Unit	n	PCDDs
Dongthanh, Hochiminh City	2001	Soil, paddy field, control sites for DongThanh open dumping site	$pg\,g^{-1}$ dry wt	3	160
			TEQ ($pg\,g^{-1}$ dry wt)		0.8
Foodstuff/Biota					
Hanoi	1985–87	Pork	TEQ ($pg\,g^{-1}$ wet wt)	2	—
Hanoi	1985–87	Pork fat	TEQ ($pg\,g^{-1}$ wet wt)	1	—
Hanoi	1985–87	Catfish	TEQ ($pg\,g^{-1}$ wet wt)	1	—
Hanoi	1985–87	Carp	TEQ ($pg\,g^{-1}$ wet wt)	1	—
Hanoi	1985–87	Eel	TEQ ($pg\,g^{-1}$ wet wt)	1	—
Hanoi	1985–87	Cow fat	TEQ ($pg\,g^{-1}$ wet wt)	1	—
Hanoi	1985–87	Chicken fat	TEQ ($pg\,g^{-1}$ wet wt)	1	—
Southern Vietnam	1985–87	Pork fat	TEQ ($pg\,g^{-1}$ wet wt)	1	—
Southern Vietnam	1985–87	Pork fat	TEQ ($pg\,g^{-1}$ wet wt)	1	—
Song Be, southern Vietnam	1985–87	Pork fat	TEQ ($pg\,g^{-1}$ wet wt)	1	—
Song Be, southern Vietnam	1985–87	Chicken fat	TEQ ($pg\,g^{-1}$ wet wt)	1	—
Tan Uyen, southern Vietnam	1985–87	Chicken egg	TEQ ($pg\,g^{-1}$ wet wt)	1	—
Tan Uyen, southern Vietnam	1985–87	Fish	TEQ ($pg\,g^{-1}$ wet wt)	1	—
Tan Than, southern Vietnam	1985–87	Fish	TEQ ($pg\,g^{-1}$ wet wt)	1	—
Southern Vietnam	1985–87	Channa fish muscle	TEQ ($pg\,g^{-1}$ wet wt)	1	—
Southern Vietnam	1985–87	Channa fish muscle	TEQ ($pg\,g^{-1}$ wet wt)	1	—
Southern Vietnam	1985–87	Channa fish liver	TEQ ($pg\,g^{-1}$ wet wt)	1	—
Southern Vietnam	1985–87	Clarius fish muscle	TEQ ($pg\,g^{-1}$ wet wt)	1	—
Binh Tri Thien, middle Vietnam	1985–87	Catfish	TEQ ($pg\,g^{-1}$ wet wt)	1	—

PCDFs	Total PCDD/Fs	Dioxin-like PCBs	Reference	Remark
23	180 (130–260)	160 (46–300)	Minh et al. 2003	
0.35	1.15 (0.36–1.2)	0.12 (0.027–0.19)	Minh et al. 2003	
—	2.1 (1.9–2.2)	—	Schecter et al. 1989c	
—	0.58	—	Schecter et al. 1989c	
—	0.26	—	Schecter et al. 1989c	
—	0.64	—	Schecter et al. 1989c	
—	0.35	—	Schecter et al. 1989c	
—	3.5	—	Schecter et al. 1989c, 1989d	
—	3.3	—	Schecter et al. 1989c, 1989d	
—	10.9	—	Schecter et al. 1989c	
—	4.1	—	Schecter et al. 1989c	
—	3.5	—	Schecter et al. 1989c, 1989d	
—	32	—	Schecter et al. 1989c, 1989d	
—	1.6	—	Schecter et al. 1989c	
—	0.53	—	Schecter et al. 1989c	
—	0.82	—	Schecter et al. 1989c	
—	0.17	—	Schecter et al. 1989c	
—	0.26	—	Schecter et al. 1989c	
—	4.2	—	Schecter et al. 1989c	
—	0.36	—	Schecter et al. 1989c	
—	5.7	—	Schecter et al. 1989c	

Table 5. (*cont.*)

Location	Year	Sample/sampling site description	Unit	n	PCDDs
Hochiminh City	1985–87	Eel	TEQ ($pg\,g^{-1}$ wet wt)	2	—
Hochiminh City	1985–87	Fish	TEQ ($pg\,g^{-1}$ wet wt)	1	—
Hochiminh City	1985–87	Fish	TEQ ($pg\,g^{-1}$ wet wt)	1	—
Hochiminh City	1985–87	Fish	TEQ ($pg\,g^{-1}$ wet wt)	1	—
Hochiminh City	1985–87	Crab	TEQ ($pg\,g^{-1}$ wet wt)	1	—
Hochiminh City	1985–87	Crab meat	TEQ ($pg\,g^{-1}$ wet wt)	1	—
Hochiminh City	1985–87	Chicken egg	TEQ ($pg\,g^{-1}$ wet wt)	1	—
Hochiminh City	1985–87	Duck egg	TEQ ($pg\,g^{-1}$ wet wt)	1	—
Southern Vietnam	1985–87	Turtle fat	TEQ ($pg\,g^{-1}$ wet wt)	1	—
Southern Vietnam	1985–87	Turtle muscle	TEQ ($pg\,g^{-1}$ wet wt)	1	—
Southern Vietnam	1985–87	Turtle liver	TEQ ($pg\,g^{-1}$ wet wt)	1	—
Southern Vietnam	1985–87	Turtle gall baldder	TEQ ($pg\,g^{-1}$ wet wt)	1	—
Southern Vietnam	1985–87	Turtle ovaries	TEQ ($pg\,g^{-1}$ wet wt)	1	—
Southern Vietnam	1985–87	Snake	TEQ ($pg\,g^{-1}$ wet wt)	1	—
Hochiminh City	1988/89	Pork stick	$pg\,g^{-1}$ lipid	1	39
Hochiminh City	1988/89	Pork fat	$pg\,g^{-1}$ lipid	1	41
Hochiminh City	1988/89	Chicken liver	$pg\,g^{-1}$ lipid	1	94
Hochiminh City	1988/89	Chicken fat	$pg\,g^{-1}$ lipid	1	44
Bienhoa City, southern Vietnam	2002	Snake head (channa striata), nearby Orange Agent Air Base	$pg\,g^{-1}$ lipid	1	16,000
Bienhoa City	2002	Climbing perch (anabas testudineus), nearby Orange Agent Air Base	$pg\,g^{-1}$ lipid	1	13

PCDFs	Total PCDD/Fs	Dioxin-like PCBs	Reference	Remark
—	0.05 (0.04–0.07)	—	Schecter et al. 1989c	
—	0.05	—	Schecter et al. 1989c	
—	0.9	—	Schecter et al. 1989c	
—	0.28	—	Schecter et al. 1989c	
—	2.6	—	Schecter et al. 1989c	
—	0.52	—	Schecter et al. 1989c	
—	0.55	—	Schecter et al. 1989c	
—	ND	—	Schecter et al. 1989c	
—	1.8	—	Schecter et al. 1989c	
—	2.6	—	Schecter et al. 1989c, 1989d	
—	22	—	Schecter et al. 1989c, 1989d	
—	6.5	—	Schecter et al. 1989c, 1989d	
—	86	—	Schecter et al. 1989c,1989d	
—	13	—	Schecter et al. 1989c	
4.3	43	—	Schecter et al. 1990a	
3.5	45	—	Schecter et al. 1990a	
58	150	—	Schecter et al. 1990a	
5.1	49	—	Schecter et al. 1990a	
200	16,200	119,000	Schecter et al. 2003	
15	28	4,500	Schecter et al. 2003	

Table 5. (*cont.*)

Location	Year	Sample/sampling site description	Unit	n	PCDDs
Bienhoa City	2002	Catfish (clarias fuscus), nearby Orange Agent Air Base	$pg\,g^{-1}$ lipid	1	4.4
Bienhoa City	2002	Catfish (clarias fuscus), nearby Orange Agent Air Base	$pg\,g^{-1}$ lipid	1	6.3
Bienhoa City	2002	Carp (ostechilus hasselti), nearby Orange Agent Air Base	$pg\,g^{-1}$ lipid	1	41
Bienhoa City	2002	Duck, nearby Orange Agent Air Base	$pg\,g^{-1}$ lipid	1	560
Bienhoa City	2002	Duck, nearby Orange Agent Air Base	$pg\,g^{-1}$ lipid	1	550
Bienhoa City	2002	Toad, nearby Orange Agent Air Base	$pg\,g^{-1}$ lipid	1	22,000
Bienhoa City	2002	pork, nearby Orange Agent Air Base	$pg\,g^{-1}$ lipid	1	2.5
Bienhoa City	2002	pork nearby Orange Agent Air Base	$pg\,g^{-1}$ lipid	1	3.6
Bienhoa City	2002	Beef, nearby Orange Agent Air Base	$pg\,g^{-1}$ lipid	1	6.7
Bienhoa City	2002	Beef, nearby Orange Agent Air Base	$pg\,g^{-1}$ lipid	1	7.2
Bienhoa City	2002	Chicken, nearby Orange Agent Air Base	$pg\,g^{-1}$ lipid	1	400
Bienhoa City	2002	Chicken, nearby Orange Agent Air Base	$pg\,g^{-1}$ lipid	1	5.9
Bienhoa City	2002	Chicken, nearby Orange Agent Air Base	$pg\,g^{-1}$ lipid	1	11
Bienhoa City	2002	Chicken, nearby Orange Agent Air Base	$pg\,g^{-1}$ lipid	1	450
Human samples					
Nine hospital in Hanoi	1984	Adipose tissues	$pg\,g^{-1}$ lipid	9	150 (9.2–240)

PCDFs	Total PCDD/Fs	Dioxin-like PCBs	Reference	Remark
5.1	9.5	1,200	Schecter et al. 2003	
17	23	3,300	Schecter et al. 2003	
13	54	7,200	Schecter et al. 2003	
40	600	3,300	Schecter et al. 2003	
41	590	2,800	Schecter et al. 2003	
2,600	25,000	960,000	Schecter et al. 2003	
0.21	2.7	1,300	Schecter et al. 2003	
11	15	450	Schecter et al. 2003	
1.5	8.2	580	Schecter et al. 2003	
1.3	8.5	4.6	Schecter et al. 2003	
45	450	40,000	Schecter et al. 2003	
26	32	34	Schecter et al. 2003	
1.4	12	51	Schecter et al. 2003	
42	490	39,000	Schecter et al. 2003	
35 (5.3–47)	190 (15–290)	—	Schecter et al. 1986	2,3,7,8-substituted congeners only

Table 5. (*cont.*)

Location	Year	Sample/sampling site description	Unit	n	PCDDs
Hochiminh City hospital	1984	Adipose tissues	$pg\,g^{-1}$ lipid	15	1,600 (190–4,700)
Seven areas in south Vietnam	1973	Breast milk, Orange Agent sprayed sites, Vietnam War has not ended	$pg\,g^{-1}$ lipid	9	680 (220–1,500)
Hochiminh City	1985	Breast milk, Children Hospital	$pg\,g^{-1}$ lipid	4	12 (10–13)
Hochiminh City	1985	Breast milk, Obstetrical and Gynecological Hospital	$pg\,g^{-1}$ lipid	3	<3–12
Hanoi	1985	Breast milk, Swedish Hospital	$pg\,g^{-1}$ lipid	2	<2.0
Dongnai River, southern Vietnam	1985	Breast milk, near Tan Uyen village, Dongnai River	$pg\,g^{-1}$ lipid	2	<23
Tan Than village, southern Vietnam	1985–87	Breast milk, rural area	$pg\,g^{-1}$ lipid	2	510
Song Be Hospital, southern Vietnam	1985–87	Breast milk, rural area	$pg\,g^{-1}$ lipid	3	580
Pediatric Hospital, Hochiminh City	1985–87	Breast milk, urban area	$pg\,g^{-1}$ lipid	7	760
Obstetrical and Gynecological Hospital, Hochiminh City	1985–87	Adipose tissue	$pg\,g^{-1}$ lipid	3	6.4 (4.4–9.6)
Binh Dan Hospital, Hochiminh City	1985–87	Adipose tissue	$pg\,g^{-1}$ lipid	5	8.8 (1.2–19)
Dong Nai Provincial Hospital, southern Vietnam	1985–87	Adipose tissue	$pg\,g^{-1}$ lipid	2	10 (7.6–13)
Hue provincial Hospital, middle Vietnam	1985–87	Adipose tissue	$pg\,g^{-1}$ lipid	3	16 (9.6–26)

PCDFs	Total PCDD/Fs	Dioxin-like PCBs	Reference	Remark
110 (30–280)	1,700 (220–5,000)	—	Schecter et al. 1986	2,3,7,8-substituted congeners only
62 (<2–190)	740 (220–1,700)	—	Schecter et al. 1987	
—	12 (10–13)	—	Schecter et al. 1987	2,3,7,8-TCDD only
—	<3–12	—	Schecter et al. 1987	2,3,7,8-TCDD only
—	<2.0	—	Schecter et al. 1987	2,3,7,8-TCDD only
—	<23	—	Schecter et al. 1987	2,3,7,8-TCDD only
140	650	—	Schecter et al. 1989a	
100	680	—	Schecter et al. 1989a	
260	1,020	—	Schecter et al. 1989a	
—	6.4 (4.4–9.6)	—	Schecter et al. 1989e	2,3,7,8-TCDD only
—	8.8 (1.2–19)	—	Schecter et al. 1989e	2,3,7,8-TCDD only
—	10 (7.6–13)	—	Schecter et al. 1989e	2,3,7,8-TCDD only
—	16 (9.6–26)	—	Schecter et al. 1989e	2,3,7,8-TCDD only

Table 5. (*cont.*)

Location	Year	Sample/sampling site description	Unit	n	PCDDs
Hochiminh City	1984–85	Adipose tissue	$pg\,g^{-1}$ lipid	2	1,045 (390–1,700)
Soc Trang, southern Vietnam	1984–85	Adipose tissue	$pg\,g^{-1}$ lipid	1	1,600
Vung Tau, southern Vietnam	1984–85	Adipose tissue	$pg\,g^{-1}$ lipid	1	4,700
Tayninh,southern Vietnam	1984–85	Adipose tissue, resident close to Agent Orange spray areas	$pg\,g^{-1}$ lipid	1	4,400
Nha Be, southern Vietnam	1984–85	Adipose tissue	$pg\,g^{-1}$ lipid	1	13
Long An, southern Vietnam	1984–85	Adipose tissue	$pg\,g^{-1}$ lipid	1	7
O Mon, Hau Giang, southern Vietnam	1984–85	Adipose tissue	$pg\,g^{-1}$ lipid	1	4
Cuu Long, southern Vietnam	1984–88	Adipose tissue	$pg\,g^{-1}$ lipid	1	410
Dong Nai, southern Vietnam	1984–88	Adipose tissue	$pg\,g^{-1}$ lipid	1	190
Lam Dong, southern Vietnam	1984–88	Adipose tissue	$pg\,g^{-1}$ lipid	1	1,150
Binh Tri Thien, middle Vietnam	1984–88	Adipose tissue	$pg\,g^{-1}$ lipid	1	300
Tay Ninh, southern Vietnam	1984–88	Adipose tissue	$pg\,g^{-1}$ lipid	1	1,310
Tien Giang, southern Vietnam	1984–88	Adipose tissue	$pg\,g^{-1}$ lipid	1	930
Song Be, southern Vietnam	1984–88	Adipose tissue	$pg\,g^{-1}$ lipid	1	2,010
Phu Khanh, southern Vietnam	1984–88	Adipose tissue	$pg\,g^{-1}$ lipid	1	3,000
Hanoi	1980s	Breast milk	$pg\,g^{-1}$ lipid	1 (28 samples pooled)	93

PCDFs	Total PCDD/Fs	Dioxin-like PCBs	Reference	Remark
49 (39–58)	1,100 (430–1,760)	—	Phuong et al. 1989	
130	1,730	—	Phuong et al. 1989	
31	4,730	—	Phuong et al. 1989	
160	4,560	—	Phuong et al. 1989	
—	13	—	Phuong et al. 1989	2,3,7,8-TCDD only
—	7	—	Phuong et al. 1989	2,3,7,8-TCDD only
—	4	—	Phuong et al. 1989	2,3,7,8-TCDD only
71	480	—	Phiet et al. 1989	
36	230	—	Phiet et al. 1989	
195	1,350	—	Phiet et al. 1989	
100	400	—	Phiet et al. 1989	
88	1,400	—	Phiet et al. 1989	
76	1,000	—	Phiet et al. 1989	
150	2,160	—	Phiet et al. 1989	
280	3,300	—	Phiet et al. 1989	
23	120	—	Schecter et al. 1989f	

Table 5. (*cont.*)

Location	Year	Sample/sampling site description	Unit	n	PCDDs
Hochiminh City	1980s	Breast milk	pg g^{-1} lipid	1 (38 samples pooled)	300
Song Be, southern Vietnam	1980s	Breast milk	pg g^{-1} lipid	1 (12 samples pooled)	280
Tan Uyen, southern Vietnam	1980s	Breast milk	pg g^{-1} lipid	3 pooled	640 (160–310)
Can Gio, southern Vietnam	1980s	Breast milk	pg g^{-1} lipid	1 (3 samples pooled)	94
Long Xuyen, southern Vietnam	1980s	Breast milk	pg g^{-1} lipid	1 (2 samples pooled)	48
Hochiminh City	1980s	Breast milk	pg g^{-1} lipid	1 (15 samples pooled)	390
Hochiminh City	1980s	Breast milk	pg g^{-1} lipid	1 (8 samples pooled)	240
Song Be, southern Vietnam	1980s	Breast milk	pg g^{-1} lipid	1 (12 samples pooled)	280
Various provinces in Mekong River Delta, southern Vietnam	1980s	Adipose tissues	pg g^{-1} lipid	25	19 (7.7–36)
Binh Long, southern Vietnam	1988–89	Breast milk	pg g^{-1} lipid	1 (4 samples pooled)	215
Vung Tau, southern Vietnam	1988–89	Breast milk	pg g^{-1} lipid	1 (5 samples pooled)	280
Tay Ninh, southern Vietnam	1988–89	Breast milk	pg g^{-1} lipid	1 (4 samples pooled)	580
Song Be, southern Vietnam	1988–89	Breast milk	pg g^{-1} lipid	1 (4 samples pooled)	124
Hanoi	1988–89	Adipose tissue	pg g^{-1} lipid	1 (10 samples pooled)	274
Various provinces, southern Vietnam	1988–89	Adipose tissue	pg g^{-1} lipid	13	596

PCDFs	Total PCDD/Fs	Dioxin-like PCBs	Reference	Remark
34	330	—	Schecter et al. 1989f	
47	330	—	Schecter et al. 1989f	
72 (11–190)	710 (170–500)	—	Schecter et al. 1989f	
10	104	—	Schecter et al. 1989f	
8.2	56	—	Schecter et al. 1989f	
32	420	—	Schecter et al. 1989f	
32	270	—	Schecter et al. 1989f	
48	330	—	Schecter et al. 1989f	
7.2 (2.1–17)		—	Schecter et al. 1989g	2,3,7,8-TCDD & TCDF only
40	255	—	Schecter et al. 1990b	
42	320	—	Schecter et al. 1990b	
48	630	—	Schecter et al. 1990b	
32	156	—	Schecter et al. 1990b	
32	306	—	Schecter et al. 1990b	
36	632	—	Schecter et al. 1990b	

Table 5. (*cont.*)

Location	Year	Sample/sampling site description	Unit	n	PCDDs
Hochiminh City	early 1990s	Blood	$pg\,g^{-1}$ lipid	1 (50 samples pooled)	1,090
Dong Nai, southern Vietnam	early 1990s	Blood	$pg\,g^{-1}$ lipid	1 (33 samples pooled)	1,940
Hanoi	early 1990s	Blood	$pg\,g^{-1}$ lipid	1 (32 samples pooled)	126
Hanoi	early 1990s	Breast milk	$pg\,g^{-1}$ lipid	30	104
Da Nang, middle Vietnam	early 1990s	Breast milk	$pg\,g^{-1}$ lipid	1 (11 samples pooled)	406
Dong Nai, southern Vietnam	early 1990s	Breast milk	$pg\,g^{-1}$ lipid	1 (11 samples pooled)	180
Hanoi	1987–91	Adipose tissues, general population	$pg\,g^{-1}$ lipid	1 (10 samples pooled)	270
Hanoi	1987–91	Adipose tissues, soldiers exposed to Agent Orange during the War	$pg\,g^{-1}$ lipid	1 (10 samples pooled)	685
Song Be, southern Vietnam	1987–91	Adipose tissues, locations in proximity of Agent Orange spraying	$pg\,g^{-1}$ lipid	1	1,050
Hau Giang, southern Vietnam	1987–91	Adipose tissues, locations in proximity of Agent Orange spraying	$pg\,g^{-1}$ lipid	1	800
Dong Nai	1987–91	Adipose tissues, locations in proximity of Agent Orange spraying	$pg\,g^{-1}$ lipid	1	226
Various locations from northern Vietnam	1980–91	Whole blood	$pg\,g^{-1}$ lipid	82 (pooled)	190
Various locations from southern Vietnam	1980–91	Whole blood	$pg\,g^{-1}$ lipid	383 (pooled)	860
Various location from northern Vietnam	1991–92	Blood	$pg\,g^{-1}$ lipid	82 (pooled)	190

PCDFs	Total PCDD/Fs	Dioxin-like PCBs	Reference	Remark
82	1,170	—	Schecter et al. 1991	
130	2,070	—	Schecter et al. 1991	
41	167	—	Schecter et al. 1991	
23	127	—	Schecter et al. 1991	
130	536	—	Schecter et al. 1991	
55	235	—	Schecter et al. 1991	
32	300	—	Schecter et al. 1992a	
56	740	—	Schecter et al. 1992a	
68	1,120	—	Schecter et al. 1992a	
62	860	—	Schecter et al. 1992a	
23	249	—	Schecter et al. 1992a	
100	290	—	Schecter et al. 1992b	
110	970	—	Schecter et al. 1992b	
100	290	—	Schecter et al. 1995	

Table 5. (*cont.*)

Location	Year	Sample/sampling site description	Unit	n	PCDDs
Various location from central Vietnam	1991–92	Blood	$pg\,g^{-1}$ lipid	183 (pooled)	930
Various locations from southern Vietnam	1991–92	Blood	$pg\,g^{-1}$ lipid	433 (pooled)	760
Hanoi	1999	Whole blood, general population	$pg\,g^{-1}$ lipid	1 (100 samples pooled)	150
Bien Hoa, Dong Nai	1999	Whole blood, residents in proximity of Agent Orange Air Base	$pg\,g^{-1}$ lipid	20	590 (140–1,410)
Aso, Aluoi Valley, central Vietnam	1999	Whole blood, males, residents in proximity of Agent Orange Air Base	$pg\,g^{-1}$ lipid	78 (pooled)	104
Aso, Aluoi Valley	1999	Whole blood, females, residents in proximity of Agent Orange Air Base	$pg\,g^{-1}$ lipid	85 (pooled)	94
Huong Lam, Aluoi Valley	1999	Whole blood, males, residents in proximity of Agent Orange Air Base	$pg\,g^{-1}$ lipid	64 (pooled)	150
Huong Lam, Aluoi Valley	1999	Whole blood, females, residents in proximity of Agent Orange Air Base	$pg\,g^{-1}$ lipid	56 (pooled)	160
Hong Thuong, Aluoi Valley	1999	Whole blood, males, residents in proximity of Agent Orange Air Base	$pg\,g^{-1}$ lipid	70 (pooled)	220
Hong Thuong, Aluoi Valley	1999	Whole blood, females, residents in proximity of Agent Orange Air Base	$pg\,g^{-1}$ lipid	62 (pooled)	180

PCDFs	Total PCDD/Fs	Dioxin-like PCBs	Reference	Remark
210	1,140	—	Schecter et al. 1995	
94	850	—	Schecter et al. 1995	
40	190	29	Schecter et al. 2001	
67 (28–184)	660 (190–1,520)	60 (20–156)	Schecter et al. 2001	
31	135	—	Dwernychuk et al. 2002	
33	127	—	Dwernychuk et al. 2002	
110	260	—	Dwernychuk et al. 2002	
130	290	—	Dwernychuk et al. 2002	
84	304	—	Dwernychuk et al. 2002	
83	260	—	Dwernychuk et al. 2002	

Table 5. (*cont.*)

Location	Year	Sample/sampling site description	Unit	n	PCDDs
Hong Van, Aluoi Valley	1999	Whole blood, males, residents in proximity of Agent Orange Air Base	$pg\,g^{-1}$ lipid	77 (pooled)	88
Hong Van, Aluoi Valley	1999	Whole blood, females, residents in proximity of Agent Orange Air Base	$pg\,g^{-1}$ lipid	64 (pooled)	110
Aso, Aluoi Valley	1999	Breast milk, resident in proximity of Agent Organe Air Base	$pg\,g^{-1}$ lipid	4	25 (8.8–36)
Huong Lam, Aluoi Valley	1999	Breast milk, resident in proximity of Agent Organe Air Base	$pg\,g^{-1}$ lipid	4	42 (30–68)
Hong Thuong, Aluoi Valley	1999	Breast milk, resident in proximity of Agent Organe Air Base	$pg\,g^{-1}$ lipid	4	54 (36–63)
Hong Van, Aluoi Valley	1999	Breast milk, resident in proximity of Agent Organe Air Base	$pg\,g^{-1}$ lipid	4	34 (19–80)
Tay Mo, suburb Hanoi	1999	Breast milk, residents near dumping site for municipal wastes	$pg\,g^{-1}$ lipid	8	32 (10–81)
			TEQ ($pg\,g^{-1}$ lipid)		2.7 (1.4–4.8)
Tay Mo, suburb Hanoi	1999	Breast milk, residents living far from dumping site for municipal wastes	$pg\,g^{-1}$ lipid	10	27 (14–41)
			TEQ ($pg\,g^{-1}$ lipid)		2.9 (1.7–4.0)

PCDFs	Total PCDD/Fs	Dioxin-like PCBs	Reference	Remark
48	136	—	Dwernychuk et al. 2002	
22	130	—	Dwernychuk et al. 2002	
7.6 (4.1–16)	33 (13–45)	—	Dwernychuk et al. 2002	
24 (11–54)	66 (46–92)	—	Dwernychuk et al. 2002	
32 (11–47)	86 (47–110)	—	Dwernychuk et al. 2002	
16 (4.4–42)	50 (20–120)	—	Dwernychuk et al. 2002	
20 (7.3–42)	51 (18–120)	24,000 (4,200–46,000)	Kunisue et al. 2004	
3.3 (1.3–7.2)	6.0 (2.9–9.3)	7.5 (1.5–15)	Kunisue et al. 2004	
20 (9.6–45)	47 (26–67)	15,000 (2,800–22,000)	Kunisue et al. 2004	
3.4 (2.0–5.6)	6.3 (3.6–8.1)	5.3 (1.9–13)	Kunisue et al. 2004	

In spite of the dioxin problems caused by Agent Orange in southern Vietnam, as just discussed, there have not been many studies of contamination status, bioaccumulation characteristics, and the fate and toxic implications of dioxins in a tropical environment such as Vietnam, and data were not available until our laboratory reported dioxin contamination in dumping sites for municipal wastes in 2003 (Minh NH et al. 2003). In recent years, public media have voiced concern regarding the open dumping sites in Asian developing countries where large amounts of municipal solid wastes are dumped. Unfortunately, these open dumping sites are usually located near human habitats; therefore, exposure to various toxic chemicals from dumping sites is a matter of serious concern because of the effects on human health, wildlife, and environmental quality. Natural burning by the generation of methane gas under anaerobic conditions and combustion by waste pickers scavenging in dumping sites are favorable factors for the formation of PCDD/Fs in dumping sites. To find an answer to the question of whether open dumping sites are potential sources of PCDD/Fs, comprehensive investigations have been conducted to examine PCDD/F concentrations in soils and human breast milk from open dumping sites from the Philippines, Cambodia, India, and Vietnam (Minh NH et al. 2003; Kunisue et al. 2004).

Concentrations of PCDD/Fs in soils from dumping sites in Hanoi were significantly higher than those in Hochiminh. Mean and range concentrations in dumping sites from Hanoi were 6100 pg/g dry wt [95 pg/g toxicity equivalent (TEQ)] and 125–50,500 pg/g dry wt (0.4–850 pg/g TEQ), respectively (Minh NH et al. 2003) (Fig. 12). However, in general, dioxin residues in soils from Vietnam were less than those in other Asian countries examined in this study, indicating less dioxin contamination in Vietnam. However, one soil sample contained very high concentration of PCDD/Fs (50 ng/g dry wt; 850 pg/g TEQ) (Minh NH et al. 2003). This level is greater than most of the background level soils in industrialized countries and even higher than those reported for dioxin-contaminated locations in the world (Minh

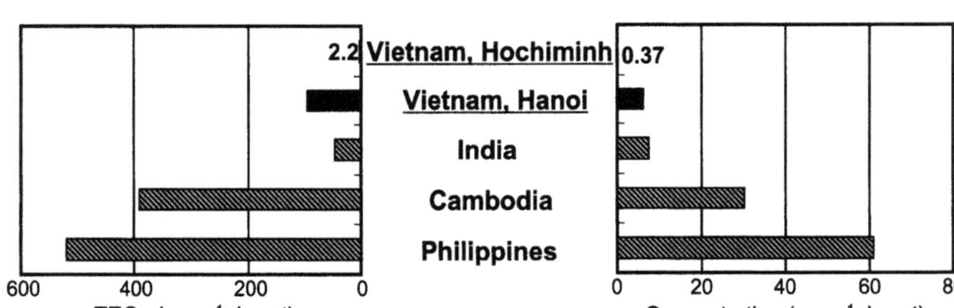

Fig. 12. Concentrations and TEQs of PCDD/Fs in soils from open dumping sites for municipal wastes in developing Asian countries (Minh NH et al. 2003).

NH et al. 2003). In addition, PCDD/F levels in soils collected from dumping sites were significantly higher than those in soils from agricultural and residential areas far from dumping sites, indicating that open dumping sites can be a potential source of dioxins in Vietnam. To understand the degree of human exposure to dioxins in open dumping sites, levels in human milk samples were also examined (Kunisue et al. 2004). Mean PCDD/Fs concentration in breast milk from women living near dumping sites in Hanoi was 52 pg/g lipid (6.0 pg/g TEQs) [range, 18–120 pg/g lipid (2.9–9.3 pg/g TEQ] (Kunisue et al. 2004). These values are comparable to those in the Philippines and Cambodia but lower than those in India. In contrast to soils, there were no significant differences in PCDD/F residues in milk of women living near dumping sites and far from dumping sites (control sites) (Kunisue et al. 2004).

On the basis of these results, it can be emphasized that although the status of contamination in Vietnam was less pronounced as compared to other countries in the East and South Asian regions, the issue of dioxin pollution in such open dumping sites should be considered as one of the priority research targets in the future because of the sporadic evidence of elevated dioxin contamination in dumping sites from Hanoi. Control measures and legislation toward management and mitigation of dioxin emissions in open dumping sites are urgently needed.

IV. Environmental Behavior and Bioaccumulation

A. Transport Behavior in Tropical Environments

Multimedia monitoring studies have not only helped increase knowledge of the status of contamination but also provided in-depth insights into the transport behavior and fate of POPs in tropical ecosystems in developing Asian countries. On a global scale, it has been made clear that the fate of persistent OCs in the open ocean is different from that in the coastal environment (Iwata et al. 1993, 1994; Tanabe et al. 1994). The open ocean plays a role as a final sink for POPs, and cold waters at high latitudes such as the Arctic Ocean, which are located far from pollution sources, serve as significant reservoirs (Iwata et al. 1993). On the other hand, pronounced contamination by OC insecticides was observed in coastal waters of Asian countries as a result of extensive usage for agriculture and public health purposes (Iwata et al. 1994). On a smaller scale, latitudinal distribution of persistent OCs such as PCBs, DDTs, and HCHs in sediments from the southern Vietnamese estuary were also examined (Iwata et al. 1994). The authors suggested using the relationship between the concentration ratio of a compound in the water phase (concentration in water sample) and the particulate phase (organic carbon normalized concentration in sediment) (S/W ratio) and latitude. The degree of correlation and the ratio (S/W) are indications of the efficiency of their transport.

Analytical results showed relatively uniform distribution of HCHs along the latitudes, and the ratio (S/W) showed the highest degree of positive correlation with latitude. This phenomenon indicates that HCH isomers pose higher potency for atmospheric transport and global redistribution. Other compounds such as DDTs, CHLs, and PCBs with lower correlations and gradients exhibited less efficiency for atmospheric transport.

In general, the degree of usage, the climate, and the physicochemical properties of persistent organic compounds are the major factors controlling the magnitude of contamination. When comparing the distribution of OCs in different environmental media in the Asia-Pacific region, we found that distribution patterns in fish and sediment were relatively similar and showed less spatial variation. The pattern in air and water showed greater geographic variability. A recent extensive survey of OC insecticides from the Red and Duong Rivers, the two largest rivers flowing through the northern delta region of Vietnam, and various lakes in Hanoi has shown that DDT and HCH levels in rivers is apparently greater than those in lakes (Hung and Thiemann 2002). This observation suggests different behavior of semivolatile compounds in tropical lakes and in rivers, showing shorter retention time in lakes than in rivers. Elevated temperatures may enhance volatilization of such compounds, leading to shorter residence time in the tropical water phase, lower levels, and relatively uniform distribution in sediments and aquatic biota. In this context, the role of the tropics in Asia-Pacific region as potential emission sources for higher-latitude areas deserves further attention.

As mentioned earlier, very little information is available regarding contamination by PCDD/Fs in developing Asian countries. Therefore, the fate and behavior of these compounds in developing countries are still obscure. On the basis of residues in soils from dumping sites, flux of PCDD/Fs to soils can be estimated (Fig. 13). The estimated fluxes to soils in dumping sites in developing Asian countries including Vietnam were surprisingly higher. Fluxes to soils in Asian dumping sites were higher than those of some other locations in the world, including contaminated areas in the U.S. and Hong Kong (see Fig. 13). Loads of PCDD/Fs to the dumping sites were also estimated (Fig. 14). Results indicated that these sites in the Philippines and India, with a huge area of approximately 23 and 140 ha, could receive the highest annual amounts of 3900 and 1400 mg/yr PCDD/Fs (35 and 8.8 mg TEQ/yr), respectively. A dumping site in Hochiminh, Vietnam, had the lowest loading rate because of less contamination of PCDD/Fs in soils. For comparison, total annual fluxes to the Kanto region, Japan, one of the most polluted areas in the world, were found to range from 50 to 900 g TEQ in a total area of 32,000 km^2 (approximately 3 million ha) (Ogura et al. 2001). The area of dumping sites in India is 140 ha, which is 21,000 times smaller than that of Kanto region; this area was estimated to receive 8.8 mg TEQ every year. These data suggest that dumping sites

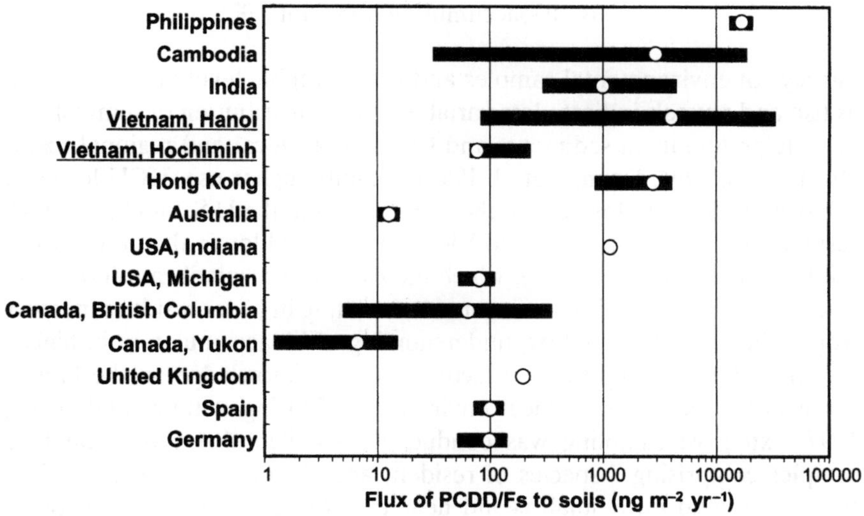

Fig. 13. Comparison of fluxes to soils from open dumping sites in Asian countries with those from other locations in the world (Minh NH et al. 2003; Wagrowski and Hites 2000).

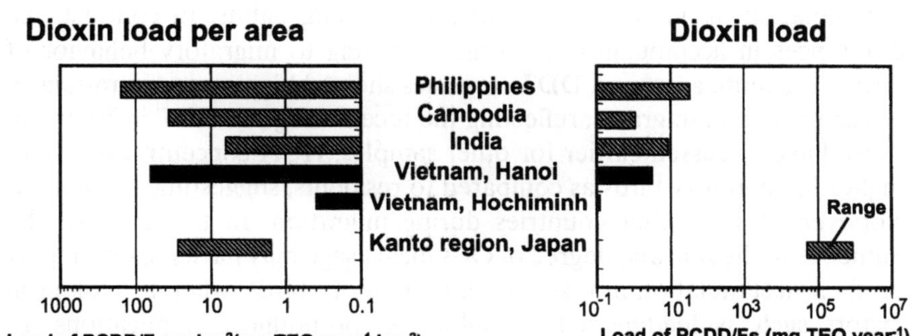

Fig. 14. Comparison of the load of PCDD/Fs to dumping sites in Asian countries with those in Kanto region, an extensive industrial area in Japan (Minh NH et al. 2003).

in India and the Philippines may be significant reservoirs of PCDD/Fs. Possible impacts on human health and wildlife living near dumping sites are of great concern and warrant further comprehensive studies. On the basis of the results of this study, it is important to note that despite decrease in global pollution by POPs in the future, developing countries may continue to be a potential source of certain compounds, particularly PCDD/Fs.

B. Bioaccumulation in Biota

Surveys of environmental samples and lower trophic level organisms such as fish and mussels indicate less variable spatial distribution of semivolatile organic pollutants in sediments and fish on both local and regional scales (Iwata et al. 1994; Kannan et al. 1995). Despite higher use, HCH levels in fish from India were lower than those in Japan and the U.S. Similarly, DDT residues in fishes from India and Vietnam were still lower than those from the U.S. (Kannan et al. 1995), which indicates rapid volatilization of these insecticides in the tropical environment, resulting in lower residues in lower trophic biota. In this context, understanding of bioaccumulation in higher trophic animals is important to acquire further knowledge on the fate of persistent OCs in the tropical environment. During autumn and spring 1997, extensive sampling was conducted to collect 101 individual bird samples comprising 7 species of resident and 17 species of migrant birds from a wetland on a small island near the Red River estuary, northern Vietnam (Minh TB et al. 2002). Concentrations of OC insecticides and PCBs were analyzed in these resident and migratory birds to understand the magnitude of contamination, bioaccumulation, and metabolism in birds. Interestingly, an in-depth analysis of the variation of OC residues according to the feeding habits and migratory behaviors revealed different pictures of accumulation. In particular, there were no clear differences in OC accumulation in birds according to their feeding habits. In contrast, the differences in accumulation patterns according to migratory behavior of birds were quite apparent. DDTs residues showed higher concentrations in residents than in migrants, reflecting the recent usage of DDT in Vietnam, as we have discussed earlier for other samples. HCH concentrations were higher in migratory birds as compared to residents, suggesting exposure at stopover sites in other countries during migration. In this context, the influence of the rate and degree of OCs local usage may have greater impact on their accumulation in resident and migratory birds, whereas biological factors such as feeding habit exhibit less pronounced implications. In general, results observed in residents and migratory birds from North Vietnam clearly suggest the suitability of using birds as a bioindicator for monitoring toxic contaminants.

On the basis of isomer-specific analysis of PCB congeners, it is possible to estimate the metabolism capacity of contaminants in birds. Tanabe et al. (1988) suggested the concept of using a "metabolic index" (MI) value to estimate enzyme activities in higher trophic animals including birds. Higher MI values imply greater capability to metabolize persistent OCs. Estimated enzyme activities in some species of resident and migratory birds from North Vietnam compared with those in other species of birds and mammals are given in Fig. 15. Estimated PB-type enzyme activities of some species such as black-capped kingfisher and whiskered tern were comparable to those in the common cormorant but higher than those in other higher-

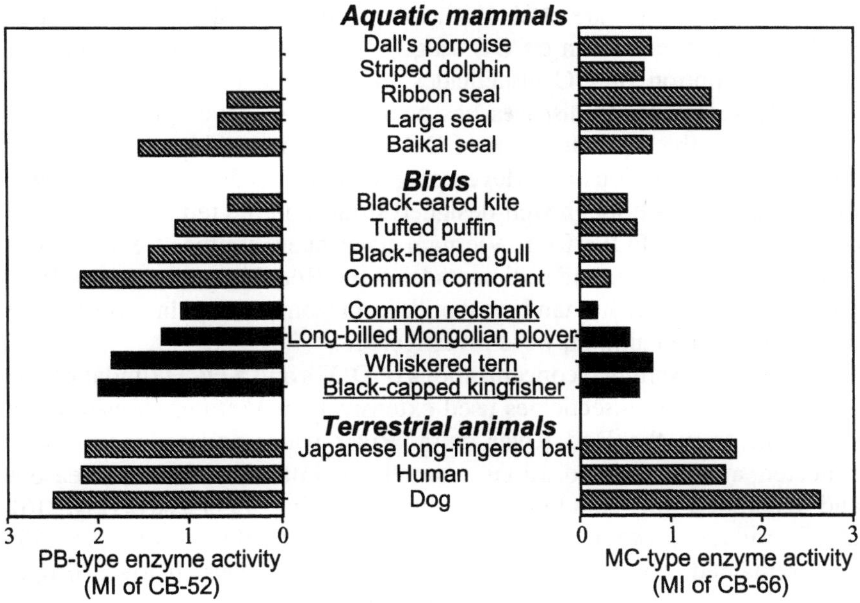

Fig. 15. Comparison of estimated phenobarbital (PB)- and 3-methylcholanthrene (MC)-type enzyme activities in higher trophic animals by metabolic indices of chlorobiphenyl (CB)-52 and CB-66. *Black bars* represent enzyme activities of Vietnamese birds (Minh TB et al. 2002).

trophic fish-eating bird species such as kite, puffin, and gull, while MC-type enzyme activities were comparable or slightly higher in Vietnamese birds. Higher PB- and, to some extent, MC-type enzyme activities in shorebirds from Vietnam suggest that these species may have a higher capacity to metabolize PCB congeners; this may explain relatively lower levels of PCBs and other OCs in shorebirds from Vietnam compared to high-trophic top predator species (Minh TB et al. 2002).

V. Temporal Trends

Well-designed studies on temporal trends of POPs contamination in developing countries are generally limited. This is a common issue in underdeveloped nations, where advanced knowledge and state-of-the-art analytical equipment are still substantially lacking. Therefore, reliable and reproducible data for long-term trends of POP contamination, a key requirement in monitoring programs, are limited in these countries, including Vietnam. Nevertheless, studies on temporal trends of contamination are very important, particularly in tropical developing countries. Few reports have suggested the role of the southern Asian region as a possible emission source of toxic contaminants for pristine areas such as the Arctic and the Antarctic (Iwata et al. 1993, 1994; Kannan et al. 1995; Kunisue et al. 2002). Despite the ban on OC usage in most of the developed nations, high consumption of OC insecticides to enhance food production and eradicate vector-borne diseases has been a fact in developing countries. Thus, despite the rapid decrease of OC residues in developed nations, the status of contamination in the developing world seems different, with slower rates of decline. Although well-designed studies on the temporal trends of contamination in POPs from Vietnam have been limited, trends of OC residues in river water and sediments from the Red River estuary and from Hochiminh City, and human breast milk from women living in suburb areas of Hanoi and Hochiminh, were investigated.

Viet (2002) examined concentrations of DDTs and γ-HCH (lindane), the two most common insecticides used extensively in Vietnam, in water and sediments from the Red River delta. River water and sediments were collected at the same locations annually in both dry and wet seasons and examined for contamination trends during 1995–2001 (Fig. 16). DDT residues in water have declined relatively rapidly during 1995–1998 and remained constant until recent years at levels below 10ng/L. Concentrations in sediments also exhibited a reducing trend but to a lesser extent. DDT residues in sediments dropped by a factor of 2 during 1997 and 2000. DDT was officially banned in Vietnam in 1995 (Sinh et al. 1999). The reduction of DDT concentrations in both water and sediments from Red River, northern Vietnam, indicate the effect of legislative action to reduce the degree of pollution. An interesting result was observed on trends of lindane in sediment, showing peak concentrations in 1997 and lower levels during 1995–1996 and 1999–2001. Recent studies examined HCH residues in sediments from different sites in the Red River delta and

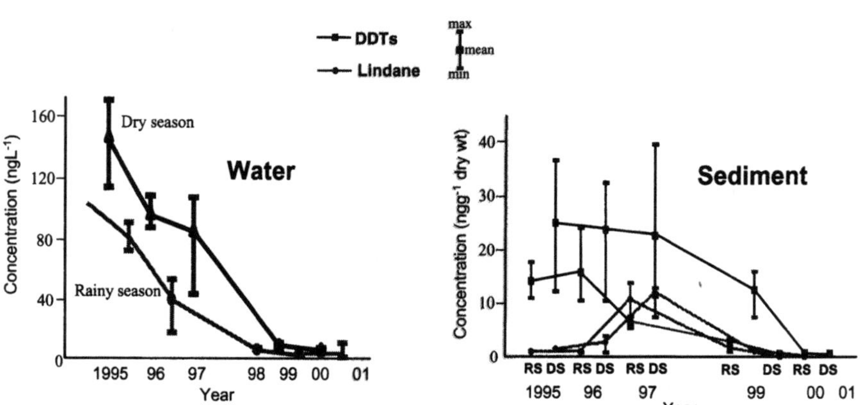

Fig. 16. Temporal trends of DDTs and lindane in surface water and sediment from Red River Delta, northern Vietnam during 1995–2001 (Viet et al. 2002).

estuary and found that HCH concentrations in the 1997 survey were higher than those analyzed in the 1995 sampling survey (Nhan et al. 1998, 1999). Such a fluctuating trend of HCH contamination suggests sporadic input of this insecticide into the Red River watershed. In general, results in water and sediments in recent years indicated a rapid decline of DDTs and HCHs in surface water, but a slow reduction in sediment levels.

In southern Vietnam, several studies reported levels of DDTs and PCBs in urban sediments from Hochiminh metropolitan area (Iwata et al. 1994; Phuong et al. 1998; Minh NH et al. 2007a). Results of these investigations in similar sampling areas were compiled to evaluate contamination trends. PCBs levels decreased from 310 ng/g dry wt in 1990 (Iwata et al. 1994) to 220 ng/g in 1996 (Phuong et al. 1998) and to 82 ng/g in 2004 (Minh et al. 2007a). In the same period, DDTs decreased from 240 ng/g dry wt to 80 and 37 ng/g, respectively.

It is well known that decrease of POP residues in environmental matrices often follows first-order kinetics, and thus their residue level data can be fitted in simple log-linear regression for examination of temporal trends (Bignert et al. 1998; Noren and Meironyte 2000). Using this approach, data from the above studies (Iwata et al. 1994; Phuong et al. 1998; Minh et al. 2007b) were combined to estimate half-life (defined as the time needed for a compound to decrease to one-half of its original concentration) of DDTs and PCBs in sediments from canals in the metropolitan area of Hochiminh (Fig. 17). The estimated half-lives were approximately 5 yr for DDTs and 7 yr for PCBs.

The half-life of DDTs in this study is in agreement with those observed in various environmental matrices from northern Europe (Bignert et al. 1998) as well as those reported for human breast milk in Hochiminh (Minh et al. 2004a). The estimated half-life for PCBs in sediments from Hochiminh

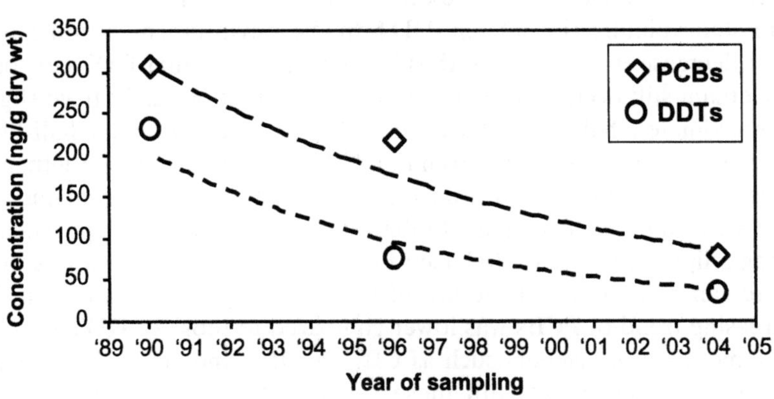

Fig. 17. Temporal trends of DDTs and PCBs in sediments collected from Hochiminh City during the last 14 years (Minh NH et al. 2007a).

was relatively short compared to 11–17 yr reported in the literature (Bignert et al. 1998). Different methods for quantification of PCBs may have led to variations in reported total PCBs and thus caused variation in the estimation of PCB half-life. In general, results of previous monitoring studies suggest there was a somewhat general trend of POP contamination observed in various ecosystems, i.e., the relatively rapid decline of DDTs and slower decrease of PCBs (Tanabe et al. 2003). Temporal trends of POP contamination in the Mekong River delta remain a critical issue that needs long-term and comprehensive monitoring.

On the other hand, monitoring concentrations in biota should provide more realistic data to understand trends of environmental contamination because the impact of any change in the environmental input of persistent OCs can be realized relatively slowly in biota samples compared to those observed in environmental abiotic samples. A similar phenomenon was also seen in the trends of POP pollution in lower and higher trophic animals (Tanabe et al. 2003).

Minh NH et al. (2004a) assessed the declining rates of human exposure to DDTs and PCBs over 10 yr (1989 and 2001). A first-order kinetics approach has been used to estimate the declining rates of DDTs and PCBs in human breast milk from Vietnam (Fig. 18). The decrease of persistent DDTs, PCBs, and HCHs in human breast milk was suggested to follow first-order kinetics (Noren and Meironyte 2000). Another important assessment is half-life ($t_{dec1/2}$). Based on concentrations of OCs in 1989 reported by Schecter et al. (1989a) and those of 2001 (Minh NH et al. 2004a), the rate constant and $t_{dec1/2}$ were estimated.

Levels of *p,p'*-DDT have decreased from 4700 to 2700 ng/g lipid over a 10-yr period with $t_{dec1/2}$ of ≈3 yr. On the other hand, *p,p'*-DDE decreased rather slowly with a $t_{dec1/2}$ of 6 yr. These results are somewhat in agreement with those of Sweden showing half-lives of 4 and 6 yr for *p,p'*-DDT and *p,p'*-DDE (Noren and Meironyte 2000). The slightly shorter half-life observed here could be because the tropical climate in Vietnam may have facilitated volatilization of *p,p'*-DDT in the environment, leading to its faster decrease in food chains, thus in humans. Given that DDTs exposure in Vietnam still occurs to some extent, the calculation might have slightly underestimated half-life values. Nevertheless, the estimated half-life is fairly useful for assessing the trend of DDT contamination of Vietnamese breast milk. Assuming that the decrease in the trend of DDTs remains more or less constant, we can estimate the DDT levels reaching approximately 700 ng/g lipid in the year 2011 (see Fig. 18). However, lower levels in the future can be expected if the use of DDT is completely phased out. The decreasing trend of PCBs was lower compared to those of DDT (11–18 yr for some major congeners such as CB-138, -153, and -180). This result is somewhat in agreement with those reported from Sweden showing the half-life for some PCB congeners varying from 11 to 17 yr (Noren and Meironyte 2000).

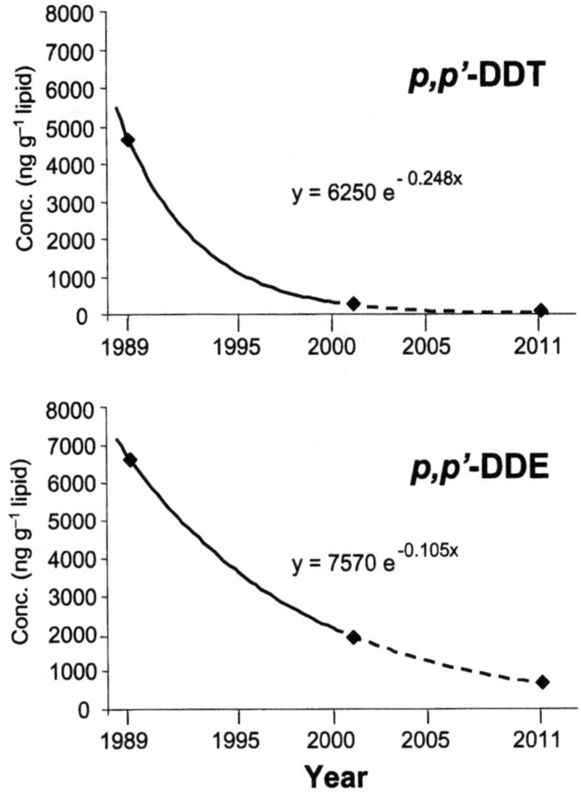

Fig. 18. Estimation of time-trend curve of *p,p'*-DDT and *p,p'*-DDE residues in human breast milk in Vietnam (Minh NH et al. 2004a).

VI. Environmental and Human Health Implications

Widespread contamination by OC insecticides, particularly DDTs, in different environmental samples of Vietnam has been apparent, as indicated in our survey in the early 1990s. In an evaluation of the estuarine sediments collected from various locations from the northern to southern part of the country, high concentrations of DDTs were observed (Iwata et al. 1994). Environment Canada has recently updated the sediment quality guidelines for protection of aquatic life. The Interim Fresh Water Sediment Quality Guidelines (ISQGs) and the Probable Effect Levels (PELs) for *p,p'*-DDE are 1.42 and 6.75 ng/g dry wt, respectively, whereas these values for *p,p'*-DDT are 1.19 and 4.77 ng/g dry wt (Canadian Environmental Quality Guideline 2002). Among the 18 locations examined throughout Vietnam, about half of the sediment samples contained *p,p'*-DDE and *p,p'*-DDT greater than the ISQG values. Some samples collected from the municipal sewage canal contained elevated levels of DDTs, far exceeding the probable effect levels (PELs). PCB concentrations in Vietnamese sediments in these locations were also above the PEL level. Likewise, concentrations of DDTs in soils collected from some locations from north, middle, and south Vietnam (Thao et al. 1993a,b) approached or exceeded the guideline level (700 ng/g dry wt) proposed by Environment Canada and the level of 1000 ng/g dry wt recommended by the Japanese government.

Results of a 2004 survey indicated that concentrations of DDTs in all sediment samples from Hochiminh canals are higher than the ISQG, and those in more than half of the samples still exceeded the PEL (Minh TB et al. 2005; Minh NH et al. 2007a). Nearly half of sediment samples from the Sai Gon-Dong Nai River had concentrations above the ISQG and 20% of the samples were above the PEL; only 10% of the estuarine and coastal samples showed values over the ISQG and no samples over the PEL (Fig. 19). On the other hand, PCBs levels in all sediments were lower than the PEL (277 ng/g dry wt; data not shown) and only sediments from Hochiminh City canals contained PCBs higher than the ISQG (34 ng/g). The results demonstrate high toxic potential of sediment in the city canals and therefore suggest a need for appropriate management. In this context, it is important to note that despite the decrease in POPs pollution, present levels in many sediment samples still exceed the PELs. Input of OCs to the

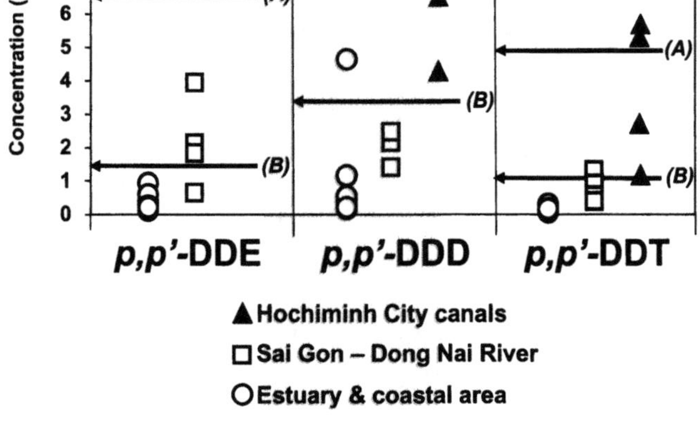

Fig. 19. Concentrations of DDT compounds in sediments from Sai Gon–Dong Nai River, southern Vietnam in comparison with the Canadian Environmental Quality Guideline for Sediment (Minh NH et al. 2007a).

aquatic environment is likely the result of the discharge of untreated sewage from municipal areas, and thus more efficient management on discharge is needed to prevent discharge of toxic chemicals into the aquatic environment. Taking into account all these facts, the magnitude of contamination by DDTs in Vietnam is of concern and warrants further studies. From the environmental health and global contamination point of view, the role of Vietnam as a potential source of DDTs for other countries in the region as well as in higher latitudes should be considered as a priority for future research.

As for PCDD/Fs, the formation of these contaminants in open dumping sites in Asian developing countries raises human health concerns, not only for communities living near the dumping sites but also for people who live far away, because PCDD/Fs may undergo atmospheric transport and deposition in distant areas. For risk assessment of soils contaminated by dioxins and related compounds, the Agency for Toxic Substances and Disease Registry (ATSDR) U.S. Department of Health, proposed guidelines recommending that areas having soil concentrations within the range 50–1000 pg TEQ/g should be evaluated for bioavailability, ingestion rates, community concerns, etc., and that soils with levels >1000 pg TEQs/g should be considered for stronger actions such as health studies, exposure investigations, etc. (ATSDR guideline, 1997). The Japanese Government recently issued new standards for dioxins in soil, establishing 1000 pg TEQ/g as the maximum acceptable level and those within 250–1000 pg TEQ/g be kept under surveillance. Many soil samples in dumping sites from Asian countries including Vietnam contained TEQ concentrations exceeding 250 pg/g TEQs (Fig. 20; Minh NH et al. 2003), suggesting the necessity of continuous monitoring. Particularly, some soils from Cambodia and Hanoi dumping sites contained TEQ concentrations >1000 pg/g, suggesting their potential for causing adverse health risks for humans and wildlife.

In the perspective of human health implication, surveys conducted in early 1990s on OCs in foodstuffs provided useful information regarding dietary intake of these compounds by Vietnamese (Kannan et al. 1992). The estimated average daily intakes based on exposure through foodstuffs to PCBs in Vietnam were higher than India and Thailand, comparable to those reported for developed nations such as the U.S. and Germany. Particularly, average daily intake of DDTs by Vietnamese was estimated to be 19 μg/person/d, and this value was the highest of countries in the region and in developed nations (Kannan et al. 1992). Although the data used for this estimation were reported a decade earlier, this fact clearly suggests elevated exposure to DDTs and PCBs by Vietnamese and that the usage of DDT has been extensive during the past 10 yr.

Surveys in the framework of the recent Asia-Pacific Mussel Watch Program indicated that dietary intakes of DDTs and PCBs from fish in Vietnam were higher than those in Cambodia and Thailand, but still lower than those in industrialized nations such as Australia, Japan, and Hong

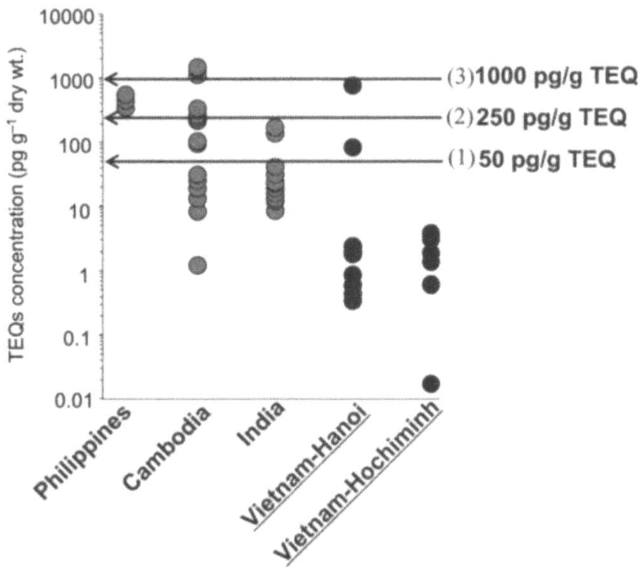

Fig. 20. Concentrations of PCDD/Fs in soils from dumping sites in developing Asian countries compared with various environmental guideline values (Minh NH et al. 2003).

Kong (Monirith et al. 2000). On the basis of data of average seafood consumption reported by the Food and Agriculture Organization of the United Nations, the average daily intakes of PCBs and DDTs from seafood for different countries in the Asia-Pacific region were estimated (Table 6). Interestingly, results again showed that intakes of DDTs by Vietnamese were apparently higher than those reported in other countries examined.

In addition to the elevated exposure of DDT via seafood to the Vietnamese general population, certain cohorts living near municipal dumping sites may be at a higher risk from dioxins and dibenzofurans. A methodical approach has been developed to evaluate the risk of exposure to PCDD/Fs via soil ingestion and dermal absorption (Minh NH et al. 2003). Human exposure to PCDD/Fs in soil is considered to be different for children and adults because of differences in ingestion rates as well as body weights of children and adults. Intakes of dioxins were estimated to be the highest in people of the Philippines, followed by Cambodia, India, Hanoi (North Vietnam), and Hochiminh (South Vietnam). Intakes of PCDD/Fs by those living near dumping sites in Vietnam were about 2- to 200 fold greater than those of residents in control sites, thus emphasizing greater health risks. In addition, it is important to note that the estimated intakes of dioxins via soil ingestion and dermal exposure by children were higher than those for adults, suggesting greater risk of dioxin exposure for

Table 6. Estimated Daily Intakes of Persistent Organochlorines Via Mussels by Different Populations in Asia-Pacific Region.

Country	Survey year	Seafood consumption[a] (g $person^{-1}$ d^{-1})	Intake of PCBs[b] (ng $person^{-1}$ d^{-1})	Intake of DDTs (ng $person^{-1}$ d^{-1})	Intake of HCHs (ng $person^{-1}$ d^{-1})
Cambodia	1998	20	15	6.6	<0.2
China	1999–2001	71	180	17,000	57
Hong Kong	1998–1999	69	260	8,300	14
India	1998	13	49	55	26
Indonesia	1998	52	68	52	2.1
Japan	1994	196	5,900	690	63
South Korea	1998	114	420	400	30
Malaysia	1998	156	160	220	<1.6
Philippines	1998	77	440	31	2.3
Russia	1999	54	3,400	650	54
Vietnam	1997	47	66	1,900	2.8

[a]Seafood consumption data were cited from FAO Food Balance Sheets, FAO Statistics Divison (FAO 2006).

[b]Intakes were estimated on the basis of residue concentrations in mussels (Asia-Pacific Mussel Watch Program) reported by Monirith et al. (2003).

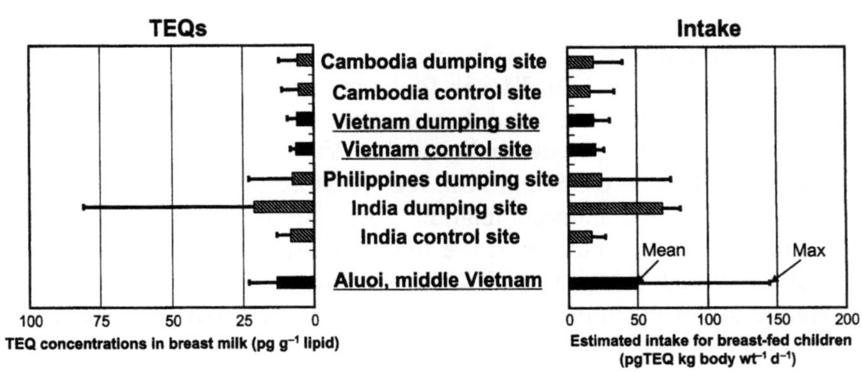

Fig. 21. TEQs concentrations and estimated intakes of PCDD/Fs for breast-fed children from open dumping sites and control sites in Asian countries and Aluoi Valley, a former Agent Orange spraying target during the Vietnam War (Kunisue et al. 2004; Dwernychuk et al. 2002).

children near dumping sites (Minh NH et al., 2003). Further investigations should be focused on children and infants, as they are the most susceptible group and have greater exposure to dioxins. Breast-fed children intakes of PCDD/Fs estimated on the basis of residues in breast milk of women living near open dumping sites in Asian countries are given in Fig. 21. The intakes estimated for Vietnamese were comparable to those in Cambodia but lower than in the Philippines and India. In addition, intakes estimated for children living near the hot spot of dioxin contamination from Agent Orange in southern Vietnam are still very high, even after spraying ended almost three decades ago (see Fig. 21; Dwernychuk et al. 2002). Thus, Vietnam could serve as a suitable location for future research on possible toxic effects of dioxins on wildlife and humans as this is a place where a unique situation of both current and historical dioxin contamination exists.

VII. Conclusions and Recommendations

Multimedia monitoring studies have made substantial contributions to the discovery, description, and understanding of the environmental problems caused by persistent man-made chemicals. Over the past two decades, comprehensive surveys were conducted in the Asia-Pacific region, including Vietnam, to make clear various issues of POPs contamination such as extent of pollution, transport behavior and fate, spatial and temporal trends, and toxic implications for the environment, wildlife, and human health. The following conclusions can be made on the basis of these monitoring studies.

- POPs contamination is still widespread in various environmental media in Vietnam. In particular, DDT, a well-known insecticide that is economic and effective for both agricultural crop production and eradication of vector-borne diseases in developing countries, is the prime contaminant in Vietnam. Elevated DDT levels found in various environmental media such as water, soils, sediment, fish, bivalves, and birds in both past and recent monitoring surveys indicate that Vietnam may be a potential source of DDT in the South Asia region. More importantly, this insecticide poses high bioaccumulation potential in the food chain and eventually bioaccumulates in the human body to a substantial extent. Relatively high levels of DDTs found in human breast milk from both Hanoi and Hochiminh warrant relatively higher risk for breast-fed children.
- In general, the hot spots of POPs pollution were linked closely with extensive industrial and human activities. Elevated levels were observed in the metropolitan areas in Hanoi and Hochiminh as well as urban sites along the Mekong River. This fact suggests that the fate of persistent organic pollutants such as DDTs, HCHs, PCBs, and PBDEs in the Vietnamese environment seems to be governed by human activities rather than their physicochemical properties such as "long-range atmospheric transport potency." In addition, our results clearly demonstrate that open dumping sites in developing Asian countries, including Vietnam, are new potential hot spots of pollution by various toxic chemicals, particularly dioxins and DDTs.

- Several studies have been conducted to understand the levels and impacts of dioxins in specific hot spots from southern Vietnam as a result of Agent Orange spraying during the war. Surveys conducted during the late 1980s/early 1990s indicate elevated levels of dioxins from these sites. In addition, our recent studies have raised a concern regarding dioxin contamination in open dumping sites in developing Asian countries, suggesting their role as a potential source of dioxins and related compounds. Although dioxin residues in soils and human milk collected from dumping sites in Vietnam were generally lower than those in other Asian countries, the status of dioxin pollution in these dumping sites was much greater than those in control sites. The role of dumping sites as a source of dioxins, therefore, is an important environmental issue for future research.
- For temporal trends of POPs pollution, available data demonstrate a relatively rapid decline in OC insecticide residues such as DDTs and HCHs in the Vietnamese environment, while PCBs show slower decreasing trends. In this context, it is important to note that despite the fact that DDT residues have declined over the last decade, the present levels are still high and well exceed the environmental quality guideline values.
- Open dumping sites for municipal wastes in developing Asian countries are potential sources of many chemical contaminants. Long-term health impacts of POPs on humans living in and around these dumping sites are probably a major concern for future research in Vietnam.
- With rapid growth of industrial activities, the use of modern chemicals such as PBDEs as flame retardants is another issue that needs particular attention. In this context, the role of "e-waste," the specific dumping sites of used electronic devices and appliances and computers, as a significant source of modern and toxic chemicals such as polybrominated compounds, should be elucidated in developing Asian countries including Vietnam.
- Considering these critical issues of POPs pollution, it is clear that continued and constant efforts should be made to deal with environmental problems. Despite the possible decrease in global contamination by POPs in the future, developing countries in the Asia-Pacific region may continue to be a potential sources for certain contaminants such as DDTs and PCDD/Fs. Comprehensive and long-term monitoring programs are still urgently needed. In this context, well-designed monitoring of temporal trends of POP residues in developing countries, including Vietnam, over an extended period is crucial for tracing the unrevealed sources and predicting future prospects of their pollution. In addition, the issue of toxic chemicals exposure to humans, particularly those currently residing in extremely poor living conditions, such as near open dumping sites for municipal wastes in developing countries, should be the major strategy for future research.

Summary

This review provides a comprehensive overview of the contamination by persistent organic pollutants (POPs) in Vietnam based on the results of extensive monitoring studies conducted during the last two decades. Available data on POP levels in various environmental media and humans are compiled and discussed to help provide in-depth insights into their distributions and sources, transport behavior and fate, bioaccumulation features, temporal trends, and the potential impacts on ecosystems and human health. Surveys conducted in the framework of the Asia-Pacific Mussel Watch Program during early 1990s indicated widespread contamination by PCBs and organochlorine insecticides, particularly DDT and HCH in various environmental compartments such as air, water, soils, sediments, and fish collected from different parts of Vietnam. Recent studies continued to reveal elevated contamination by DDTs in fish, mussels, and birds from Vietnam. DDT concentrations in fish and birds from Vietnam are among the highest reported for countries in the Asia-Pacific region, suggesting the role of Vietnamese environment as a potential emission source of DDTs in this region. Although residues of this insecticide in Vietnam have declined relatively rapidly during the past decade, recent levels still exceed the proposed guideline values proposed by Canada and the U.S. Open dumping sites for municipal wastes in some major cities such as Hanoi and Hochiminh are a matter of concern with regard to environmental pollution, particularly contamination by DDTs and PCDD/Fs. Daily intakes of DDTs via seafood estimated for the Vietnamese general population were among the highest reported for East Asian countries. In the open dumping sites, intakes of dioxins by residents were significantly greater than those living far from dumping sites. Particularly, the estimated intakes of dioxins via soil ingestion and dermal exposure for children were higher than those for adults, suggesting greater risk of dioxin exposure for children near dumping sites. Future studies should be focused on the temporal trends of POPs in biota in Vietnam to predict the future trends of contamination and to understand possible toxic impacts on organisms. In addition, human exposure and possible toxic effects, particularly on children, should be considered as priority research, because they are the most susceptible group and have greater exposure to POPs.

Acknowledgments

We thank Annamalai Subramanian, CMES, Ehime University, Japan for his critical reading of this manuscript. This study was supported by grants from the Environmental Science and Technology in the Core University Program between Japan Society for the Promotion of Science (JSPS) and National Center for Natural Science and Technology, Vietnam (NCST), Research Revolution 2002 (RR2002) of Project for Sustainable Coexistence of Human, Nature and the Earth (FY2002) and "21st Century Center of Excellence (COE) Program" from the Ministry of Education, Culture, Sports, Science and Technology, Japan.

References

Agency for Toxic Substances and Disease Registry (ATSDR) (1997) Interim policy guideline: dioxins and dioxin-like compounds in soil. U.S. Department of Health and Human Services, Public Health Services, Atlanta, GA.

Agency for Toxic Substances and Disease Registry (ATSDR) (1998) Toxicological Profile for Chlorinated Dibenzo-*p*-dioxins (CDDs). Department of Health and Human Services, Public Health Services, Atlanta, GA.

Ahmed MT, Ismail SMM, Mabrouk SS (1998) Residues of some chlorinated hydrocarbon pesticides in rain water, soil and ground water and their influence on some soil microorganisms. Environ Int 24:665–670.

Bignert A, Olsson M, Persson W, Jensen S, Zakrisson S, Litzén K, Eriksson U, Häggberg L, Alsberg T (1998) Temporal trends of organochlorines in Northern Europe, 1967–1995. Relation to global fractionation, leakage from sediments and international measures. Environ Pollut 99:177–198.

Canadian Council of the Ministers of the Environment (2002) Canadian Environmental Quality Guidelines 1999, updated 2001, 2002. Canadian sediment quality guideline for the protection of aquatic life. Summary table updated 2002. Winnipeg, MB, Canada.

Choi JW, Matsuda M, Kawano M, Min BY, Wakimoto T (2001) Accumulation profiles of persistent organochlorines in waterbirds from an estuary in Korea. Arch Environ Contam Toxicol 41:353–363.

Dwernychuk LW, Cau HD, Hatfield CT, Boivin, TG, Hung TM, Dung PT, Thai ND (2002) Dioxin reservoirs in southern Viet Nam: a legacy of Agent Orange. Chemosphere 47:117–137.

Food and Agriculture Organization of the United Nations (FAO). FAO Statistics Division (FAOSTAT), FAOSTAT Core Consumption Data, http://faostat.fao.org/site/346/default.aspx, accessed June 2006.

Guruge KS, Tanabe S, Fukuda M, Yamagishi S, Tatsukawa R (1997) Accumulation pattern of persistent organochlorine residues in common cormorants (*Phalacrocorax carbo*) from Japan. Mar Pollut Bull 34:186–193.

Hoshi H, Minamoto N, Iwata H, Shiraki K, Tatsukawa R, Tanabe S, Fujita S, Hirai K, Kinjo T (1998) Organochlorine pesticides and polychlorinated biphenyls congeners in wild terrestrial mammals and birds from Chubu region, Japan: interspecies comparison of the residue levels and composition. Chemosphere 36:3211–3221.

Hung DQ, Thiemann W (2002) Contamination by selected chlorinated pesticides in surface waters in Hanoi, Vietnam. Chemosphere 47:357–367.

Iwata H, Tanabe S, Sakai N, Tatsukawa R (1993) Distribution of persistent organochlorines in the oceanic air and surface seawater and the role of the ocean on their global transport and fate. Environ Sci Technol 27:1080–1098.

Iwata H, Tanabe S, Sakai N, Nishimura A, Tatsukawa R (1994) Geographical distribution of persistent organochlorines in air, water and sediments from Asia and Oceania, and their implication for global redistribution from lower latitudes. Environ Pollut 85:15–33.

Iwata H, Tanabe S, Ueda K, Tatsukawa R (1995) Persistent organochlorine residues in air, water, sediments and soils from the lake Baikal region, Russia. Environ Sci Technol 29:792–801.

Iwata H, Agusa T, Inoue S, Kubota R, Minh NH, Minh TB, Tu NPC, Kajiwara N, Kunisue T, Subramanian A, Tanabe S, Viet PH, Tuyen BC (2004) Contamination of trace elements in groundwater and persistent organochlorines in sediment from Mekong Delta, South Vietnam. In: Proceedings of the International Symposium on the Development of Water Resource Management System in Mekong Watershed, December 3–4, 2004, Hanoi, Vietnam, pp 25–31.

Kannan K, Tanabe S, Quynh HT, Hue ND, Tatsukawa R (1992) Residue pattern and dietary intake of persistent organochlorine compounds in foodstuffs from Vietnam. Arch Environ Contam Toxicol 22:367–374.

Kannan K, Tanabe S, Tatsukawa R (1995) Geographical distribution and accumulation features of organochlorine residues in fish from tropical Asia and Oceania. Environ Sci Technol 29:2673–2683.

Kim JH, Smith A (2001) Distribution of organochlorine pesticides in soils from South Korea. Chemosphere 43:137–140.

Kunisue T, Minh TB, Fukuda K, Wantanabe M, Tanabe S, Titenko A (2002) Seasonal variation of persistent organochlorines accumulation in birds from Lake Baikal, Russia and the role of the South Asian Region as a source of pollution for wintering migrants. Environ Sci Technol 36:1396–1404.

Kunisue T, Watanabe M, Subramanian A, Sethuraman A, Titenko AM, Qui V, Prudente M, Tanabe S (2003) Accumulation features of persistent organochlorines in resident and migratory birds from Asia. Environ Pollut 125:157–172.

Kunisue T, Watanabe M, Iwata H, Subramanian A, Monirith I, Minh TB, Baburajendran R, Tana TS, Viet PH, Prudente M, Tanabe S (2004) Dioxins and related compounds in human breast milk collected around open dumping sites in Asian developing countries: bovine milk as a potential source. Arch Environ Contam Toxicol 47:414–426.

Li YF (1999) Global technical hexachlorocyclohexane usage and its contamination consequences in the environment: from 1948 to 1997. Sci Total Environ 232: 121–158.

Marta V, Viktor P, Jana K, Ján U (1997) Analytical methods for the determination of organochlorine compounds: application to environmental samples in the Slovak Republic. J Chromatogr A 774:333–347.

McGrath D (1995) Organic micropollutant and trace element pollution of Irish soils. Sci Total Environ 164:125–133.

Meijer SN, Ockenden WA, Sweetmen A, Breivik K, Grimalt JO, Jones KC (2003) Global distribution of PCBs and HCB in background surface soils: implications for sources and environmental processes. Environ Sci Technol 37:667–672.

Minh NH, Minh TB, Watanabe M, Kunisue T, Monirith I, Tanabe S, Sakai S, Subramanian A, Sasikumar K, Viet PH, Tuyen BC, Tana TS, Prudente M (2003) Open dumping sites in Asian developing countries: a potential source of polychlorinated dibezo-*p*-dioxins and polychlorinated dibenzofurans. Environ Sci Technol 37:1493–1502.

Minh NH, Someya M, Minh TB, Kunisue T, Watanabe M, Tanabe S, Viet PH, Tuyen BC (2004a) Persistent organochlorine residues in human breast milk from Hanoi and Hochiminh City, Vietnam: contamination, accumulation kinetics and risk assessment for infants. Environ Pollut 129:431–441.

Minh NH, Kajiwara N, Kunisue T, Subramanian A, Iwata H, Tanabe S, Viet PH, Tuyen BC (2004b) Contamination of persistent organochlorines in sediments from Mekong River Delta, South Vietnam. Organohalogen Compd 66: 3615–3620.

Minh NH, Minh TB, Kajiwara N, Kunisue T, Subramanian A, Iwata H, Tana TS, Baburajendran R, Karuppiah S, Viet PH, Tuyen BC, Tanabe S (2006a) Contamination by persistent organic pollutants in dumping sites of Asian developing countries: implications of emerging pollution sources. Arch Environ Contam Toxicol 50:474–481.

Minh NH, Minh TB, Kajiwara N, Kunisue T, Iwata H, Viet PH, Tu NPC, Tuyen BC, Tanabe S (2006b) Contamination by polybrominated diphenyl ethers and persistent organochlorines in catfish and feed from Mekong River Delta, Vietnam. Environ Toxicol Chem 25:2700–2709.

Minh NH, Minh TB, Iwata H, Kajiwara N, Kunisue T, Takahashi S, Viet PH, Tuyen BC, Tanabe S (2007a) Persistent organic pollutants in sediments from Saigon–Dong Nai River basin, Vietnam: levels and temporal trends. Arch Environ Contam Toxicol 52:458–465.

Minh NH, Minh TB, Kajiwara N, Kunisue T, Iwata H, Viet PH, Tu NPC, Tuyen BC, Tanabe S (2007b) Pollution sources and occurrences of selected persistent organic pollutants (POPs) in sediments of the Mekong River Delta, South Vietnam. Chemosphere 67:1794–1801.

Minh TB, Kunisue T, Yen NTH, Watanabe M, Tanabe S, Hue ND, Qui V (2002) Persistent organochlorine residues and their bioaccumulation profiles in resident and migratory birds from North Vietnam. Environ Toxicol Chem 21:2108–2118.

Minh TB, Iwata H, Agusa T, Minh NH, Inoue S, Kubota R, Tu NPC, Kajiwara N, Kunisue T, Subramanian A, Viet PH, Tuyen BC, Chamnan C, Tana TS, Tanabe S (2005) Contamination by arsenic and persistent organic pollutants in Mekong River: geographical distribution, patterns of accumulation and implications for environmental quality and human health. In: Proceedings of the 2nd International Symposium on the Development of Water Resource Management System in Mekong Watershed, December 7, 2005, Bangkok, Thailand, pp 15–23.

Monirith I, Nakata H, Watanabe M, Takahashi S, Tanabe S, Tana TS (2000) Organochlorine contamination in fish and mussels from Cambodia and other Asian countries. Water Sci Technol 42:241–252.

Monirith I, Ueno D, Takahashi S, Nakata H, Sudaryanto A, Subramanian A, Karuppiah S, Ismail A, Muchtar A, Zheng J, Richardson BJ, Prudente M, Hue ND, Tana TS, Tkalin AV, Tanabe S (2003) Asia-Pacific mussel watch: monitoring contamination of persistent organochlorine compounds in coastal waters of Asian countries. Mar Pollut Bull 46:281–300.

Nakata H, Tanabe S, Tatsukawa R, Amano M, Miyazaki N, Petrov E (1995) Persistent organochlorines and their bioaccumulation kinetics in Baikal seal (*Phoca sibirica*) from Lake Baikal, Russia. Environ Sci Technol 29:2877–2885.

Nhan DD, Am NM, Hoi NC, Dieu LV, Carvalho FP, Villeneuve JP, Cattini C (1998) Organochlolrine pesticides and PCBs in the Red river delta, North Vietnam. Mar Pollut Bull 36:742–749.

Nhan DD, Am NM, Carvalho FP, Villeneuve JP, Cattini C (1999) Organochlorine pesticides and PCBs along the coast of North Vietnam. Sci Total Environ 237(8):363–371.

Nhan DD, Carvalho FP, Am NM, Tuan QT, Yen NTH, Villeneuve JP, Cattini C (2001) Chlorinated pesticides and PCBs in sediments and mollusks from freshwater canals in the Hanoi region. Environ Pollut 112:311–320.

Noren K, Meironyte D (2000) Certain organochlorine and organobromine contaminants in Swedish human milk in perspective of 20–30 years. Chemosphere 40:1111–1123.

Ogura I, Masunaga S, Nakanishi J (2001) Atmospheric deposition of polychlorinated dibenzo-*p*-dioxins, polychlorinated dibenzofurans, and dioxin-like polychlorinated biphenyls in the Kanto Region, Japan. Chemosphere 44:1473–1487.

Phiet PH, Anh TK, Vu DQ, Schecter A (1989) Preliminary observations of the clinical histories, polychlorinated dibenzodioxin and dibenzofuran tissue levels and 2,3,7,8-TCDD toxic equivalents of potentially dioxin exposed patients living in the south of Vietnam. Chemosphere 19:937–940.

Phuong NTG, Hung BS, Vu DQ, Schecter A (1989) Dioxin levels in adipose tissues of hospitalized women living in the south of Vietnam in 1984–85 with a brief review of their clinical histories. Chemosphere 19:933–936.

Phuong PK, Son CPN, Sauvain J, Tarradellas J (1998) Contamination by PCBs, DDTs, and heavy metals in sediments of Ho Chi Minh City's canals, Viet Nam. Bull Environ Contam Toxicol 60:347–354.

Schecter A, Ryan JJ, Constable JD (1986) Chlorinated dibenzo-p-dioxin and dibenzofuran levels in human adipose tissue and milk samples from the north and south of Vietnam. Chemosphere 15:1613–1620.

Schecter A, Ryan JJ, Constable JD (1987) Polychlorinated dibenzo-*p*-dioxin and polychlorinated dibenzofuran levels in human breast milk from Vietnam compared with cow's milk and human breast milk from the North American continent. Chemosphere 16:2003–2016.

Schecter A, Fuerst P, Fuerst C, Meemken HA, Groebel W, Constable JD (1989a) Levels of polychlorinated dibenzofurans, dibenzodioxins, PCBs, DDT, DDE, hexachlorobenzene, dieldrin, hexachlorocyclohexanes, and oxychlordane in human breast milk from the United States, Thailand, Vietnam and Germany. Chemosphere 18:445–454.

Schecter A, Eitzer BD, Hites RA (1989b) Chlorinated dioxin and dibenzofuran levels in sediments collected from rivers in Vietnam. Chemosphere 18:831–834.

Shecter A, Kooke R, Serne P, Olie K, Huy DQ, Hue ND, Constable J (1989c) Chlorinated dioxin and dibenzofuran levels in food samples collected between 1985–87 in the north and south of Vietnam. Chemosphere 18:627–634.

Schecter A, Kooke R, Serne P, Olie K, Huy DQ, Hue ND, Constable J (1989d) Chlorinated dioxin and dibenzofuran levels in food samples collected between 1985–87 in the north and south of Vietnam. Chemosphere 18:627–634.

Schecter A, Tong HY, Monson SJ, Gross ML, Constable J (1989e) Adipose tissue levels of 2,3,7,8-TCDD in Vietnamese adults living in Vietnam, 1984–87. Chemosphere 18:1057–1062.

Schecter A, Furst P, Ryan JJ, Furst C, Meemken H, Groebel W, Constable J, Vu D (1989f) Polychlorinated dioxin and dibenzofuran levels from human milk from several locations in the United States, Germany and Vietnam. Chemosphere 19:979–984.

Schecter A, Vu D, Tong HY, Monson SJ, Gross ML, Constable JD (1989g) Levels of 2,3,7,8-TCDD and 2,3,7,8-TCDF in human adipose tissue from hospitalized persons in the north and south of Vietnam 1984–88. Chemosphere 19: 1001–1004.

Schecter A, Furst P, Furst C, Groebel W, Constable JD, Kolesnikov S, Beim A, Boldonov A, Trubitsun E, Vlasov B, Cau HD, Dai LC, Quynh HT (1990a) Levels of chlorinated dioxins, dibenzofurans and other chlorinated xenobiotics in food from the Soviet Union and the South of Vietnam. Chemosphere 20:799–806.

Schecter A, Startin JR, Rose M, Wright C, Parker I, Woods D, Hansen H (1990b) Chlorinated dioxin and dibenzofuran levels in human milk from Africa, Pakistan, Southern Vietnam, the southern US and England. Chemosphere 20:919–925.

Schecter A, Furst P, Furst C, Papke O, Ball M, Dai LC, Quynh HT, Phuong NTG, Beim A, Vlasov B, Chongchet V, Constable JD, Charles K (1991) Dioxins, dibenzofurans and selected chlorinated organic compounds in human milk and blood from Cambodia, Germany, Thailand, the USA, the USSR, and Vietnam. Chemosphere 23:1903–1912.

Schecter A, Papke O, Ball M, Cau HD, Dai LC, Minh NQ, Quynh HT, Phuong NNT, Phiet PH, Chi HK, Vo DT, Constable JD, Spencer J (1992a) Dioxin and dibenzofuran levels in blood and adipose tissue of Vietnamese from various locations in Vietnam in proximity to Agent Orange Spraying. Chemosphere 25:1123–1128.

Schecter A, Papke O, Ball M, Grachev M, Beim A, Koptug V, Cau HD, Dai LC, Quynh HT, Phuong NGT, Chi HK (1992b) Dioxin and dibenzofuran levels in human blood samples from Guam, Russia, Germany, Vietnam and the USA. Chemosphere 25:1129–1134.

Schecter A, Dai LC, Thuy LTB, Quynh HT, Minh DQ, Cau HD, Phiet PH, Phuong NTN, Constable JD, Baughman R, Papke O, Ryan JJ, Furst P, Ralsanen S (1995) Agent Orange and the Vietnamese: The persistence of elevated dioxin levels in human tissues. Am J Public Health 85:516–522.

Schecter A, Toniolo P, Dai LC, Thuy LTB, Wolff MS (1997) Blood levels of DDT and breast cancer risk among women living in the North of Vietnam. Arch Environ Contam Toxicol 33:453–456.

Schecter A, Dai LC, Papke O, Prange J, Constable JD, Matsuda M, Thao VD, Piskac AL (2001) Recent dioxin contamination from Agent Orange in residents from a southern Vietnam city. J Occup Environ Med 43:435–443.

Schecter A, Quynh HT, Pavuk M, Papke O, Malisch R, Constable JD (2003) Food as a source of dioxin exposure in the residents of Bien Hoa City, Vietnam. J Occup Environ Med 45:781–788.

Sinh NN, Thuy LTB, Kinh NK, Thang LB (1999) The persistent organic pollutants and their management in Vietnam. In: Proceedings of the Regional Workshop on the Management of Persistent Organic Pollutants (POPs), United Nations Environment Program, March 16–19, 1999, Hanoi, Vietnam, pp 385–406.

Stellman JM, Stellman SD, Christian R, Weber T, Tomasallo C (2003) The extent and patterns of usage of Agent Orange and other herbicides in Vietnam. Nature (Lond) 422:681–687.

Strandberg B, van Bavel B, Bergqvist P, Broman D, Ishaq R, Pettersen H, Rappe C (1998) Occurrence, sedimentation and spatial variation of organochlorine contaminants on settling particulate matter and sediments in the northern part of the Baltic Sea. Environ Sci Technol 32:1754–1759.

Tanabe S (2000) Asian developing regions: persistent organic pollutants in the seas. In: Sheppard CRC (ed) Sea at the millennium: an environmental evaluation. Elsevier, Amsterdam, pp 447–462.

Tanabe S (2002) Contamination and toxic effects of persistent endocrine disrupters in marine mammals and birds. Mar Pollut Bull 45:69–77.

Tanabe S, Subramanian A (2006) Bioindicators of POPs. Monitoring in developing countries. Kyoto University Press and Trans Pacific Press, Kyoto, Japan.

Tanabe S, Watanabe S, Kan H, Tatsukawa R (1988) Capacity and mode of PCB metabolism in small cetaceans. Mar Mamm Sci 4:103–124.

Tanabe S, Iwata H, Tatsukawa R (1994) Global contamination by persistent organochlorines and their ecotoxicological impacts in marine mammals. Sci Total Environ 154:163–177.

Tanabe S, Senthilkumar K, Kannan K, Subramanian A (1998) Accumulation features of polychlorinated biphenyls and organochlorine pesticides in residents and migratory birds from South India. Arch Environ Contam Toxicol 34: 387–397.

Tanabe S, Niimi S, Minh TB, Miyazaki N, Petrov EA (2003) Temporal trends of persistent organochlorine contamination in Russia: a case study of Baikal and Caspian seal. Arch Environ Contam Toxicol 44:533–545.

Thao VD, Kawano M, Tatsukawa R (1993a) Persistent organochlorine residues in soils from tropical and subtropical Asian countries. Environ Pollut 81:61–71.

Thao VD, Kawano M, Matsuda M, Wakimoto T, Tatsukawa R (1993b) Chlorinated hydrocarbon insecticide and polychlorinated biphenyl residues in soils from southern provinces of Vietnam. Int J Environ Anal Chem 50:147–159.

Viet PH (2002) Persistent organic pollutants in Vietnam—an overview. In: Proceedings of the Workshop on Environmental Monitoring of Persistent Organic Pollutants (POPs) in East Asian Countries, December 2–4, 2002, Tokyo/Tsukuba, Japan, pp 113–121.

Viet PH, Hoai PM, Minh NH, Ngoc NT, Hung PT (2000) Persistent organochlorine pesticides and polychlorinated biphenyls in some agricultural and industrial areas in Northern Vietnam. Water Sci Technol 42:223–229.

Wagrowski DM, Hites RA (2000) Insights into the global distribution of polychlorinated dibenzo-*p*-dioxins and dibenzofurans. Environ Sci Technol 34:2952–2958.

Manuscript received April 25; accepted May 1, 2007.

Index

MIX
Papier aus verantwortungsvollen Quellen
Paper from responsible sources
FSC® C105338

If you have any concerns about our products,
you can contact us on
ProductSafety@springernature.com

In case Publisher is established outside the EU,
the EU authorized representative is:
Springer Nature Customer Service Center GmbH
Europaplatz 3, 69115 Heidelberg, Germany

Printed by Libri Plureos GmbH
in Hamburg, Germany